Machine Learning Applications in the Water Domain

Machine Learning Applications in the Water Domain

Guest Editor

Yonggui Wang

MDPI

Basel • Beijing • Wuhan • Barcelona • Belgrade • Novi Sad • Cluj • Manchester

Guest Editor
Yonggui Wang
School of Geography and
Information Engineering
China University
of Geosciences
Wuhan
China

Editorial Office
MDPI AG
Grosspeteranlage 5
4052 Basel, Switzerland

This is a reprint of the Special Issue, published open access by the journal *Water* (ISSN 2073-4441), freely accessible at: https://www.mdpi.com/journal/water/special_issues/248W1J7QLT.

For citation purposes, cite each article independently as indicated on the article page online and as indicated below:

Lastname, A.A.; Lastname, B.B. Article Title. *Journal Name* **Year**, *Volume Number*, Page Range.

ISBN 978-3-7258-8563-3 (Hbk)
ISBN 978-3-7258-8564-0 (PDF)
https://doi.org/10.3390/books978-3-7258-8564-0

Contents

About the Editor

Yonggui Wang

Yonggui Wang is currently the Director of the Wuhan Advanced Technology Institute and an Associate Professor at the School of Geography and Information Engineering, China University of Geosciences, where he also holds the qualification of Senior Technology Manager. Dr. Wang's primary research interests lie at the intersection of environmental numerical simulation, big data, and artificial intelligence, with a focus on applying data-driven approaches to address complex water, environmental and ecological challenges. He has led 16 national and provincial/ministerial research projects, published 48 papers as first or corresponding author, been granted 29 invention patents, and edited 3 books. His work has earned him the Special Prize for Science and Technology Progress from the Chinese Dam Engineering Society, as well as an Excellence Award in the Hubei Provincial Staff Innovation and Entrepreneurship Competition. With extensive experience in multidisciplinary integration, technology transfer, and commercialization, he is dedicated to translating cutting-edge informatics and geoscience research into practical solutions for environmental management and societal benefit.

water

MDPI

Editorial

Machine Learning Applications in the Water Domain

Yonggui Wang [1,2]

[1] Hubei Key Laboratory of Regional Ecology and Environmental Change, School of Geography and Information Engineering, China University of Geosciences, Wuhan 430074, China; wangyg@cug.edu.cn
[2] Wuhan Institute of Advanced Technology, Wuhan 430070, China

1. Introduction

Water is an indispensable resource for life and nature. However, under the combined pressures of global climate change, continuous population growth, and accelerating urbanization, global water security and river ecosystems are facing multiple challenges [1]. The IPCC report indicates that global warming will intensify the hydrological cycle and increase the frequency of extreme weather events, posing severe challenges to regional water resource security. Meanwhile, human activities, such as land use change, urbanization, and agricultural irrigation, are also profoundly altering watershed hydrological processes. Previous studies have shown that two-thirds of the global population, approximately 4 billion people, experience severe water scarcity for at least one month each year [2] and water pollution may further aggravate future water scarcity at the basin scale [3]. Therefore, how to quantitatively separate and assess the relative contributions of climate change and human activities to water resources, and how to predict the evolution of regional water resources under future climate change scenarios have become important scientific questions in current hydrological research.

In recent years, the remarkable adaptability of machine learning has demonstrated its potential as a powerful tool in environmental science and engineering, bringing new opportunities to water resource research [4]. Compared with traditional methods, machine learning can automatically learn complex patterns from large volumes of data without explicitly programming physical mechanisms. It has been widely applied in water-related subfields, including floods, weather, land use and soil, water quality, surface water, water resources, and groundwater [5], showing great potential in hydrological process prediction, water quality monitoring, and watershed simulation. However, despite these advances, there remains a lack of systematic integration of emerging topics such as machine vision for water body identification, knowledge–data coupling, and model interpretability. It is precisely this gap that motivated the organization of this Special Issue, to provide a dedicated platform for showcasing cutting-edge research on machine learning applications in hydrology and water quality—especially in time-series prediction, remote sensing recognition, hybrid modeling, and performance evaluation. The value and significance of this Special Issue lie in its ability to bridge the gap between algorithmic innovation and practical water management, offering insights into how ML can resolve long-standing challenges in pollution source apportionment, ungauged basin simulation, and real-time monitoring network design. Importantly, the collected studies demonstrate that machine learning is not merely an auxiliary tool but a transformative methodology capable of reshaping our understanding of water systems.

Since its launch, this Special Issue has received an enthusiastic and broad response from the global research community, attracting high-quality contributions from diverse

 https://doi.org/10.3390/w18111351

fields such as hydrology, water quality modeling, remote sensing, and environmental informatics. The number and quality of submissions reflect that this topic has become a major research hotspot in contemporary water science. Looking forward, it is evident that artificial intelligence, particularly machine learning, will inevitably change the research paradigm in the water domain—from data-driven discovery to intelligent decision support. It will enable more accurate predictions, deeper causal inference, and more adaptive management strategies. We believe that continued efforts in this direction will yield significant new achievements, ultimately providing novel insights, innovative monitoring methods, and intelligent management solutions for humanity's understanding, monitoring, and stewardship of water resources.

2. An Overview of Published Articles

Contribution 1: The growing risks posed by climate change and extreme flood events increase susceptibility to erosion and structural instability under elevated hydraulic loading conditions. This study proposes a spillway erosion prediction framework that integrates machine learning with Bayesian probability analysis, using readily available spillway width and stream power as core variables to classify erosion potential and support preliminary screening for dam safety management when detailed site-specific information is limited. Based on the ASDSO Dam Failures and Incidents Database and the National Inventory of Dams, the study constructed a dataset of 500 samples and applied Logistic Regression, support vector machine, and Random Forest models to classify spillways as eroded or non-eroded cases. Bayesian probability analysis was further used to quantify the posterior reliability of model predictions. The Random Forest model demonstrated the best predictive performance, achieving 82.7% accuracy and 92.5% specificity on the test dataset, and identified spillway width as a more influential predictor of erosion risk than stream power. The Bayesian analysis further indicated that when the model predicts erosion, there is approximately a 59% probability that the spillway will actually experience erosion.

Contribution 2: Water quality prediction requires capturing not only the long-term dependencies of water quality time-series data, but also the spatial connectivity between upstream and downstream monitoring sites, unidirectional pollutant transport, and extreme pollution events. To address the limitation that traditional deep learning models mainly focus on temporal features while neglecting real river network topology, this study developed an enhanced spatiotemporal relationship-guided deep learning method. A Gated Recurrent Unit (GRU) model was used as the baseline, a Graph Attention Network–Gated Recurrent Unit (GAT-GRU) model was constructed by incorporating a graph attention mechanism, and an Enhanced Spatiotemporal Relation-Guided Gated Recurrent Unit (ESRG-GRU) model was further developed by reconstructing the adjacency matrix and improving the extremum-sensitive loss function. Using 40 monitoring stations in the Tuojiang River Basin, Sichuan Province, China, the study compared model training efficiency for 1-day, 3-day, and 7-day forecast horizons and evaluated 7-day total phosphorus (TP) prediction performance. The ESRG-GRU model improved average NSE by 0.1034 and 0.0347 compared with GRU and GAT-GRU, respectively, while reducing average RMSE by 15.676% and 7.143%, with only a limited increase in training time. This method effectively integrates spatiotemporal features and river network topology, improving the accuracy, physical consistency, and interpretability of long-term, multi-station water quality prediction, and providing support for basin water quality management and early warning.

Contribution 3: Reservoir operation requires a dynamic balance between flood risk mitigation, water supply reliability, and operational stability. This study applies deep reinforcement learning (DRL) models, including Deep Q-Network (DQN), Proximal Policy

 https://doi.org/10.3390/w18111351

Optimization (PPO), and Deep Deterministic Policy Gradient (DDPG), to optimize reservoir operations at the Soyang River Dam in South Korea using 30 years of daily data from 1993 to 2022. By integrating observed and remote-sensing hydrometeorological inputs, such as precipitation, temperature, soil moisture, and evaporation, the DRL framework supports adaptive storage and release decisions and helps overcome the limitations of traditional Dynamic Programming (DP) and Stochastic Dynamic Programming (SDP) under high-dimensional, multi-variable conditions. Discharge was derived through a mass balance formulation from optimized storage, and model performance was evaluated using cumulative reward, action stability, and counts of total capacity and flood control violations. The results show that all three DRL strategies maintained larger storage buffers, generally reduced peak discharge on high-inflow days, and decreased flood threshold exceedances under the same inflow conditions. Among them, PPO achieved the highest cumulative reward and the most stable actions, DQN prioritized flood control compliance, and DDPG provided continuous control with intermediate responsiveness. The SHAP analysis further demonstrated that the DRL agents learned hydrologically consistent and interpretable operation strategies, characterized by retaining water during the rising limb, moderating the crest, and releasing during the recession. Overall, this study highlights the ability of DRL to integrate multi-source hydrometeorological data, capture nonlinear reservoir operation dynamics, and provide an adaptive, scalable, and explainable decision framework for optimized reservoir operation and flood risk mitigation under hydrological extremes.

Contribution 4: This study addresses the need to monitor vegetation dynamics on the Baishazhou sandbar in the Wuhan reach of the Changjiang River Basin, especially the spatial and temporal variability of sandbar roughness for improving hydrodynamic flood modeling during flooding events. It proposes a repeatable, cost-effective, and fine-scale workflow for estimating vegetation height, providing a methodological basis for deriving dynamic vegetation roughness from UAV and satellite imagery. The study used a DJI Mavic 3 Multispectral UAV (SZ DJI Technology Co., Ltd., Shenzhen, China) to collect multispectral images and generate DSM/DEM data, calculated vegetation indexes (VIs) including NDVI, GNDVI, NDRE, LCI, and OSAVI, and employed a Fully Connected Neural Network (FCNN) to establish the relationship between vegetation height (h) and VIs. The results show that linear and polynomial regression models had limited performance in estimating vegetation height, while the trained FCNN model performed rather well, with R^2 of about 0.85–0.88, RMSE = 1.10 m, and a maximum relative error of 17.2%.

Contribution 5: To address the limitations of traditional optimization-based models with high computational demands and physical-based models that rely on static operation policies under rapidly changing flood conditions, this study combines an IPOPT optimization model with a GRU-based reservoir operation model and introduces data integration (DI) to explicitly incorporate lagged observations for real-time flood control. Applied to the Namgang Dam in South Korea, the study used hourly reservoir inflow, storage, release, and precipitation data from 42 significant flood events during 2000–2024. The Interior Point Optimizer (IPOPT) was first used to generate optimized hydrographs by minimizing peak outflows, and the resulting optimal reservoir storage and release values were then used to train the Gated Recurrent Unit (GRU) model. Six input scenarios were designed to assess the effects of input variable selection, DI, and forecasted inputs on model performance. Results show that the GRU model with DI using historical reservoir inflow and storage achieved the best performance, with RMSE = 202.2, NSE = 0.93, and R^2 = 0.89, consistently outperforming models without DI; adding precipitation or forecasted inflows did not improve model performance. This study demonstrates that GRU-DI can accurately

https://doi.org/10.3390/w18111351

replicate optimized reservoir operations, supporting timely and informed decision-making for real-time reservoir operation and flood risk mitigation.

Contribution 6: Reservoir evaporation is a vital variable for water resource management in semi-arid regions, but the complex and nonlinear relationships among meteorological variables limit the prediction capability of traditional empirical methods. This study used daily hydrometeorological data from 2004 to 2011 at the Atatürk Dam Reservoir in Turkey, applied the ReliefF algorithm to select key input variables such as sunshine duration, temperature, precipitation, and humidity, and compared the Multilayer Perceptron (MLP) and Adaptive Neuro-Fuzzy Inference System (ANFIS) under different data partitioning strategies, training algorithms, and membership functions. The results show that ANFIS had strong fitting capacity during the training phase but showed a tendency toward overfitting, whereas MLP demonstrated better generalization ability on the test sets, with the best model achieving $R^2 = 0.8069$ and RMSE = 0.0923. The study indicates that MLP can effectively capture the nonlinear characteristics of reservoir evaporation and support evaporation forecasting, reservoir operation, and climate-resilient water resource planning in semi-arid regions.

Contribution 7: Groundwater levels in the Yellow River Basin are influenced by long-term hydrological variations, topographic and geomorphic heterogeneity, and human activities, leading to complex spatiotemporal fluctuations, while the limited spatial and temporal coverage of monitoring wells constrains large-scale continuous assessment. This study proposes an enhanced Transformer deep learning model that integrates groundwater monitoring well data with GRACE/GRACE-FO terrestrial water storage anomalies and GLDAS soil moisture data. Groundwater storage was derived using the mass balance equation, and the data were matched to 1277 pixels at a $0.25^\circ \times 0.25^\circ$ grid scale. The model incorporates Key Feature Enhancement Architecture (KFEA) and Dynamic–Static Feature Fusion (DSFF) to adaptively enhance key feature channels and integrate surrounding spatial contextual information, thereby improving the capture of long-range temporal dependencies and robustness to anomalous data. Results show that the enhanced Transformer model significantly outperformed LSTM, with R^2 increasing by approximately 17.5%, while RMSE and MAE decreased by approximately 12.4% and 10.9%, respectively. The model also revealed a fluctuating downward trend in groundwater levels in the Yellow River Basin, providing a quantitative method for large-scale groundwater time-series monitoring, variation assessment, and water resource management.

Contribution 8: Existing water distribution networks (WDNs) may experience damaging hydraulic transients and water hammer events during sudden demand increase, but previous studies often relied on a single worst transient case and considered only pressure at pipe end nodes, which may not ensure full protection along pipes. This study developed a comprehensive optWDN model to identify the minimum set of worst transient cases and optimize the diameters of additional duplicate pipes using a genetic algorithm (GA), while satisfying both steady-state and transient pressure constraints. The existing New York Tunnel (NYT) network was used for model validation. The model introduced two versions of the Surge Damage Potential Factor (SDPF): SDPFN, based on pressure at pipe end nodes, and SDPFP, based on pressure at all computational nodes along pipes. Results show that achieving SDPFN = 0 does not guarantee full protection, because pressure heads along pipes can be more extreme than those at pipe end nodes. For concurrent sudden demand increase at two nodes, optWDN identified eight critical transient cases and optimized duplicate pipe diameters that achieved SDPFN = 0 and SDPFP = 0 for all 171 possible cases. Although the final costs were 46% and 74% higher than previous studies for SDPFN = 0 and SDPFP = 0, respectively, the additional cost was necessary because previous solutions failed

https://doi.org/10.3390/w18111351

to achieve zero SDPF in many transient loading cases. The study also demonstrated that regulating the duration of demand increase can reduce surge protection costs, providing a more reliable and cost-effective framework for full protection of existing WDNs.

Contribution 9: The multi-objective optimization operation of a cascade reservoir system must coordinate water release for hydropower generation in the mainstream and water release for regional water supply in tributaries, but it is challenged by the high-dimensional decision-making process, nonlinear system dynamics, stochasticity of runoff, and the curse of dimensionality. Based on the synchronicity and substitutability properties of storage distribution, this study derives the first-order derivative relationship between individual reservoir water release decisions and the water availability of the cascade system or a specific individual reservoir. The theoretical analysis identifies a non-decreasing monotonic relationship between the cascade reservoir system's overall water availability, optimal carryover storage, and water release for tributary use, as well as monotonic relationships between water release for mainstream use and individual reservoir water availability. On this basis, an improved reinforcement learning (IRL) method with a reduced searching region is developed and applied to the WB-Xll-Xjb-Sx reservoir system in the upper reaches of the Yangtze River in China. Using historical inflow runoff from 1959 to 2023, the comparison with stochastic dynamic programming (SDP) and traditional reinforcement learning (RL) shows that IRL keeps relative errors within 10%, while its computation time is consistently less than 0.10% of SDP and 5% of RL. The method also preserves highly interpretable optimization results following the synchronic storage distribution property, demonstrating high computational efficiency and reasonable interpretability.

Contribution 10: Water quality management in California's Sacramento–San Joaquin Delta must balance agricultural operations, drinking water treatment, and ecosystem management, while traditional sampling and parametric regression methods are limited in providing accurate, continuous, and spatially distributed ion constituent simulations. To improve simulations of chloride, bromide, and sulfate concentrations, this study developed artificial neural network models for three critical water intake locations: the Harvey O. Banks Pumping Plant, Tracy Pumping Plant, and O'Neill Forebay at the Gianelli Pumping Plant. Using Electrical Conductivity (EC), Sacramento X2, Water Year Type (WYT), Month, and Location as predictors, the models achieved R^2 values above 0.96, with standard deviation in $R^2 \leq 0.006$ in cross-validation. Compared with existing methods, the models reduced MAE by 60.7–74.0% for chloride, 63.3–72.5% for bromide, and 70.4–87.9% for sulfate. The study further developed four complementary interactive dashboards, including a Model Interpretability Dashboard to enhance transparency and build stakeholder trust. For the first time, spatial coverage limitations were addressed through hybrid modeling that combines DSM2 hydrodynamic simulation with machine learning, enabling continuous prediction of ion distributions across 162 points in the Interior Delta and providing accessible tools for informed water management decisions.

Contribution 11: Although long short-term memory (LSTM) networks have been widely used for water quality prediction, the confidence analysis of their prediction accuracy and the underlying causes of variability across different water quality indicators and basins remain insufficiently explored. This study selected the Yangtze River Basin (YRB), Haihe River Basin (HRB), and Huangshui River Basin (HSB), and developed LSTM models to forecast key water quality parameters, including ammonia nitrogen (AN), biochemical oxygen demand (BOD), chemical oxygen demand (COD), dissolved oxygen (DO), hydrogen ion concentration (pH), and total phosphorus (TP). A novel confidence evaluation method was introduced by combining Nash–Sutcliffe Efficiency (NSE), confidence intervals, and a confidence coefficient. The results showed that LSTM exhibited higher credibility

 https://doi.org/10.3390/w18111351

in the Haihe River Basin, acceptable reliability in the Huangshui River Basin, and weaker applicability in the Yangtze River Basin except for TP. LSTM demonstrated greater accuracy and stability in predicting total phosphorus (TP), with median NSE values of 0.71 in the HRB and 0.73 in the YRB. The study further revealed a linear relationship between LSTM prediction performance and the temporal autocorrelation and cross-correlation coefficients of water quality parameters, with coefficients of determination (R^2) ranging from 0.59 to 0.85. These findings emphasize that the reliability of LSTM-based predictions depends strongly on basin-specific data characteristics and correlation structures, providing a basis for confidence assessment and practical model selection in water quality forecasting.

Contribution 12: Predicting freshwater quality, including surface and groundwater, is essential for the sustainable management of water resources, while the rapid development of AI/ML/DL techniques has created a complex methodological landscape that requires a systematic understanding of trends, gaps, and future directions. This study collected 1822 articles from the Scopus database covering 2000–2024 and applied Topic Modeling together with the bibliometric–systematic literature review (B-SLR) methodology to analyze water quality prediction studies. The results identified exponential growth in scientific publications since 2020, indicating that this field has reached a stage of maturity. The predominant techniques include ensemble models, especially Bagging and Boosting, deep neural networks such as long short-term memory (LSTM) and convolutional neural networks (CNNs), and traditional algorithms such as artificial neural networks (ANNs), support vector machines (SVMs), and decision trees (DTs). The study further shows that hybrid and explainable architectures, explainable artificial intelligence (XAI), SHAP, LIME, Transfer Learning, Generative Adversarial Networks (GANs), Group Method of Data Handling (GMDH), and Transformer architectures are becoming important research directions. Among them, hybrid and interpretable models are the most promising alternatives for improving prediction accuracy, transparency, and informed decision-making in AI/ML/DL-based freshwater quality prediction.

3. Concluding Remarks

This Special Issue presents representative advances in machine learning applications in the water domain, showing that the field is moving from single data-driven prediction toward an integrated paradigm that combines multi-source data fusion, process knowledge, interpretable prediction, and intelligent decision support. The main advances can be summarized in three aspects:

1. Machine learning has substantially improved the prediction and simulation of complex hydrological, water quality, and water engineering processes. The studies in this Special Issue applied LSTM, GRUs, Transformers, graph neural networks, Random Forest, MLP, ANFIS, and deep reinforcement learning to water quality prediction, groundwater level estimation, reservoir evaporation, reservoir operation, spillway erosion, and transient protection in water distribution networks. These methods effectively capture nonlinear relationships, long-term temporal dependencies, high-dimensional interactions, and responses to extreme events, showing strong prediction accuracy, computational efficiency, and adaptability across different water-related applications.
2. Multi-source data fusion and process knowledge have promoted the transition from black-box prediction to physically consistent prediction. The collected studies integrated monitoring station records, meteorological and hydrological data, remote sensing products, UAV imagery, groundwater monitoring well data, and engineering operation data. They also incorporated river network topology, hydrodynamic simula-

 https://doi.org/10.3390/w18111351

tion, mass balance equations, monotonic property constraints, dynamic–static feature fusion, and reservoir operation rules. This coupling of knowledge and data enhances the ability of machine learning models to represent real hydrological, hydraulic, and operational processes, improving their reliability and transferability for basin-scale analysis, multi-station prediction, and engineering operation.

3. Interpretability, risk assessment, and intelligent optimization have expanded the decision-support value of machine learning in water resource management. Methods such as SHAP interpretation, Bayesian probability analysis, confidence evaluation, extremum-sensitive loss functions, action stability metrics, and reinforcement learning-based optimization improved model transparency, reliability, and operational relevance. These approaches not only improve prediction accuracy, but also identify key driving factors, quantify prediction confidence, optimize operation strategies, and support water quality warning, groundwater management, reservoir flood control, and hydraulic infrastructure safety assessment.

Looking forward, this Special Issue points to several promising directions: the emergence of foundation models for ungauged basins and emerging water issues; digital twins with real-time ML for adaptive monitoring and early warning; physics-guided and causal ML to move beyond correlation toward intervention; and lightweight edge models deployed on drones, buoys, and satellites for ubiquitous sensing. In summary, ML in the water domain is evolving toward accurate sensing, process-informed modeling, interpretable prediction, and intelligent decision-making. It will not remain a mere analytical tool but become a core driver of a new paradigm—intelligent water science—offering transformative capabilities for water security, environmental protection, and climate adaptation.

Funding: This research received no external funding.

Conflicts of Interest: The author declares no conflicts of interest.

List of Contributions

1. Ghimire, S.N.; Schulenberg, J.; Flynn, S. Extreme Events and Dam Safety: Machine Learning Approach to Predict Spillway Erosion. *Water* **2026**, *18*, 373. https://doi.org/10.3390/w18030373.
2. Chen, R.; Wang, Y.; Wang, H.; Wang, S.; Yang, J. Enhanced Spatiotemporal Relationship-Guided Deep Learning for Water Quality Prediction. *Water* **2026**, *18*, 185. https://doi.org/10.3390/w18020185.
3. Sseguya, F.; Jun, K.S. Deep Reinforcement Learning for Optimized Reservoir Operation and Flood Risk Mitigation. *Water* **2025**, *17*, 3226. https://doi.org/10.3390/w17223226.
4. Guo, Y.; Nones, M.; Zhou, Y.; Zhu, R.; Ding, W. Combination of UAV Imagery and Deep Learning to Estimate Vegetation Height over Fluvial Sandbars. *Water* **2025**, *17*, 3160. https://doi.org/10.3390/w17213160.
5. Li, L.; Jun, K.S. GRU-Based Reservoir Operation with Data Integration for Real-Time Flood Control. *Water* **2025**, *17*, 3039. https://doi.org/10.3390/w17213039.
6. Dökmen, F.; Coşkun Dilcan, Ç.; Ahi, Y. Daily Reservoir Evaporation Estimation Using MLP and ANFIS: A Comparative Study for Sustainable Water Management. *Water* **2025**, *17*, 2623. https://doi.org/10.3390/w17172623.
7. Zhou, T.; Fu, C.; Liu, Y.; Xiang, L. Groundwater Level Estimation Using Improved Transformer Model: A Case Study of the Yellow River Basin. *Water* **2025**, *17*, 2318. https://doi.org/10.3390/w17152318.
8. Ahmed, H.M.; Imam, Y.E.; El-Ghandour, H.A.; Elansary, A.S. Identifying Worst Transient Cases and Optimizing Surge Protection for Existing Water Networks. *Water* **2025**, *17*, 1816. https://doi.org/10.3390/w17121816.

https://doi.org/10.3390/w18111351

9. Li, X.; Ma, H.; Chen, S.; Xu, Y.; Zeng, X. Improved Reinforcement Learning for Multi-Objective Optimization Operation of Cascade Reservoir System Based on Monotonic Property. *Water* **2025**, *17*, 1681. https://doi.org/10.3390/w17111681.
10. Namadi, P.; He, M.; Sandhu, P. Advancing Ion Constituent Simulations in California's Sacramento–San Joaquin Delta Using Machine Learning Tools. *Water* **2025**, *17*, 1511. https://doi.org/10.3390/w17101511.
11. Fang, P.; Wang, Y.; Zhao, Y.; Kang, J. Analysis of Prediction Confidence in Water Quality Forecasting Employing LSTM. *Water* **2025**, *17*, 1050. https://doi.org/10.3390/w17071050.
12. Muñoz-Alegría, J.A.; Núñez, J.; Oyarzún, R.; Chávez, C.A.; Arumí, J.L.; Rodríguez-López, L. A Bibliometric-Systematic Literature Review (B-SLR) of Machine Learning-Based Water Quality Prediction: Trends, Gaps, and Future Directions. *Water* **2025**, *17*, 2994. https://doi.org/10.3390/w17202994.

References

1. Vörösmarty, C.J.; McIntyre, P.B.; Gessner, M.O.; Dudgeon, D.; Prusevich, A.; Green, P.; Glidden, S.; Bunn, S.E.; Sullivan, C.A.; Liermann, C.R.; et al. Global Threats to Human Water Security and River Biodiversity. *Nature* **2010**, *467*, 555–561. [CrossRef] [PubMed]
2. Mekonnen, M.M.; Hoekstra, A.Y. Four Billion People Facing Severe Water Scarcity. *Sci. Adv.* **2016**, *2*, e1500323. [CrossRef] [PubMed]
3. Wang, M.; Bodirsky, B.L.; Rijneveld, R.; Beier, F.; Bak, M.P.; Batool, M.; Droppers, B.; Popp, A.; van Vliet, M.T.H.; Strokal, M. A Triple Increase in Global River Basins with Water Scarcity Due to Future Pollution. *Nat. Commun.* **2024**, *15*, 880. [CrossRef] [PubMed]
4. Zhu, M.; Wang, J.; Yang, X.; Zhang, Y.; Zhang, L.; Ren, H.; Wu, B.; Ye, L. A Review of the Application of Machine Learning in Water Quality Evaluation. *Eco-Environ. Health* **2022**, *1*, 107–116. [CrossRef] [PubMed]
5. Sit, M.; Demiray, B.Z.; Xiang, Z.; Ewing, G.J.; Sermet, Y.; Demir, I. A Comprehensive Review of Deep Learning Applications in Hydrology and Water Resources. *Water Sci. Technol.* **2020**, *82*, 2635–2670. [CrossRef] [PubMed]

water

MDPI

Article

Extreme Events and Dam Safety: Machine Learning Approach to Predict Spillway Erosion

Sanjeeta N. Ghimire [1,*], Joseph Schulenberg [2] and Stefan Flynn [3]

1 Department of Civil and Environmental Engineering, Florida Polytechnic University, 4700 Research Way, Lakeland, FL 33805, USA
2 Department of Civil, Materials, and Environmental Engineering, University of Illinois at Chicago, Chicago, IL 60607, USA
3 US Army Corps of Engineers, 1500 Rock Island District, Rock Island, IL 61201, USA
* Correspondence: sghimire@floridapoly.edu

Abstract

This study examines the erosion potential of earthen spillways under the growing risks posed by changing climate and extreme flood events, which threaten the stability and safety of dam infrastructure. Specifically, it employs a machine learning approach to evaluate how readily available spillway width and stream power can predict erosion potential. Site-specific erosion prediction methods are often costly and time-consuming because they rely on extensive field investigations and physical modeling. To address these challenges, this research employs multiple machine learning algorithms, including logistic regression, Support Vector Machine, and Random Forest, on existing data to classify spillways as erodible or non-erodible cases. The Random Forest model demonstrated the best predictive performance, achieving 82.7% accuracy on the test dataset. To further interpret the reliability of model predictions, a Bayesian probability analysis was performed, revealing that when the model predicts erosion, there is a 59% probability that the dam will actually experience erosion. These results highlight how integrating existing datasets with machine learning and probabilistic reasoning can enhance dam safety assessment by considering the accuracy, efficiency, and reliability of spillway erosion predictions.

Keywords: extreme flood events; spillway erosion; dam safety; stream power; spillway width; machine learning

1. Introduction

Embankment erosion is a critical concern in hydraulic engineering, particularly in auxiliary spillways, where high water velocities and discharge rates can cause rapid and extensive surface damage, threatening the structural integrity of dams. With the escalating impacts of climate change and the increasing frequency of extreme flood events, rising reservoir water levels pose significant risks to dam safety, as demonstrated by the catastrophic failures of the Edenville and Sanford Dams in Michigan [1]. These events underscore the urgent need for reliable and efficient methods to predict embankment erosion potential for earthen dam and spillway design and risk management.

Erosion prediction techniques are often time-consuming, costly, and limited in scalability, requiring extensive site-specific field and laboratory testing [2–5]. To address these limitations, Ghimire and Schulenberg [6] developed the Windows Dam Analysis Modules

(WinDAM) C screening model, which enables the identification of multiple at-risk spillways within a single simulation. Their study revealed that approximately 75 percent of the analyzed dams were at severe risk of erosion, primarily linked to narrower spillway widths. However, several other factors influence the embankment breach, such as embankment material, slope, dam height, and reservoir stage [7]. Among the parameters previously examined, spillway width and stream power have emerged as the most dominant predictors of erosion [8], surpassing other indicators such as the headcut erodibility index, representative soil diameter, and spillway channel slope ratio. Our goal in this paper is to evaluate whether these two parameters aid the prediction of spillway erosion using a machine learning framework.

Spillway width represents a geometric design parameter, whereas stream power is a hydraulic parameter that quantifies the energy per unit time and area exerted by flowing water on the spillway channel, a process closely tied to extreme flood dynamics. Stream power, expressed as the rate of energy dissipation, acts as a hydraulic stressor that initiates and accelerates erosion [9]. Given their interdependence, variations in spillway width and stream power jointly influence overall spillway performance and erosion susceptibility. Yet, despite their importance, the combined predictive role of these parameters remains under-researched in the literature, and our paper fills this gap by using an existing dataset and employing various machine learning methods.

Methodologically, the application of machine learning in civil engineering and infrastructure has expanded rapidly in recent years [10–12]. However, studies on dams and spillway performance, particularly auxiliary spillways, have primarily relied on traditional statistical methods, such as linear regression models, deterministic threshold-based criteria, and parametric reliability analyses, without emphasizing model-building and inherent nonlinearities. This study contributes to the growing body of literature on machine learning techniques by employing and extending beyond a logistic regression framework, a widely used statistical approach for predicting dam erosion and other engineering events [9,13] to identify key parameters influencing spillway erosion. While earlier studies focused primarily on the physical mechanisms of spillway erosion and hydraulic loading [9,13], more recent work emphasizes that dam failures arise from interactions among technical, organizational, and human factors [14].

In this context, the present study adopts a machine learning approach to model and predict the erosion potential of auxiliary spillways using the two most sensitive variables identified in prior research: spillway width and stream power. This is a more focused approach compared to prior studies that emphasize spillway erosion and embankment performance as a broad set of geotechnical, hydraulic, and operational factors, including soil type and material properties, degree of compaction, spillway geometry and slope, duration and intensity of overtopping events, and maintenance and operational history [15–17]. While our study primarily targets auxiliary spillway erosion, the model applies broadly to earthen dams, embankments, and spillways. Accordingly, these terms are used interchangeably throughout the remainder of the paper.

To address the research question, we focus on a categorical variable with a binary outcome. Based on the sample of available dams and the information collected, a value of 1 is assigned to spillways with erosion and 0 to those without erosion (additional details are provided in the data description section). Accordingly, logistic regression is an appropriate statistical model for predicting the likelihood of occurrence between two mutually exclusive events, namely, whether erosion will occur, based on the given parameters. This methodological choice is consistent with existing literature on related applications. For example, a general linear model combined with a novel machine learning

https://doi.org/10.3390/w18030373

approach has been applied to model the gully erosion susceptibility of Golestan Dam in Iran using training and test samples [18]. Similarly, piping erosion in the Zarandieh watershed of Markazi Province, Iran, has been evaluated using Bayesian generalized linear models with machine learning applications [19].

Three supervised learning algorithms are applied to classify spillways as eroded or non-eroded cases: logistic regression, Support Vector Machine, and Random Forest. Model performance is evaluated using accuracy and receiver operating characteristic (ROC) metrics, and the best-fit model is further assessed through Bayesian probability analysis to quantify post-prediction reliability. By integrating data-driven algorithms with probabilistic reasoning, this study aims to improve the predictive accuracy, efficiency, and interpretability of erosion risk assessment under elevated stream power conditions, which serves as a proxy associated with extreme hydrologic events. The findings contribute to improved dam safety management by offering a scalable and evidence-based framework for spillway design and operation. As such, this paper advances spillway erosion prediction by demonstrating the application of machine learning and Bayesian probability within a decision science framework, offering a practical data-driven approach for assessing erosion potential when detailed site-specific information is limited, particularly under elevated hydraulic loading conditions. To the best of our knowledge, existing studies have not applied the specific machine learning framework adopted in this paper to evaluate whether readily available, pre-existing data on stream power and spillway width can be used to predict spillway erosion in the absence of detailed field observations.

2. Theoretical Framework

Existing studies have identified multiple factors contributing to dam and spillway erosion, including hydraulic forces, soil erodibility, and geometric design parameters. The logistic regression model has traditionally been employed to estimate the probability of spillway failure and to serve as a screening tool for erosion risk assessment [9,13]. In these studies, a threshold line was applied between stream power (y-axis) and one or more of the spillway features identified above, as illustrated in Figure 1. This threshold line visually separates spillways with a higher likelihood of erosion (above the line) from those considered non-erodible (below the line).

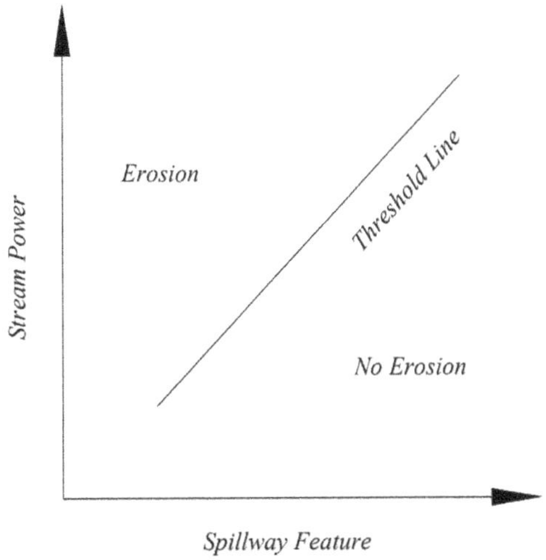

Figure 1. Threshold line between erodible and non-erodible spillways.

 https://doi.org/10.3390/w18030373

The threshold line in this paper is defined using an approach similar to that of Wibowo et al. [9]. In their study, the threshold is established based on stream power and an erodibility index (a measure of spillway "feature"), where a higher erodibility index indicates greater resistance to erosion. By analogy, in our model, a larger spillway width serves as the corresponding feature that provides greater resistance to erosion for a given level of stream power.

Two main approaches have been used in the literature to define this erosion threshold: the United States Department of Agriculture (USDA) threshold line and Annandale's threshold line [20]. The USDA method, derived from empirical field observations, associates allowable stress limits with soil type and vegetative cover, providing a qualitative guideline for erosion potential [21,22]. In contrast, Annandale's approach offers a more quantitative, physically based criterion, incorporating parameters such as stream power, erodibility index, and energy dissipation within a mathematical framework.

While both approaches are valuable for identifying critical thresholds, they rely on empirical or semi-analytical formulations that can be data-intensive and site-specific, limiting their generalizability. In the context of climate-induced extreme events, where hydrologic and hydraulic parameters can change rapidly, such static threshold methods may not fully capture nonlinear or interacting effects. This limitation motivates the adoption of a machine learning framework that can learn complex relationships between variables, such as spillway width and stream power, directly from data. Building on Annandale's energy-based rationale, this study extends the theoretical framework toward a data-driven, machine-learning approach that can dynamically predict erosion potential and enhance dam safety assessment under extreme conditions.

2.1. Stream Power Based on Annandale's Threshold Model

Stream power is the rate of energy dissipation per unit width of flow, which is calculated as:

$$P = \gamma\, q\, \Delta E \tag{1}$$

where P is the stream power per unit width, γ is the unit weight of water (9.81 kN/m^3), q is the unit discharge, which means the discharge per unit width, and is derived as $q = Q/b$, and its unit is = $\frac{\mathrm{m^3/s}}{\mathrm{m}}$ = m^2/s. ΔE is the energy loss per unit weight of water, and is further calculated by using the following formula:

$$\Delta E = \Delta z + (3/2)\, y_c - y_1 - {v_m}^2/2g \tag{2}$$

where, Δz = dam height (m); y_c = critical flow depth (m), used here to represent energy conditions at control points in steep or transitional flow regions; y_1 = downstream flow depth/tailwater depth (m), v_m = flow velocity of jet downstream of a drop after entrainment of water (m/s); g = acceleration due to gravity (m/s^2). Critical depth is used instead of normal depth because it better characterizes energy conditions at spillway chutes and hydraulic drop structures, where supercritical flow is expected. This modeling approach follows Annandale's model [20], who emphasized the use of critical depth in his formulation for evaluating erosion potential. Figure 2 shows the discharge at the drop structure, along with the parameters explained.

https://doi.org/10.3390/w18030373

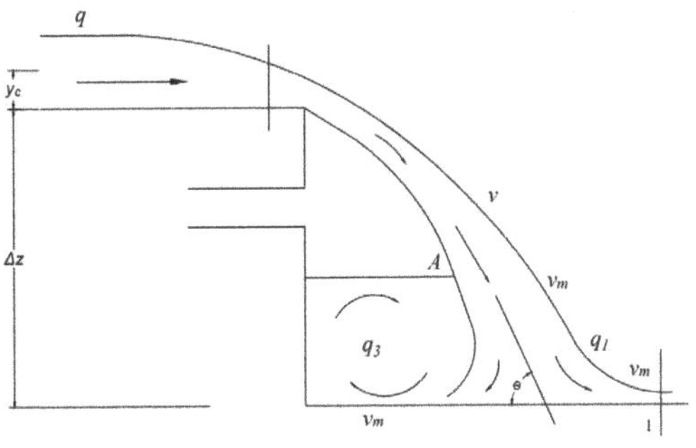

Figure 2. Discharge at a drop structure.

2.2. Unit Discharge (q)

The amount of fluid passing a section of a stream in a unit of time is called unit discharge. Unit discharge in an open channel rectangular flow is computed as follows:

$$q = Q/b \tag{3}$$

where,

q = inflow unit discharge or discharge per unit width in m^2/s
Q = maximum discharge in m^3/s
b = spillway width in m

2.3. Critical Flow Depth (y_c)

As shown in Figure 2, y_c denotes the critical flow depth, and it is calculated as:

$$y_c = (Q^2/(b^2g))^{1/3} = (q^2/g)^{1/3} \tag{4}$$

where,

y_c = critical flow depth in m
g = acceleration due to gravity = 9.81 m/s^2

2.4. Flow Velocity of Jet in Downstream (v_m)

Annandale provides the value of v_m as:

$$v_m = (v/2)\,(1 + cos\theta) \tag{5}$$

where,

v_m = flow velocity of jet in the downstream in m/s
v = velocity of jet downstream of a drop just before entrainment of water, which is:

$$v = Q/(b\,y_c) \tag{6}$$

$$cos\theta = \frac{1.06}{\sqrt{(\Delta z/y_c + 3/2)}} \tag{7}$$

where,

θ = angle of the channel bed (or drop face) relative to the horizontal
Δz = dam height or vertical drop height in m

https://doi.org/10.3390/w18030373

Overall, Equation (6) gives the velocity at a critical flow, and Equation (7) relates the slope geometry to the critical depth.

We use the above parameters to calculate stream power using Annandale's approach. Other parameters included in Figure 2, q_1 (downstream unit discharge), q_3 (trapped unit discharge due to recirculation), and A (point of velocity change) are not directly used in our analysis but are included in the figure for completeness, as originally presented in Annandale [20].

3. Data and Methodology

3.1. Data Description

To achieve the objective of this study, a comprehensive dataset was constructed by collecting information on dams that experienced erosion-related failures, as well as those that did not. The erosion information (our binary outcome variable, coded 1 for erosion cases and 0 for no erosion cases) was obtained from the Association of State Dam Safety Officials (ASDSO) Dam Failures and Incidents Database. The remaining information needed for further analysis (spillway width, dam height, and maximum discharge) was obtained from the National Inventory of Dams (NID). Based on these measures, we calculated the following parameters: Unit discharge (q) as shown in Equation (3); Critical depth (y_c) as shown in Equation (4); Flow velocity of jet in the downstream (v_m) as shown in Equation (5); Velocity of jet downstream of a drop just before entrainment of water (v) as shown in Equation (6); and $cos\theta$ as shown in Equation (7). All these parameters were used to calculate Energy Loss (ΔE) as shown in Equation (2), and ΔE was then substituted into Equation (1) to calculate Stream Power (P) = $\gamma\ q\ \Delta E$. Table 1 summarizes the key variables included in the analysis.

Table 1. Sample of dams with failure incidents with eroded data.

Variable	Erosion Status	N	Mean	Std. Dev.
Spillway Width (m)	Erosion Cases	100	32.39	53.00
	No Erosion Cases	400	88.97	155.79
Stream Power (W/m)	Erosion Cases	100	174,983.90	598,670.00
	No Erosion Cases	400	605,914.00	4,304,426.00

Although the NID enlists over 91,000 dams across the United States, it does not provide the spillway failure information. Therefore, the ASDSO database served as the source for a sample of dams that failed due to erosion-induced processes, such as spillway headcutting or surface degradation. As of May 2020, the ASDSO database documented approximately 1074 dam failure cases, from which a sample of 100 dams was selected, based on data availability. The selected cases specifically included failures attributed to spillway erosion, headcutting, spillway chute failure, or inadequate spillway capacity, aligning with the study's emphasis on embankment erosion mechanisms. These erosion cases were intentionally chosen to represent a range of failure types associated with hydraulic stress during extreme flood events.

For comparison, a control sample of 400 dams was drawn to represent non-eroded spillways. While the selection was not randomized, efforts were made to include a diverse set of dams varying in size, type, and geographic distribution, thereby ensuring heterogeneity within the non-eroded group. The 1:4 ratio of eroded to non-eroded samples (100:400) was selected to balance statistical power with the relatively limited availability of

https://doi.org/10.3390/w18030373

well-documented erosion cases. The resulting dataset of 500 samples provided a balanced and representative basis for predictive modeling.

3.2. *Supervised Machine Learning Models*

To predict the likelihood of spillway erosion under extreme hydrologic conditions, this study applies supervised machine learning algorithms that classify samples as eroded or non-eroded based on physical and hydraulic characteristics. The 500 datasets were split into training (70%) and test (30%) subsets for model training, validation, and comparison. The training set was used to develop the predictive models, while the test set was used to evaluate each model's generalization performance on unseen data.

Three machine learning algorithms were implemented: Logistic Regression, Support Vector Machine (SVM), and Random Forest (RF). We implemented these models in R using the 'caret' package. Specifically, we used createDataPartition() for the train–test split. Model training and tuning were conducted using repeated k-fold cross-validation via trainControl (method = "repeatedcv", number = 10, repeats = 5), and we enabled class probabilities (classProbs = TRUE) and saved cross-validated predictions (savePredictions = "final"). Following standard practice for binary classification, we optimized model performance using the area under the ROC curve (AUC) by specifying summaryFunction = twoClass-Summary and metric = "ROC". We estimated (i) a baseline logistic regression using train (…, method = "glm", family = binomial), (ii) a random forest using train (…, method = "rf", tuneLength = 5), and (iii) an SVM with radial kernel using train (…, method = "svmRadial", preProcess = c ("center", "scale"), tuneLength = 5), where SVM predictors were centered and scaled before model fitting.

3.2.1. Logistic Regression

The logistic regression model provides a probabilistic framework for classifying the binary outcome of erosion. It is particularly effective when the dependent variable is categorical, as in this study, where the outcome is either 1 (eroded) or 0 (non-eroded). The model estimates the probability $p(y)$ that a spillway will erode based on predictor variables—spillway width (b) and stream power (P)—and is expressed as:

$$y = \begin{cases} 1, \ with\ probability\ p \\ 0, \ with\ probability\ 1 - p \end{cases} \tag{8}$$

The linear form of logistic regression occurs in the natural log of odds of the event and is expressed as:

$$ln\ (odds) = \beta_0 + \beta_1 b + \beta_2 P \tag{9}$$

In the above expression, b = spillway width and P = stream power, with corresponding coefficients β_1, β_2 and the regression intercept β_0. The left-hand side is the natural log of the odds, also known as logit. It is defined as the ratio of the probability of an occurrence $p(y)$. An odds ratio represents the probability of a spillway being stable and is expressed as follows:

$$odds = \frac{p(y)}{1 - p(y)} \tag{10}$$

In this case, the probability of the event is calculated as:

$$ln \frac{p(y)}{1 - p(y)} = \beta_0 + \beta_1 b + \beta_2 P \tag{11}$$

https://doi.org/10.3390/w18030373

where,

$$p(y) = \frac{1}{1 + exp\left[-(\beta_0 + \beta_1 b + \beta_2 P)\right]} \tag{12}$$

This framework allows for direct interpretation of the relationship between spillway width, stream power, and erosion probability, where negative coefficients indicate that an increase in width or a reduction in hydraulic energy reduces erosion risk.

3.2.2. Support Vector Machine (SVM)

The SVM algorithm constructs an optimal decision boundary (hyperplane) that separates eroded and non-eroded spillways in a multidimensional space defined by the predictor variables [23]. Using a radial basis function (RBF) kernel, SVM captures nonlinear relationships between spillway width and stream power—relationships that often emerge during extreme flood events when flow conditions deviate from normal operating regimes. Model tuning was performed through repeated 10-fold cross-validation, optimizing the penalty parameter (C) and (σ) to maximize classification accuracy and the area under the ROC curve (AUC).

3.2.3. Random Forest (RF)

The Random Forest model is a combination of tree predictors such that each tree depends on the values of a random vector sampled independently and with the same distribution for all trees in the forest [24]. Each tree is trained on a random subset of data and predictor variables, and the final classification is determined by majority voting across trees. This nonparametric method is particularly advantageous for erosion modeling because it can capture complex, nonlinear, and interacting effects between spillway width and stream power without assuming a specific functional form. Variable importance was computed based on the Mean Decrease in Gini impurity, which quantifies each predictor's contribution to reducing model uncertainty.

Following model training, predictions were generated for the 30% test sample to evaluate classification performance. Accuracy, sensitivity, specificity, and AUC were used as evaluation metrics. Among the three models, the Random Forest algorithm achieved the highest predictive accuracy (82.7%) and the most balanced performance across evaluation metrics. To further assess model reliability beyond standard classification metrics, a Bayesian probability analysis was applied to estimate the posterior probability of actual erosion given a predicted erosion outcome. This integrated approach enhances technical rigor and its applicability to dam safety decision-making under extreme hydrologic stress.

3.2.4. Bayesian Probability Method for Decision Making

The Bayesian probability method provides a probabilistic framework for evaluating the reliability of model predictions [25]. Rooted in Bayes' theorem, it allows the probability of an event to be updated based on new evidence—an approach particularly useful when assessing dam safety under uncertainty. In this study, Bayesian inference is used to estimate the probability that an embankment will actually experience erosion, given that the machine learning model predicts erosion. This post-prediction analysis complements the accuracy measures by incorporating prior and conditional probabilities to evaluate model reliability.

Mathematically, Bayes' theorem is expressed as:

$$p(y|\hat{y}) = \frac{p(\hat{y}|y) \cdot p(y)}{p(\hat{y})} \tag{13}$$

https://doi.org/10.3390/w18030373

where, y represents the samples that actually eroded, $\hat{y}$ represents the event that the samples are predicted to erode, $p(y)$ is the prior probability of erosion in the dataset, $p(\hat{y}|y)$ is the sensitivity, or the probability that the model predicts erosion given that erosion occurred, and $p(\hat{y})$ is the overall probability that the model predicts erosion.

This formulation provides a posterior probability $p(y|\hat{y})$, representing the chance that a spillway truly eroded, given its predicted classification. The Bayesian framework is particularly valuable for identifying false positives, cases where erosion is predicted but does not occur, and for quantifying the likelihood that these predictions correspond to actual erosion events. By integrating probabilistic reasoning with machine learning models, the study provides a more robust assessment of model reliability, prediction confidence, and the real-world risk of spillway erosion under extreme hydrologic events.

4. Results and Discussion

4.1. Model Comparison and Performance Evaluation

Three supervised machine learning algorithms—Logistic Regression, Support Vector Machine (SVM), and Random Forest (RF)—were applied to classify spillways as eroded or non-eroded based on their geometric and hydraulic properties, that is, spillway width and stream power.

As shown in Table 2, the Random Forest model demonstrates the strongest predictive performance among the evaluated approaches, achieving an accuracy of 82.7% and the most balanced classification outcomes, with a sensitivity of 92.5%, specificity of 43.3%, and an AUC of 0.82. The SVM model exhibits moderate discriminative ability (accuracy = 78.7%; AUC = 0.72), whereas the logistic regression model shows relatively low predictive power (accuracy = 75.3%; AUC = 0.54). The comparatively lower AUC values for logistic regression and SVM suggest limitations in capturing nonlinear and interaction effects between spillway width and stream power. In contrast, the Random Forest model more effectively represents these complex relationships, resulting in superior overall performance and justifying its use for subsequent analysis.

Table 2. Comparison of machine learning model performance metrics.

Model	Accuracy	Sensitivity	Specificity	AUC
Logistic Regression	0.753	0.000	0.942	0.542
Support Vector Machine (SVM)	0.787	0.100	0.958	0.721
Random Forest	0.827	0.433	0.925	0.817

4.2. Random Forest Model Assessment Using Confusion Matrix

The confusion matrix for the Random Forest model on the test dataset is presented in Table 3.

Table 3. Confusion matrix for random forest model (test dataset).

Predicted	Actual	
	No Erosion	Erosion
No Erosion	111 (*TN*)	17 (*FN*)
Erosion	9 (*FP*)	13 (*TP*)

The model correctly classified 111 true non-eroded (*TN*, non-eroded cases predicted as non-eroded) and 13 true eroded (*TP*, eroded cases predicted as eroded) spillways, with

https://doi.org/10.3390/w18030373

17 false negatives (*FN*, eroded cases predicted as non-eroded) and 9 false positives (*FP*, non-eroded cases predicted as eroded). We can use information from the confusion to calculate the following model performance measures:

$$Accuracy = \frac{TP + TN}{Total} = \frac{13\ +\ 111}{13\ +\ 111\ +\ 9\ +\ 17} = 0.827 = 82.7\%$$

$$Sensitivity = \frac{TP}{TP + FN} = \frac{13}{13 + 17} = 0.433 = 43.3\%$$

$$Specificity = \frac{TN}{TN + FP} = \frac{111}{111 + 9} = 0.925 = 92.5\%$$

While the model demonstrated a high ability to identify non-eroded cases, as reflected in its specificity of 0.925, its sensitivity of 0.433 indicates limited success in detecting eroded spillways. This imbalance in classification performance is consistent with the relatively small number of erosion cases in the dataset.

This outcome highlights a common challenge in dam safety prediction: predicting a non-eroded spillway as eroded (*FP*) affects operational efficiency, whereas predicting an eroded spillway as non-eroded (*FN*) poses substantial safety risks. Given the model's high specificity (92.5%) and lower sensitivity (43.3%), accepting some false positives is a reasonable trade-off because it allows the model to be more cautious. In practical terms, this means the model may occasionally flag a spillway as eroded when it is not, but it is less likely to miss a truly eroded spillway. In dam safety management, this cautious approach is important because identifying a potential erosion problem early is far better than overlooking a dangerous condition.

4.3. Bayesian Probability Analysis

To further interpret the Random Forest model's predictive performance, Bayesian inference was used to estimate the probability that an embankment will actually experience erosion given that the model predicts erosion. This provides a probabilistic interpretation of model reliability that goes beyond traditional classification accuracy. Although this same information can be calculated from the confusion matrix as *precision* ($TP/(TP + FP)$), we present it within a Bayesian probability framework, integrating traditional statistical reasoning with modern machine learning evaluation.

Using Bayes' Theorem, we define: $p(y)$ as the prior probability of erosion in the test sample (30 out of 150 samples = 0.20), $p(1 - y)$ as the prior probability of no erosion (120 out of 150 samples = 0.80), $p(\hat{y} \mid y)$ as the probability that the model predicts erosion when it has actually occurred (13 out of 30 = 0.43, representing sensitivity), and $p(\hat{y} \mid 1 - y)$ as the probability that the model incorrectly predicts erosion when erosion did not occur (9 out of 120 = 0.075, representing the false positives).

Using these values, the overall probability that the model predicts erosion is calculated as:

$$p(\hat{y}) = p(\hat{y}|y)\ .\ p(y) + p(\hat{y}|1 - y)\ .\ p(1 - y)$$

$$p(\hat{y}) = (0.43)(0.20) + (0.075)(0.80) = 0.147$$

Substituting these values into Equation (13), we obtain:

$$p(y|\hat{y}) = \frac{p(\hat{y}|y)\ .\ p(y)}{p(\hat{y})} = \frac{(0.43)(0.20)}{0.147} \approx 0.59$$

https://doi.org/10.3390/w18030373

Therefore, when the Random Forest model predicts erosion, there is approximately a 59 percent probability that the spillway will actually experience erosion. This posterior probability, derived from the 30% test sample, provides an out-of-sample measure of the model's reliability and demonstrates how Bayesian analysis can complement machine learning classification metrics to better inform erosion risk assessments for dam professionals and practitioners.

4.4. Variable Importance and Physical Interpretation

To evaluate the relative contribution of the predictor variables, variable importance was assessed using the Mean Decrease in Gini index in the Random Forest model. As shown in Figure 3, spillway width had a higher importance score (73.07) than stream power (38.38), indicating that geometric characteristics play a more dominant role in predicting spillway erosion than hydraulic intensity alone.

This finding aligns with hydraulic engineering theory, where narrower spillways tend to concentrate flow, increase velocity, and elevate shear stress on spillway surfaces, conditions that accelerate embankment erosion during extreme flood events. The Random Forest model thus confirms that spillway width is the most influential predictor of erosion risk, consistent with the physical mechanisms observed in previous studies [8,9].

4.5. Regression Validation and Direction of Effects

To further verify the direction and statistical significance of the predictors, we rely on the results from the logistic regression (although this is not the best model, we want to analyze the coefficients to evaluate the direction of the impact) based on log-transformed variables for width and stream power. Table 4 shows a negative, statistically significant coefficient for spillway width ($\beta = -0.459$, $p < 0.001$, controlling for stream power), indicating that erosion probability decreases with increasing spillway width. The coefficient for stream power ($\beta = -0.06$, $p = 0.06$) is not statistically significant at the 5% significance level, suggesting a weaker relationship after controlling for width, consistent with the feature importance analysis presented above. Hence, we highlight the impact of the spillway width in the following sections.

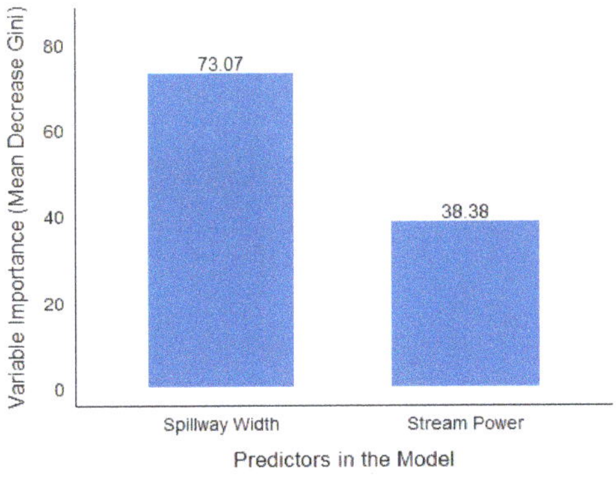

Figure 3. Variable importance from random forest.

https://doi.org/10.3390/w18030373

Table 4. Logistic regression results.

	Dependent Variable: Erosion (0/1)
ln (Spillway Width)	−0.459 ***
	(0.100)
ln (Stream Power)	−0.059
	(0.032)
Constant	0.669
	(0.413)
Observations	500
Log Likelihood	−234.774
Akaike Inf. Criteria	475.547

Note: *** ⇒ $p < 0.01$; standard errors in parentheses

Focusing on the coefficient for spillway width from the full model, the odds ratio for spillway width ($e^{-0.459} = 0.63$) implies that a 1-unit increase in the natural log of width, or approximately a 2.7-fold increase in physical width (since $e^1 = 2.718$), reduces the odds of erosion by 37% (that is, $1 - 0.63 = 0.37$), holding stream power constant. This result empirically supports the conclusion that narrower spillways are more susceptible to erosion because they constrain flow and amplify energy concentration during high-discharge events.

4.6. Decrease in Erosion Rates by Spillway Width

A descriptive analysis of observed erosion rates further validates the model results. When spillways were grouped into quartiles based on width, erosion rates decreased systematically from 30.4% in the smallest (narrowest spillway width) quartile to 6.4% in the largest (widest spillway width) quartile, as shown in Table 5 and Figure 4. This monotonic decline demonstrates a clear geometric control on erosion susceptibility and corroborates the negative relationship between spillway width and erosion probability derived from the machine learning models.

Table 5. Erosion rate by spillway width group.

Width Quartile	Mean Width (m)	Erosion Rate (%)	N
1 (Narrowest Spillway Width)	9.6	30.4	125
2	18.2	29.6	125
3	45.8	13.6	125
4 (Widest Spillway Width)	237.0	6.4	125

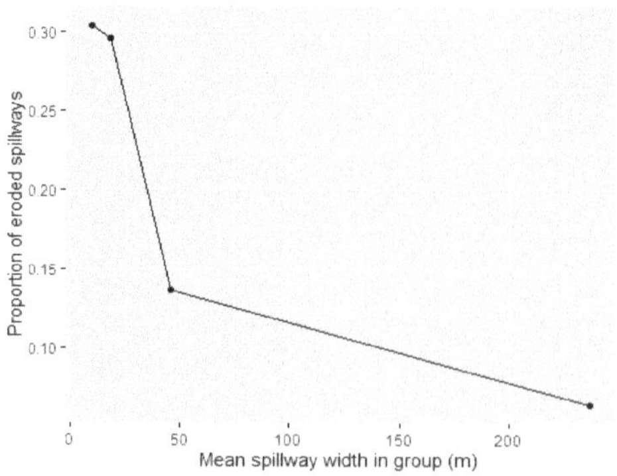

Figure 4. Decrease in erosion rate by spillway width group.

https://doi.org/10.3390/w18030373

4.7. Discussion

In sum, the results demonstrate that machine learning methods, particularly the Random Forest algorithm, can effectively predict spillway erosion potential using minimal yet physically meaningful parameters: spillway width and stream power. Compared with traditional deterministic or threshold-based models [20,22], machine learning approaches offer improved flexibility in handling nonlinear relationships and complex variable interactions.

The integration of Bayesian inference further strengthens model interpretability by translating classification outcomes into conditional probabilities that convey real-world predictive confidence [25]. From a dam safety perspective, the ability to quantify both prediction accuracy and posterior reliability is critical for early warning systems and for prioritizing maintenance resources under climate-driven extreme storm and flood scenarios.

While machine learning models provide accurate classification of erosion events, they primarily describe how well the model distinguishes eroded from non-eroded cases. In contrast, Bayesian probability analysis offers a complementary perspective by quantifying the posterior likelihood that a spillway truly erodes given its predicted classification. This probabilistic measure enhances interpretability, enabling engineers to evaluate the reliability of model predictions under real-world uncertainty. For instance, although the Random Forest model achieved 82.7% accuracy, Bayesian inference indicated a 59% posterior probability of actual erosion when erosion was predicted, offering a more nuanced assessment of prediction confidence for dam safety decision-making.

Ultimately, this study demonstrates that combining data-driven machine learning with probabilistic reasoning can enhance the predictive accuracy and operational relevance of erosion risk assessment, contributing to the broader goal of resilient dam and spillway design in an era of increasing hydrologic extremes.

4.8. Limitations and Future Work

Although this study focuses on hydraulic and geometric predictors, that is, spillway width and stream power, spillway erosion is fundamentally driven by the magnitude and duration of hydraulic events, and erosion may be continuous and progressive and not readily observable. Flood hydrology is, therefore, strongly influenced by rainfall intensity, storm duration, antecedent soil moisture, and catchment characteristics [26,27]. Likewise, rainfall–runoff processes directly control the hydrographs that produce erosive shear stresses on spillway surfaces [14,28]. Incorporating explicit flood and rainfall–runoff information could further enhance erosion prediction and improve linkage to climate-driven hydrologic processes.

A comprehensive erosion assessment, therefore, requires detailed, site-specific data from extensive field investigations, including information on soil type, compaction, spillway slope, duration of overtopping, maintenance history, etc. [15–17]. Such data are not available for inclusion in our study, and collecting them is costly and time-consuming. For this reason, the present study focuses on erosion in earthen auxiliary spillways using stream power and spillway width, two variables consistently available across large datasets and found to be most influential among the erosion-related factors in previous studies, such as Ghimire et al. [8].

In the absence of more comprehensive data, the screening approach developed in this paper is intended to support agency-level decision-making by identifying auxiliary spillways that may warrant further investigation. Once potentially high-risk spillways are flagged, agencies can then conduct detailed, site-specific hydraulic and geotechnical analyses. In this context, the contribution of this study lies in demonstrating how machine

 https://doi.org/10.3390/w18030373

learning methods can be applied for large-scale screening and in providing a foundation for future research that incorporates additional variables and more detailed site information.

Additionally, the predictive performance of the machine learning models is inherently dependent on the size, diversity, and representativeness of the available erosion dataset. If the dataset is biased toward specific dam types, climatic conditions, or geographic regions, the models' generalizability to other spillways under different conditions may be limited. Therefore, model selection and evaluation should be data-driven and context-specific, rather than assuming that the Random Forest model identified in this study will be the best-performing approach in all cases. Moreover, the black-box nature of Random Forest models limits direct physical interpretability of variable effects, reinforcing the need to use such models as complementary screening tools alongside physically based engineering analyses rather than as standalone decision instruments.

5. Conclusions

This study aimed to evaluate the likelihood of earthen auxiliary spillway erosion during extreme events, such as floods, by examining the influence of spillway width and stream power through multiple machine learning techniques, including logistic regression, Random Forest, and Support Vector Machine (SVM). Among these models, the Random Forest algorithm achieved the best predictive performance, demonstrating superior accuracy and balanced classification. The Random Forest analysis identified both spillway width and stream power as significant predictors of erosion occurrence, with spillway width exerting a stronger influence. To further assess the model's predictive reliability, a Bayesian analysis was conducted to estimate the likelihood that the predicted spillways would actually erode.

The Random Forest model achieved an overall accuracy of 82.7%, correctly classifying the majority of spillways in the test sample as either eroded or non-eroded. This performance measure alone, however, reflects an incomplete predictive power of the model. In particular, the model demonstrated strong capability in identifying non-eroded spillways, as evidenced by a high specificity of 92.5%, indicating that most stable spillways were correctly classified. By contrast, the sensitivity of 43.3% suggests more limited success in detecting eroded spillways, with a substantial proportion of erosion cases remaining undetected. This imbalance is largely attributable to the relatively small number of observed erosion events in the dataset, which constrains the model's ability to learn distinctive erosion patterns. Bayesian inference further refined this analysis by indicating a 59% probability that a spillway predicted to erode would actually experience erosion, based on the model's confusion matrix and conditional probability framework. Consequently, while the model is well-suited for screening and ruling out low-risk spillways, its performance in detecting erosion should be interpreted with caution. Nonetheless, these results underscore the usefulness of combining machine learning algorithms with probabilistic reasoning, showing that the integration of physical parameters, specifically spillway width and stream power, with Bayesian post-prediction analysis can support erosion risk assessments. Overall, this research offers practical insights for dam designers and engineers by aiding agency-level decision-making by identifying auxiliary spillways that may require further evaluation, rather than relying on it as a standalone decision instrument. Once potentially high-risk spillways are identified, agencies can then undertake detailed, site-specific hydraulic and geotechnical analyses.

Author Contributions: Conceptualization, S.N.G. and J.S.; methodology, S.N.G., J.S. and S.F.; formal analysis, S.N.G. and J.S.; data curation, S.N.G.; writing—original draft preparation, S.N.G.; writing—

https://doi.org/10.3390/w18030373

review and editing, S.N.G., J.S. and S.F.; visualization, S.N.G. All authors have read and agreed to the published version of the manuscript.

Funding: This research received no external funding.

Data Availability Statement: The data presented in this study are available in [The National Inventory of Dams (NID)] dataset, available at [https://nid.sec.usace.army.mil/nid/] (accessed on 11 November 2025), as well as the [Association of State Dam Safety Officials (ASDSO)] dataset, available at [https://damsafety.org/incidents] (accessed on 11 November 2025).

Conflicts of Interest: The authors declare no conflicts of interest.

References

1. Ghimire, S.N.; Schulenberg, J.W. Impacts of Climate Change on the Environment, Increase in Reservoir Levels, and Safety Threats to Earthen Dams: Post Failure Case Study of Two Cascading Dams in Michigan. *Civ. Environ. Eng.* **2022**, *18*, 551–564. [CrossRef]
2. Powledge, G.R.; Ralston, D.C.; Miller, P.; Chen, Y.H.; Clopper, P.E.; Temple, D.M. Mechanics of Overflow Erosion on Embankments. I: Research Activities. *J. Hydraul. Eng.* **1989**, *115*, 1040–1055. [CrossRef]
3. Wahl, T.L. Prediction of Embankment Dam Breach Parameters: A Literature Review and Needs Assessment. 1998. Available online: https://www.nrc.gov/docs/ML0901/ML090150051.pdf (accessed on 11 November 2025).
4. Amaral, S.; Caldeira, L.; Viseu, T.; Ferreira, R.M.L. Designing Experiments to Study Dam Breach Hydraulic Phenomena. *J. Hydraul. Eng.* **2020**, *146*, 04020014. [CrossRef]
5. Gholami, V.; Sahour, H.; Amri, M.A.H. Soil erosion modeling using erosion pins and artificial neural networks. *Catena* **2021**, *196*, 104902. Available online: https://www.sciencedirect.com/science/article/pii/S0341816220304525?casa_token=Cl_P_ChO_NMAAAAA:ImUqwBphRgbKQofX101xxpOJ58Vjp1NPtp14r7M4W32TgTDuxgomxE8NbgO0vuSIdCwWVvIDvLqO (accessed on 18 January 2026). [CrossRef]
6. Ghimire, S.N.; Schulenberg, J.W. An Automated Screening Technique to Identify At-Risk Auxiliary Spillways: Modified Approach in WinDAM. *J. Hydraul. Eng.* **2021**, *147*, 06021007. [CrossRef]
7. Powledge, G.R.; Ralston, D.C.; Miller, P.; Chen, Y.H.; Clopper, P.E.; Temple, D.M. Mechanics of Overflow Erosion on Embankments. II: Hydraulic and Design Considerations. *J. Hydraul. Eng.* **1989**, *115*, 1056–1075. [CrossRef]
8. Ghimire, S.N.; Schulenberg, J.W.; Neilsen, M.L. Sensitivity Analysis of the Auxiliary Spillway Erosion Based on the Material and Structural Properties. In *Geo-Extreme 2021*; American Society of Civil Engineers: Savannah, GA, USA, 2021; pp. 111–126. [CrossRef]
9. Wibowo, J.L.; Yule, D.E.; Villanueva, E.; Temple, D.M. Earth and rock surface spillway erosion risk assessment. In Proceedings of the Alaska Rocks 2005, the 40th U.S. Symposium on Rock Mechanics (USRMS), Anchorage, Alaska, 25–29 June 2005; p. ARMA-05. Available online: https://onepetro.org/ARMAUSRMS/proceedings-abstract/ARMA05/All-ARMA05/ARMA-05-813/117818 (accessed on 18 January 2026).
10. Das, P.; Saha, J.; Chouhan, P. Effects of labor out-migration on socio-economic set-up at the place of origin: Evidence from rural India. *Child. Youth Serv. Rev.* **2020**, *119*, 105512. [CrossRef]
11. Flynn, S.; Zamanian, S.; Vahedifard, F.; Shafieezadeh, A.; Schaaf, D. Data-Driven Model for Estimating the Probability of Riverine Levee Breach Due to Overtopping. *J. Geotech. Geoenviron. Eng.* **2022**, *148*, 04021193. [CrossRef]
12. Kumar, R.; Rai, B.; Samui, P. Machine learning techniques for prediction of failure loads and fracture characteristics of high and ultra-high strength concrete beams. *Innov. Infrastruct. Solut.* **2023**, *8*, 219. Available online: https://idp.springer.com/authorize/casa?redirect_uri=https://link.springer.com/article/10.1007/s41062-023-01191-w&casa_token=0m9kQB_5wacAAAAA:hoT_2Q9yLgS1UnaDD821HjnYi1HHamHmZM9EFf3APZzLcu6GAUhgxYAswNbAZwV-ZujjaeUxSaWgF52eRQ (accessed on 18 January 2026).
13. Villanueva, E. Risk Assessment of Rock Surface Spillway Erosion Using Parametric Studies. 2007. Available online: https://scholarsjunction.msstate.edu/td/3826/ (accessed on 18 January 2026).
14. Alvi, I.A.; Alvi, I.S. Why dams fail: A systems perspective and case study. *Civ. Eng. Environ. Syst.* **2023**, *40*, 150–175. [CrossRef]
15. Caldeira, L. Internal erosion in dams: Studies and rehabilitation. *Int. J. Civ. Eng.* **2019**, *17*, 457–471. Available online: https://idp.springer.com/authorize/casa?redirect_uri=https://link.springer.com/article/10.1007/s40999-018-0329-5&casa_token=uHaANx3Ecy4AAAAA:Gfyu1ZS4_WBUzV6jRBTghOJ1CaE_FfS1QqLOpZH4qjX6OjnVva3WDi-baztPfd672XqTk-6lvHfW23VuVw (accessed on 18 January 2026). [CrossRef]
16. Gebremariam, F.T.; Tesfay, A.H.; Sigtryggsdóttir, F.G.; Goitom, H.; Lia, L. Overtopping-Induced Embankment Breaching Experiments: State-of-the-Art Review on Measurement and Instrumentation. *Water* **2025**, *17*, 1051. Available online: https://www.mdpi.com/2073-4441/17/7/1051 (accessed on 18 January 2026). [CrossRef]

 https://doi.org/10.3390/w18030373

17. Xiang, S.; Zhou, J.; Qin, T.; Yang, X.; Liao, H. Flume experimental study on the landslide dam breaching behavior with different spillway morphology. *Nat. Hazards* **2025**, *121*, 22211–22236. Available online: https://idp.springer.com/authorize/casa?redirect_uri=https://link.springer.com/article/10.1007/s11069-025-07617-1&casa_token=LaFxkH247roAAAAA:SbD28A1xvchxgUZaRIHu6eChNjui9NifsbX8JxcDjJrnSmqeFK1EgZayWYLN48vQ6vl-ibtxy6Dv5X-egw (accessed on 18 January 2026). [CrossRef]
18. Arabameri, A.; Nalivan, O.A.; Saha, S.; Roy, J.; Pradhan, B.; Tiefenbacher, J.P.; Ngo, P.T.T. Novel ensemble approaches of machine learning techniques in modeling the gully erosion susceptibility. *Remote Sens.* **2020**, *12*, 1890. Available online: https://www.mdpi.com/2072-4292/12/11/1890 (accessed on 18 January 2026). [CrossRef]
19. Band, S.S.; Janizadeh, S.; Saha, S.; Mukherjee, K.; Bozchaloei, S.K.; Cerdà, A.; Shokri, M.; Mosavi, A. Evaluating the efficiency of different regression, decision tree, and bayesian machine learning algorithms in spatial piping erosion susceptibility using ALOS/PALSAR Data. *Land* **2020**, *9*, 346. Available online: https://www.mdpi.com/2073-445X/9/10/346 (accessed on 18 January 2026). [CrossRef]
20. Annandale, G.W. Erodibility. *J. Hydraul. Res.* **1995**, *33*, 471–494. [CrossRef]
21. Lane, L.J.; Renard, K.G.; Foster, G.R.; Laflen, J.M. Development and application of modern soil erosion prediction technology-The USDA experience. *Aust. J. Soil Res.* **1992**, *30*, 893–912. Available online: https://connectsci.au/sr/article-abstract/30/6/893/156675 (accessed on 18 January 2026).
22. USDA. Erodibility parameter selection for soil material horizons (surface detachment coefficient and headcut erodibility index). In *Part 628 Dams-National Engineering Handbook, Appendix 52D*; USDA: Washington, DC, USA, 2011.
23. Granata, F.; Gargano, R.; De Marinis, G. Support vector regression for rainfall-runoff modeling in urban drainage: A comparison with the EPA's storm water management model. *Water* **2016**, *8*, 69. Available online: https://www.mdpi.com/2073-4441/8/3/69 (accessed on 18 January 2026). [CrossRef]
24. Breiman, L. Random Forests. *Mach. Learn.* **2001**, *45*, 5–32. [CrossRef]
25. Ferre, T.P. Being Bayesian: Discussions from the perspectives of stakeholders and hydrologists. *Water* **2020**, *12*, 461. Available online: https://www.mdpi.com/2073-4441/12/2/461 (accessed on 18 January 2026). [CrossRef]
26. Hwang, J.; Lall, U. Increasing dam failure risk in the USA due to compound rainfall clusters as climate changes. *Npj Nat. Hazards* **2024**, *1*, 27. Available online: https://www.nature.com/articles/s44304-024-00027-6 (accessed on 18 January 2026). [CrossRef]
27. Roulo, D.; Pichuka, S. Assessing the effects of extreme rainfall patterns and their impact on dam safety: A case study on Indian dam failures. *Nat. Hazards* **2024**, *120*, 12967–12987. Available online: https://idp.springer.com/authorize/casa?redirect_uri=https://link.springer.com/article/10.1007/s11069-024-06720-z&casa_token=GkNK2y-4yoYAAAAA:gOyjT6fx99mEl6myO5LJvN8awwkorwsuFYP_OSNB6CfpF0hvNb0cx8Qdi-OY1ZGKJhuzt6UB2JcbNDDFWA (accessed on 18 January 2026). [CrossRef]
28. Fell, R. *Geotechnical Engineering of Dams*; CRC Press: Boca Raton, FL, USA, 2005. Available online: https://books.google.com/books?hl=en&lr=&id=Jc7LBQAAQBAJ&oi=fnd&pg=PP1&dq=10.%09Fell,+R.+(2005).+Geotechnical+engineering+of+dams.+CRC+press.&ots=STHnwIAlu0&sig=tB9VtTPTyuXnUEF6rohzvfK4Wyk (accessed on 18 January 2026).

 https://doi.org/10.3390/w18030373

water

MDPI

Article

Enhanced Spatiotemporal Relationship-Guided Deep Learning for Water Quality Prediction

Ruikai Chen [1,†], Yonggui Wang [1,2,*], Hongjun Wang [1], Shaofei Wang [3,†] and Jun Yang [4]

[1] Hubei Key Laboratory of Regional Ecology and Environmental Change, School of Geography and Information Engineering, China University of Geosciences, Wuhan 430074, China
[2] Wuhan Institute of Advanced Technology, Wuhan 430070, China
[3] Yantai Science and Technology Innovation Promotion Center, Yantai 264003, China
[4] Lihe Technology (Hunan) Co., Ltd., Changsha 410205, China
* Correspondence: wangyg@cug.edu.cn
† These authors contributed equally to this work.

Abstract

Water quality prediction serves as a crucial basis for water environment supervision and is of great significance for water resource protection. This study utilized meteorological and water quality data from 40 monitoring stations in the Tuojiang River Basin, Sichuan Province, China. A Gated Recurrent Unit (GRU) model and a Graph Attention Network–Gated Recurrent Unit (GAT-GRU) model were constructed. Furthermore, based on the GAT-GRU framework, an Enhanced Spatio-Temporal Relation-Guided Gated Recurrent Unit (ESRG-GRU) model was developed by incorporating an explicit river network topology and a loss function that is sensitive to extreme values to strengthen spatio-temporal relationships. Water quality predictions were made for all 40 stations, and the performance of the three models was compared. The results show that, during the 7-day forecasting period, the training time of both the ESRG-GRU and the GAT-GRU models was only about 1/40 of that required for the GRU model. In terms of prediction accuracy, the average Nash–Sutcliffe efficiency (NSE) values over the 7-day forecast period were ESRG-GRU (0.7904) > GAT-GRU (0.7557) > GRU (0.6870), while the average root mean square error (RMSE) values were ESRG-GRU (0.0156) < GAT-GRU (0.0168) < GRU (0.0185). Regarding accuracy across different regions and seasons within the river basin, the ESRG-GRU model, guided by enhanced spatio-temporal deep learning, consistently outperformed both the GRU and the GAT-GRU models. This method can effectively enhance both the efficiency and accuracy of water quality prediction, thereby providing support for water environment supervision and regional water quality improvement.

Keywords: deep learning; water quality prediction; spatio-temporal relation-guided; gated recurrent unit (GRU); attention mechanism; Tuojiang River Basin

1. Introduction

Water is essential for the survival and development of human society and serves as a cornerstone of ecosystems. High-quality water is crucial for safeguarding public health, protecting aquatic ecosystems, and ensuring sustainable societal development. However, major river basins worldwide are facing intensifying pollution pressures due to industrialization, urbanization, and climate change [1,2]. Water pollution has already exacerbated water scarcity in more than 2000 sub-basins globally, and it is projected that by 2050, an additional 3 billion people could be living under water-scarce conditions [3]. Due to water pollution issues, the South China region is also confronted with the challenge of seasonal water scarcity [4]. Against this backdrop of escalating global water pollution and persistent water scarcity, the importance of water quality prediction has become increasingly evident. Effective water quality forecasting provides decision-makers with valuable references for water resource allocation and pollution control strategies, while also allowing time to implement measures that can mitigate adverse changes in water quality.

Due to their ability to learn highly complex functional relationships [5], deep learning algorithms can effectively model water quality time-series data characterized by nonlinearity, dynamics, and non-stationarity [6]. This allows them to capture intricate nonlinear interactions between water quality variables and their driving factors, leading to high prediction accuracy and demonstrating significant potential for improving global water quality management [7,8]. Currently, deep learning models are widely favored in the field of water quality prediction [9]. Among them, Recurrent Neural Networks (RNNs) are commonly used for water quality time-series forecasting owing to their regression capabilities and proficiency in handling sequential data [7,10–12]. Long Short-Term Memory (LSTM), a variant of RNN, effectively mitigates the gradient explosion problem inherent in standard RNNs and is therefore increasingly preferred in water quality prediction studies [10,13–16]. Another RNN variant, the Gated Recurrent Unit (GRU), simplifies the LSTM structure while overcoming the vanishing gradient problem, and has seen growing application in water quality prediction in recent years [16–20]. Convolutional Neural Networks (CNNs), widely employed for image recognition and classification through feature-relationship extraction [21], have also been integrated with RNNs and their variants to enhance predictive performance [9,22–24]. Moreover, attention mechanisms have been incorporated into these models to further improve prediction accuracy [25–27]. Nevertheless, most existing studies primarily focus on modeling temporal features of water quality sequences, often neglecting the spatial variability within river networks arising from complex runoff mechanisms. They also rarely account for spatial connectivity between monitoring sites or physical attributes such as river reach length, making it difficult to achieve truly collaborative spatiotemporal prediction.

The primary challenges in water quality prediction involve addressing the long-term dependencies in water quality time-series data and the spatial correlations among upstream and downstream monitoring sites [28]. To overcome the current limitation of deep learning models that often neglect both spatial and temporal features, some studies have adopted hybrid approaches using Convolutional Neural Networks (CNNs) to extract spatial features, which are then combined with temporal features to improve prediction accuracy [29]. However, spatial relationships between actual monitoring sites are not limited to Euclidean distances. Describing interactions between sites using simple pixel adjacency methods may even introduce errors during modeling [30]. In recent years, researchers have begun integrating Graph Neural Networks (GNNs) into deep learning frameworks for water quality modeling [30,31]. The incorporation of GNNs enables models to capture non-Euclidean spatial dependencies [28], which better reflect the topological relationships of upstream and downstream nodes in river networks, thereby enhancing the accuracy of water quality predictions. Several scholars have utilized Graph Attention Networks (GAT) combined with GRU models to predict hydrological and water quality processes in river basins [32,33], achieving promising results. However, most studies rely on static, undirected graphs based purely on distance or fully connected structures to define inter-station relationships [32–34]. This approach overlooks the fundamental physical principle of unidirectional and decay-driven pollutant transport in water bodies,

https://doi.org/10.3390/w18020185

which may lead the model to learn spurious correlations that contradict actual hydrological processes. Developing deep learning models based on real river network topology and hydrological connectivity, while incorporating spatiotemporal features for water quality prediction, is therefore of great significance for improving both model interpretability and forecasting accuracy.

This study focuses on the Tuojiang River Basin in Sichuan Province as the research area. Using a Gated Recurrent Unit (GRU) model as the baseline, a Graph Attention Network–Gated Recurrent Unit (GAT-GRU) model is constructed by incorporating graph attention mechanisms. Furthermore, an Enhanced Spatiotemporal Relationship-Guided Gated Recurrent Unit (ESRG-GRU) model is developed based on the GAT-GRU framework. These models are applied to predict water quality at 40 key monitoring sites across the basin, providing a methodological reference for long-term, multi-site water quality forecasting and early warning.

2. Materials and Methods

2.1. Study Area

The study area is located within the Tuojiang River Basin in Sichuan Province, China, spanning approximately 28°49′ N to 31°43′ N and 103°40′ E to 105°40′ E. The basin covers an area of about 25,800 km^2 (shown in Figure 1). The Tuojiang River originates from the southern slope of Jiuding Mountain in the middle section of the Longmen Mountains. It flows north-to-south, passing through multiple prefecture-level cities in Sichuan, including Chengdu, Deyang, Meishan, Neijiang, Yibin, Ziyang, and Zigong, before eventually joining the Yangtze River in Luzhou. The main stream extends over 600 km and is fed by several tributaries, such as the Yazihe River, Pihe River, Qingbaijiang River, Qiuxihe River, Fuxihe River, and Laixihe River. Topographically, the basin is higher in the north and lower in the east, featuring plains and mountains in the north and low mountains and hills in the south. The region experiences a subtropical humid monsoon climate, with average annual precipitation exceeding 1000 mm, concentrated mainly between June and September.

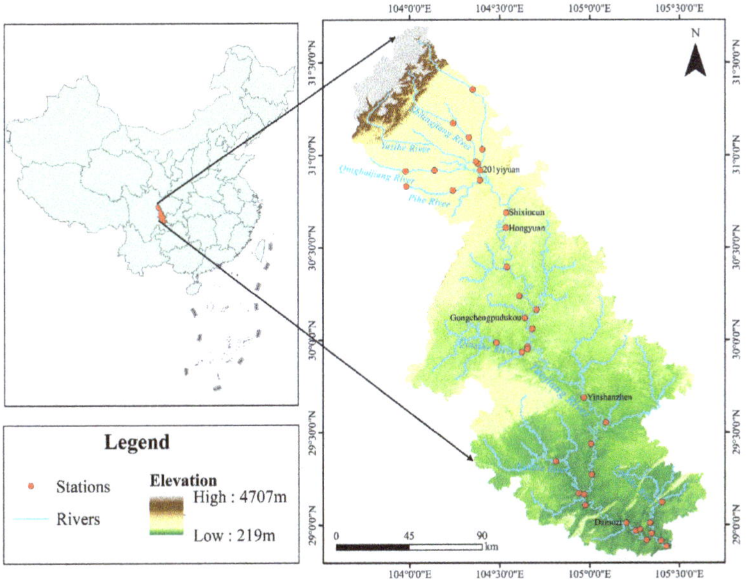

Figure 1. Overview of the study area.

https://doi.org/10.3390/w18020185

As the most densely populated and economically dynamic region in Sichuan Province, the Tuojiang River Basin contributes 30.8% of the province's total economic output and is home to 26.2% of its population [35]. However, rapid industrialization and urbanization have subjected the basin to multiple pollution pressures, including industrial discharge, municipal wastewater, and agricultural runoff. Among these, total phosphorus (TP) pollution constitutes a particularly serious threat to water quality [36].

2.2. Methods

2.2.1. Principle of Models

1. GRU Model

GRU is an optimized variant of the traditional RNN, designed primarily to address the common issues of gradient explosion and vanishing gradients in long sequence processing [37,38]. Compared with the three-gate structure of Long Short-Term Memory (LSTM), GRU reduces the number of gates and model parameters, resulting in higher computational efficiency. This makes it particularly suitable for sequence prediction or analysis tasks with high real-time requirements [39]. The structure of the GRU model is shown in Figure 2.

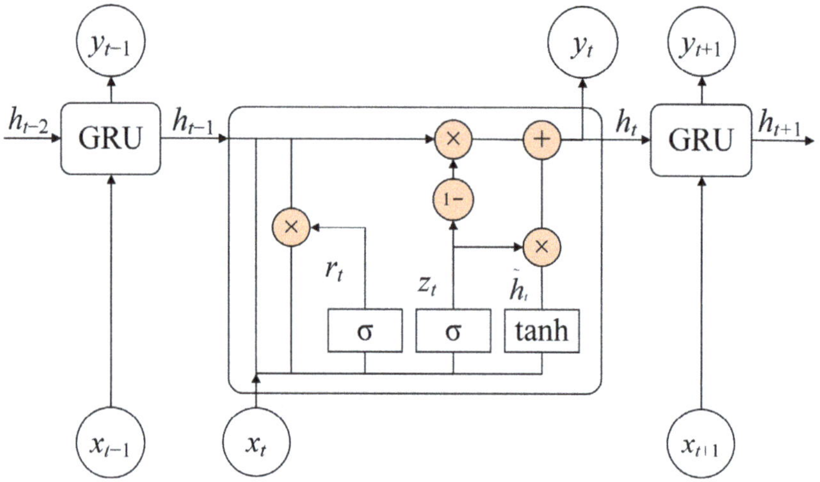

Figure 2. Internal structure diagram of the GRU.

When applying the GRU model to water quality time-series prediction, the task essentially involves multi-step ahead forecasting. Changes in water quality indicators often exhibit strong long-term dependencies and non-stationary characteristics. Predicting over long sequences may lead to the loss of key temporal features, especially making it difficult to capture the coupling relationship between sudden local fluctuations and long-term gradual trends [39]. To address these issues, this study adopts an encoder–decoder framework based on GRU and incorporates an attention mechanism. Through a dynamic weight allocation strategy, the decoder adaptively focuses on the most relevant time steps of the input sequence at each prediction step, as illustrated in Figure 3.

https://doi.org/10.3390/w18020185

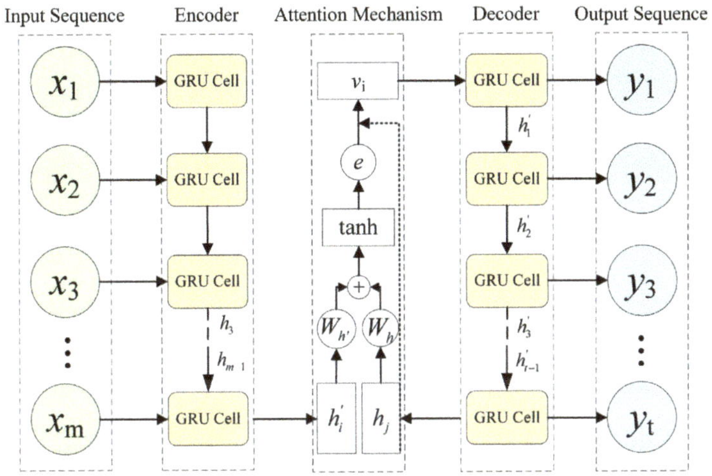

Figure 3. Framework diagram of the GRU model.

2. GAT-GRU Model

Graph Neural Networks (GNNs) have emerged as a pivotal method for spatiotemporal prediction due to their capacity to effectively model high-order dependencies among nodes in graph-structured data [40]. To address the need for modeling dynamic hydrological connectivity among river monitoring sites, this study constructs a GAT-GRU model as the foundational framework. The model adopts an encoder–decoder architecture, integrating Graph Attention Network (GAT) with GRU, with the aim of simultaneously capturing the spatial topological relationships and temporal evolution patterns of the river network.

(1) GAT Layer

The GAT layer enables adaptive feature extraction from graph-structured data by dynamically allocating interaction weights between nodes [41,42]. Unlike traditional Graph Convolutional Networks that rely on fixed-weight convolution, GAT introduces an adaptive attention mechanism, which assigns aggregation weights to neighboring nodes through masked self-attention [43]. As illustrated in Figure 4 and Equation (2), the attention coefficient e_{ij} is computed only for nodes $j \in \mathbb{N}_i$, where $\mathbb{N}_i$ denotes the neighborhood of node v_i in the graph. To eliminate scale variations among the weight coefficients of different nodes, the $soft$max function is applied to normalize the coefficients within the neighborhood (Equation (2)), ensuring they satisfy the properties of a probability distribution. Finally, the output feature $\hat{h}_i$ of node v_i is generated by aggregating the information from its neighbors in a weighted manner, as shown in Equation (3).

$$e_{ij} = a(Wx_i, Wx_j) \tag{1}$$

$$\alpha_{ij} = soft\text{max}_j(e_{ij}) = \frac{\exp(e_{ij})}{\sum_{k \in \mathbb{N}_i} \exp(e_{ik})} \tag{2}$$

$$\hat{h}_i = \sigma\Big(\sum_{j \in \mathbb{N}_i} \alpha_{ij} W x_j\Big) \tag{3}$$

Here, a denotes the shared attention mechanism; e_{ij} represents the importance of node v_j's features to node v_i; α_{ij} denotes the normalized attention weight from node v_j to node v_i; and $a \in \mathbb{R}^{2F'}$ is a learnable weight vector.

https://doi.org/10.3390/w18020185

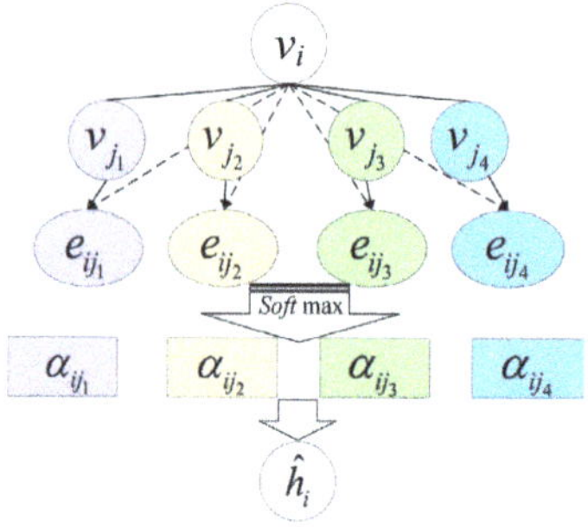

Figure 4. AT attention mechanism.

(2) GAT-GRU Model Architecture

The overall workflow of the GAT-GRU model consists of three core stages, as illustrated in Figure 5. The encoder, composed of multiple GRU layers, is used to extract temporal dynamic features from the input sequence X'. After dimension adjustment, the input data is fed into the GRUs to capture long-term historical dependencies, generating a hidden state sequence H_e. The spatio-temporal attention module enhances features through a cascaded spatial–temporal attention mechanism: the spatial attention layer is based on GAT and uses an adjacency matrix to constrain the interaction range between nodes. The aggregated spatial features S are then passed to the temporal attention layer, which employs a Query–Key–Value mechanism to focus on key historical time steps. Through weighted summation, the final spatio-temporal contextual features C are obtained and serve as the input to the decoder. The decoder adopts a GRU structure symmetric to the encoder. Its initial input is the first time step of the spatio-temporal contextual features C. Forward propagation is then performed through the GRUs (as shown in Equation (4)), computing the output for each time step. Finally, the output is passed through a fully connected layer to produce the final prediction result.

$$h_t = GRU(x_t, h_{t-1}) \tag{4}$$

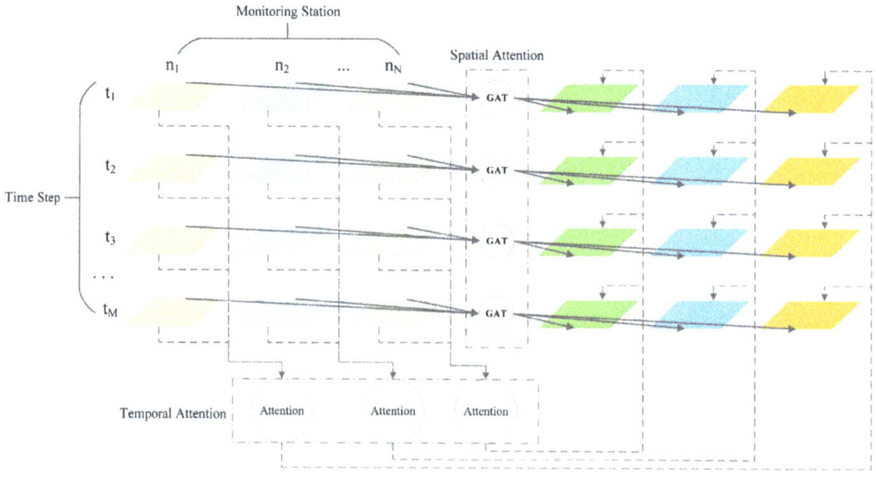

Figure 5. Structure of the GAT-GRU model.

https://doi.org/10.3390/w18020185

3. ESRG-GRU Model

While the architectures of the aforementioned foundational deep learning models, particularly the GAT-GRU model, already possess basic spatiotemporal modeling capabilities, their embedding of the physical mechanisms within hydrological systems remains insufficient. To address this limitation, this study proposes the Enhanced Spatiotemporal-Relationship-Guided Gated Recurrent Unit (ESRG-GRU). This model aims to solve two core issues: first, resolving the "black-box" weight allocation problem in the spatial dimension through explicit river network topology modeling; and second, mitigating the underestimation of peak values in the temporal dimension caused by data non-stationarity through an extremum-sensitive loss function.

(1) Physically Constrained Spatial Modeling

Traditional data-driven models often overlook the physical connectivity and lag effects of water flow. Building upon the GAT-GRU model structure, the ESRG-GRU model explicitly incorporates the hydrological physical attributes of the river network by reconstructing the adjacency matrix and refining the attention mechanism. A digital elevation model (DEM) was employed to delineate basin-wide flow directions using the D8 algorithm [44], thereby constructing a directed and weighted adjacency matrix that represents actual hydrological processes. Based on the coordinates of the monitoring stations, the upstream-downstream relationships between sites were established. As shown in Figure 6, the river reach length between stations is marked based on statistically derived river reach lengths. Finally, the directional connections and river reach lengths between all stations are summarized in an adjacency matrix denoted as $A \in \{0, dis\}^{N \times N}$. If vertices $v_i, v_j \in A$ and $(v_i, v_j) \in E$, then A_{ij} is assigned the river reach length *dis*; otherwise, it is set to 0. This formulation leverages the latent physical information of cross-site transport.

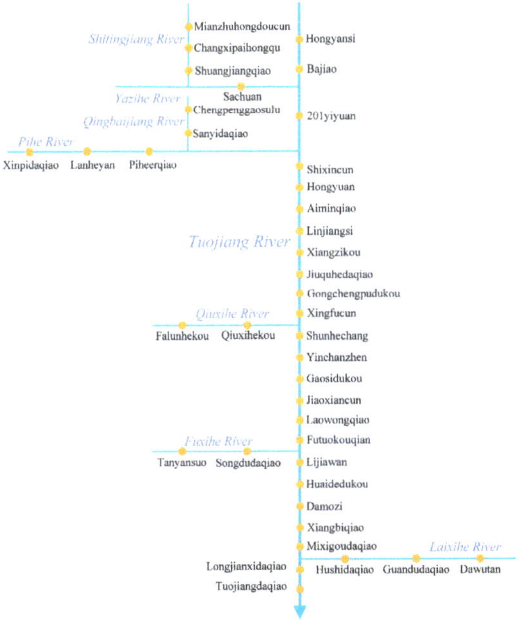

Figure 6. Generalized topological relationship of river network.

 https://doi.org/10.3390/w18020185

Building upon the GAT, a similarity weighting mechanism based on river reach length is introduced. Attention weights are constrained using a logarithmic distance decay, which preserves potential correlations between distant stations while suppressing interference from spatially proximal but hydrologically unrelated nodes (e.g., those belonging to different tributaries) through a masking mechanism. This approach enables interpretable extraction of spatial features. The calculation formula is as follows:

$$e'_{ij} = \frac{1}{1+\log(1+dis)} \tag{5}$$

(2) Extremum-Aware Temporal Optimization

Water quality time series often exhibit significant non-stationarity and right-skewed distribution characteristics (e.g., concentration surges caused by storms or pollutant discharges). Conventional loss functions such as Mean Squared Error (MSE), which assume normally distributed errors, tend to smooth out extremes, leading to systematic underestimation of sudden pollution events. To address this, the present study designs a hybrid loss function $Loss_{total}$, which combines the Nash–Sutcliffe Efficiency (NSE) coefficient with an extremum loss term (Exloss) [45], as shown in Equation (6). The NSE is adopted as the base loss term to evaluate the model's ability to simulate the overall trend of the hydrological process (Equation (7)). An Exloss term derived from generalized extreme value theory is introduced to apply asymmetric penalties according to the skewness of the data distribution (Equations (8) and (9)). Specifically, the model adaptively identifies extreme regions based on dynamic quantiles of the input data (upper quantile 80%, lower quantile 20%). For target values exceeding these thresholds, an asymmetric penalty factor is applied: when the model underestimates a high-concentration extreme, a high penalty weight λ_{up} is imposed; when it overestimates an extreme, a lower penalty weight λ_{down} is applied. Through this asymmetric dynamic adjustment, the model is forced to maintain balanced loss expectations in high-value intervals, thereby improving its ability to capture water pollution events.

$$Loss_{total} = Loss_{NSE} + Loss_{Exloss} \tag{6}$$

$$Loss_{NSE} = \frac{\Sigma(Y-\hat{Y})^2}{\Sigma\left(Y-\bar{Y}\right)^2} \tag{7}$$

$$Loss_{Ex} = \frac{L_{up} + L_{down}}{1 - up_{th} + down_{th}} \tag{8}$$

$$L_{up,down} = \lambda_{up,down}\mathrm{ReLU}\left(ReLU\left(Y, Q_{up,down}(Y)\right) - \mathrm{ReLU}\left(\hat{Y}, Q_{up,down}(\hat{Y})\right)\right) \tag{9}$$

In the formulation, Y and $\hat{Y}$ denote the observed and predicted values, respectively; $\bar{Y}$ is the mean of the observed values; the ReLU activation ensures that penalties are applied only when the prediction error is positive, avoiding unnecessary optimization of negative errors; Q_{up} and Q_{down} represent the upper (80%) and lower (20%) quantile calculations, respectively; up_{th} and $down_{th}$ are the corresponding quantile ratios; and $1 - up_{th} + down_{th}$ serves as a normalization factor to ensure the loss is independent of the chosen quantile proportions.

The training workflow of the ESRG-GRU model is illustrated in Figure 7. Preprocessed geographic data are used to construct the directed weighted adjacency matrix for the spatial attention mechanism and the similarity weights based on river-reach lengths, while the preprocessed water quality and meteorological data serve as multi-dimensional feature

https://doi.org/10.3390/w18020185

inputs to the model. The training procedure is as follows: the multi-dimensional features first undergo high-level feature extraction and fusion through an encoder composed of multiple GRU layers. The encoded features are then aggregated under the guidance of an improved spatial attention mechanism. Subsequently, a temporal attention mechanism further processes the aggregated spatial features via a Query-Key-Value mechanism to focus on key historical time steps. Finally, a weighted summation yields the spatiotemporal contextual features. These features are fed into a decoder that mirrors the GRU structure of the encoder, which then computes and outputs the water quality prediction sequence for future time steps. During the training phase, after each epoch, the value of the hybrid loss function between the model predictions and the true values is computed and recorded. Training stops when either the preset maximum number of epochs is reached or the validation loss fails to decrease for 10 consecutive epochs, at which point the trained model is obtained; otherwise, parameters are updated and the above steps continue.

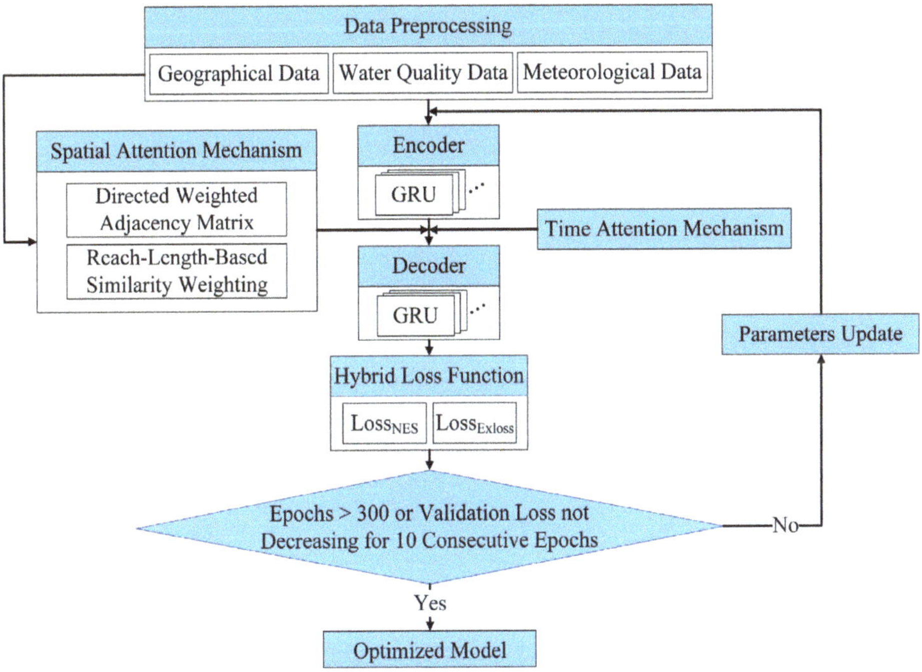

Figure 7. ESRG-GRU training process.

2.2.2. Model Comparative Evaluation Methods

1. Model Efficiency Evaluation Methods

To evaluate model efficiency, we compared the total training times of the GRU, GAT-GRU, and ESRG-GRU models under identical hardware configurations. All three models were trained on the same dataset comprising 40 monitoring stations, and their respective training durations were recorded to assess relative computational efficiency.

2. Model Accuracy Comparative Evaluation Methods

Model accuracy was evaluated using two widely adopted metrics: the Nash–Sutcliffe Efficiency (NSE) [46], commonly applied in hydrological model goodness-of-fit assessment, and the Root Mean Square Error (RMSE) [47], a standard statistical measure used

https://doi.org/10.3390/w18020185

in meteorological, air-quality, and climate studies. The NSE quantifies how well a model reproduces the overall shape of a time series by comparing the mean squared error of predictions to the variance of the observations. Its value ranges from $(-\infty,1]$, with values closer to 1 indicating higher model accuracy, as shown in Equation (10). According to established interpretation guidelines [48], an NSE $\leq$ 0.5 is considered unsatisfactory, 0.5 < NSE < 0.6 is satisfactory, 0.6 $\leq$ NSE $\leq$ 0.8 is good, and NSE > 0.8 is very good. The RMSE is derived by taking the square root of the arithmetic mean of the squared errors (Equation (11)). Its value lies within [0, +∞), and values closer to 0 indicate smaller deviations between predicted and observed values, reflecting higher model precision.

$$NSE = 1 - \frac{\sum_{i=1}^{n}(y_i - \hat{y}_i)^2}{\sum_{i=1}^{n}\left(y_i - \bar{y}\right)^2} \tag{10}$$

$$RMSE = \sqrt{\frac{1}{n}\sum_{i=1}^{n}(y_i - \hat{y}_i)^2} \tag{11}$$

In the equations, n denotes the number of samples, y_i is the observed value, $\hat{y}_i$ is the predicted value, and $\bar{y}$ is the mean of the observed values.

3. Results and Discussion

3.1. Model Construction and Training

3.1.1. Data Preprocessing

The geographic coordinates of the water quality station locations and the digital elevation model (DEM) data were standardized to the WGS 1984 datum with the UTM Zone 47 N projected coordinate system using tools from the GDAL library in Python. Subsequently, the DEM data were processed for hydrological analysis: pits were filled, a flow direction matrix was computed, and flow accumulation was calculated. A threshold of flow accumulation > 3000 was applied to extract the river network. Based on the station distribution and the river flow direction, the hydrological distance (i.e., the stream length along the extracted network) between upstream and downstream monitoring stations was calculated.

The original meteorological data—comprising six variables: precipitation, wind speed, humidity, atmospheric pressure, daily maximum temperature, and daily minimum temperature—were obtained at daily resolution from 1 January 2021 to 30 June 2024, and at hourly resolution from 14 August 2022 to 2 June 2024. Following the principle of nearest-neighbor matching, each water quality station was assigned data from the meteorological station with the shortest Euclidean distance. The hourly dataset served as the primary series, with gaps filled via nearest-neighbor interpolation from the daily data. This produced a complete meteorological time series at a 4 h resolution spanning 1 January 2021 to 30 June 2024.

Due to inconsistencies in time-stamping and missing values in the raw data, the Pandas library in Python was employed for temporal alignment. First, the time column in string format was converted to the datetime type for the period from 1 January 2021 to 30 June 2024. The data were then uniformly resampled to a 4 h interval to ensure temporal consistency. Missing values were filled using linear interpolation in Pandas, as given in Equation (12):

$$y_t = y_1 + \frac{(t - t_1)(y_2 - y_1)}{t_2 - t_1} \tag{12}$$

https://doi.org/10.3390/w18020185

where t is the missing time point, t_1 and t_2 are the preceding and following timestamps, y_1 and y_2 are the corresponding observed values, and y_t is the interpolated value.

To prevent large disparities in feature scales (e.g., total phosphorus concentration in the range 0–1 mg/L versus electrical conductivity in hundreds of μS/cm) from impeding the convergence of gradient-based optimization algorithms, all input features were normalized using min-max scaling, as shown in Equation (13):

$$\tilde{x} = \frac{x - x_{min}}{x_{max} - x_{min}} \tag{13}$$

where $\tilde{x}$ is the normalized value, x is the original value, and x_{max} and x_{min} are the maximum and minimum values of the sequence, respectively.

3.1.2. Model Training

This study focuses on 40 monitoring stations in the Tuojiang River Basin, Sichuan Province, and constructs three water quality prediction models: GRU, GAT-GRU, and ESRG-GRU. The GRU model was trained individually for each station. In contrast, by incorporating Graph Attention Networks (GAT) to capture spatial dependencies among stations, the GAT-GRU and ESRG-GRU models are capable of generating predictions for all 40 stations simultaneously, requiring only a single model to be trained. To mitigate the impact of random initialization on model performance, each model was trained independently 15 times, and the best result from these runs was selected for comparative analysis.

The input features comprise 15 meteorological and water quality indicators: water temperature (WT), pH, dissolved oxygen (DO), electrical conductivity (EC), turbidity (TUR), permanganate index (PV), ammonia nitrogen (AN), total phosphorus (TP), total nitrogen (TN), precipitation, wind speed, humidity, atmospheric pressure, daily maximum temperature, and daily minimum temperature. The input data have a temporal resolution of 4 h. After filling missing values via linear interpolation and normalizing all variables, each station contains 7662 data points. The output target is the concentration of total phosphorus (TP). The dataset was divided into training, validation, and test sets in a ratio of 6:2:2. Forecast horizons of 1, 3, and 7 days were selected, corresponding to input lengths of 12, 18, and 42 time steps, and output lengths of 6, 18, and 42 time steps, respectively.

Model training employed the Adam optimizer, which combines the advantages of AdaGrad and RMSProp, dynamically adjusts the learning rate, and offers benefits such as low memory usage, fast convergence, and high computational efficiency [49]. The initial learning rate was set to 0.001, with a decay strategy that reduces the rate if the validation loss does not decrease over several consecutive epochs. The batch size was 64, the hidden layer dimension was 32, the number of hidden layers was 2, and the maximum number of epochs was 300. To prevent overfitting, an early stopping mechanism was applied, terminating training if the validation loss showed no improvement for 10 consecutive epochs. All experiments were conducted under identical hardware conditions and with the same level of software optimization. Model training was performed on a Windows 10 Pro 22H2 operating system. The hardware environment was configured as follows: an Intel (R) Core (TM) i7-8700 CPU @ 3.20 GHz, an NVIDIA GeForce RTX 2080 GPU (8 GB), and 16 GB of DDR4 RAM. The deep learning framework used was PyTorch 1.10.1 with CUDA 11.3, and Python version 3.9. During training, no parallel computing, memory optimization, or compilation optimization measures were enabled.

https://doi.org/10.3390/w18020185

3.2. Model Training Efficiency Comparison

By comparing the training times of the three models, as shown in Figure 8, the GRU model required approximately 20 to 50 times longer to train across the 40 stations than the other two models. The training time of the GAT-GRU model for the 7-day forecast horizon was even less than 1/40 of that of the GRU model. This is because the GRU model must be trained separately for each station, whereas the GAT-GRU and ESRG-GRU models incorporate a Graph Attention Network (GAT) to capture spatial dependencies between stations, enabling parallel processing of data from all 40 stations within a single model, which significantly improves training efficiency. Compared to the GAT-GRU model, the ESRG-GRU model involves reconstructing the spatial adjacency matrix and introducing a novel hybrid loss function, resulting in increased training times of 31.567% (1-day), 6.230% (3-day), and 31.922% (7-day), respectively. Relative to the training time of the GRU model (taken as 1/40 of its actual duration), the ESRG-GRU model showed increases of 85.841% (1-day), 13.792% (3-day), and 11.837% (7-day). As the forecast horizon lengthens, the training efficiency of the ESRG-GRU model also progressively improves compared to that of the GRU model. These results demonstrate that the ESRG-GRU and GAT-GRU models, which incorporate graph structure and topological relationships, effectively utilize spatial correlations among stations, enable parallel computation across sites, and avoid the efficiency loss caused by repeated calculations in single-station models. Thus, they far outperform the GRU model in multi-station, long-term prediction efficiency, indicating that the spatiotemporally guided deep learning model incorporating graph structures holds considerable potential and value for practical applications involving long-term, synchronous multi-station forecasting.

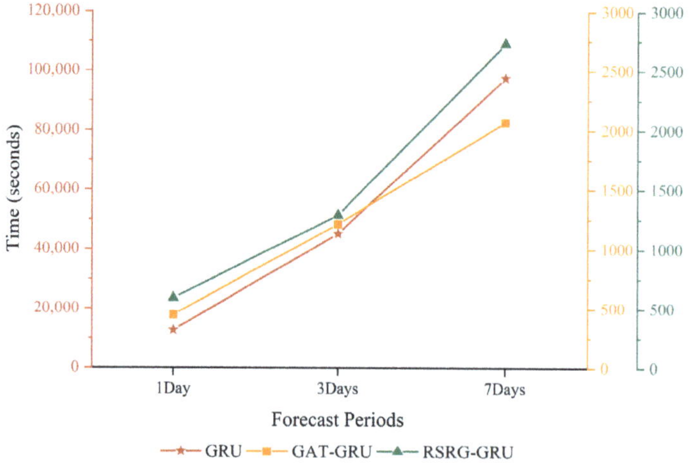

Figure 8. Training durations of models with different forecast periods.

3.3. Comparative Analysis of Model Accuracy

When comparing accuracy across different forecast horizons, our findings align with those of other studies: as the time step increases, the accuracy of the same model at the same station gradually decreases [38,50,51]. The decline in accuracy with longer forecast horizons can be attributed to two main factors. First, longer output sequences require the model to handle more extended and complex time-series data, where noise tends to accumulate more easily, thereby interfering with the model's ability to understand intricate temporal relationships and leading to reduced prediction accuracy [52,53]. Second, as the input sequence length

 https://doi.org/10.3390/w18020185

increases, the dependencies between the beginning and end of the sliding-window-processed data become weaker, making it harder for the model to capture correlations across distant time steps, which further exacerbates the decline in accuracy. To evaluate the improvement in model accuracy, we take the 7-day forecast horizon as an example for analysis.

3.3.1. Comparative Analysis of Model Accuracy: Spatial Pattern

In the 7-day forecasting period, as shown in Figure 9, the accuracy of the three models exhibits pronounced spatial heterogeneity across the study area. The 40 stations are divided into upper, middle, and lower reaches based on Shixincun and Yinshan Town as boundaries: stations above Shixincun are classified as upper reach, those at or below Yinshan Town as lower reach, and the remainder as middle reach.

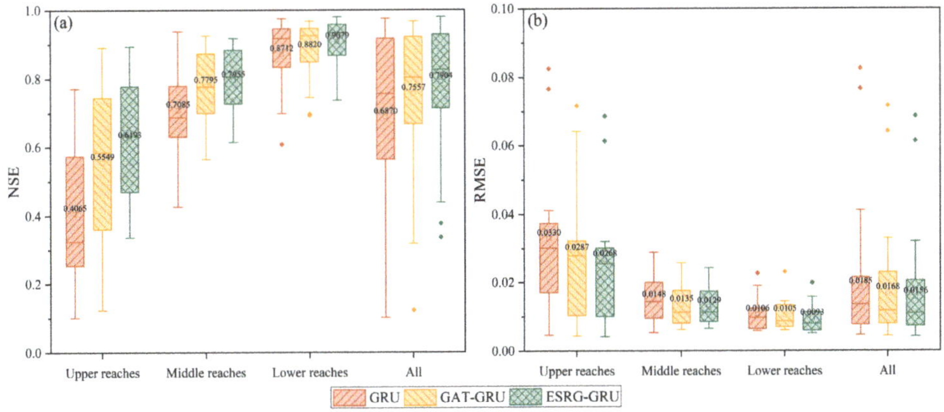

Figure 9. Comparison of the accuracy of different models in different regions: (**a**) NSE of different models in different regions; (**b**) RMSE of different models in different regions.

Across the entire basin, the average NSE values of the three models are 0.7904 (ESRG-GRU), 0.7557 (GAT-GRU), and 0.6870 (GRU), all falling within the "good" performance range. In the lower reach, the average NSE of all three models exceeds 0.8, with the ESRG-GRU model even surpassing 0.9. In contrast, in the upper reach, the average NSE of both the GAT-GRU and GRU models remains below 0.6, corresponding to "satisfactory" and "unsatisfactory" levels, respectively. The average RMSE values across the whole basin are all below 0.02—specifically, 0.0156 (ESRG-GRU), 0.0168 (GAT-GRU), and 0.0185 (GRU). However, in the upper reach, the average RMSE of all three models exceeds 0.02, with the GRU model reaching 0.033. These results indicate that incorporating physical spatiotemporal mechanisms effectively enhances the model's ability to represent long-sequence data. Related studies also suggest that spatial information can improve a model's capacity to learn complex patterns from data, as well as its generalization ability and prediction accuracy [54].

When comparing different regions using the same model, the average NSE consistently follows the order lower reach > middle reach > upper reach, while the average RMSE shows the opposite trend: lower reach < middle reach < upper reach. This indicates that, spatially, the models capture the dynamic variations in water quality more effectively in the lower reach than in the upper reach, leading to more accurate predictions downstream. The lower prediction accuracy for the upper reach may be attributed to the strongly nonlinear and abrupt characteristics of water quality parameters in that region. The more rugged mountainous terrain in the upper reach makes the area more susceptible to localized heavy rainfall and direct surface runoff, which increases input data uncertainty and makes it more challenging for

 https://doi.org/10.3390/w18020185

models to capture underlying water quality patterns, thereby limiting predictive performance. In contrast, flow conditions are more stable in the middle and lower reaches, where water quality varies less abruptly, making it easier for models to learn and forecast.

When comparing the three models within the same region, the average NSE consistently ranks as ESRG-GRU > GAT-GRU > GRU, and the average RMSE as ESRG-GRU < GAT-GRU < GRU. This indicates that the ESRG-GRU model achieves the best overall prediction performance across stations in each region. In the 7-day long-term forecast, both the GAT-GRU and ESRG-GRU models demonstrate higher accuracy than the GRU model, likely because they incorporate graph-structural features, which provide an advantage in capturing long-term temporal dependencies [30]. The ESRG-GRU model, in particular, strengthens spatiotemporal dynamic modeling through physically meaningful graph structures. When handling long sequences, its built-in extremum-sensitive penalty and spatially weighted mechanisms help filter out noise interference to some extent, thereby maintaining more stable prediction accuracy.

The distribution of NSE and RMSE values for selected representative stations is shown in Figures 10 and 11. The accuracy performance of upstream stations is consistent with the regional patterns observed across the study area. The NSE values at upstream sites generally follow the order ESRG-GRU > GAT-GRU > GRU, and within the same tributary, higher NSE values are observed at downstream stations compared to upstream ones, while RMSE values show the opposite trend. Further upstream, the increasing proportion of mountainous terrain enhances river instability and susceptibility to extreme events, which intensifies nonlinear fluctuations in water quality [55,56], posing greater challenges for model learning and prediction.

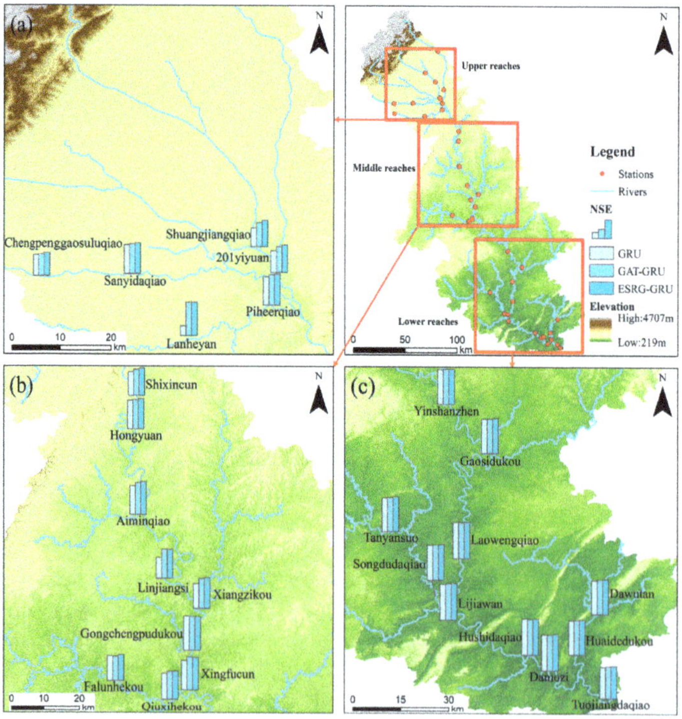

Figure 10. Comparison of NSE values at typical sites of different models: (**a**) NSE values of different models for typical upstream regional sites; (**b**) NSE values of different models for typical midstream regional sites; (**c**) NSE values of different models for typical downstream regional sites.

https://doi.org/10.3390/w18020185

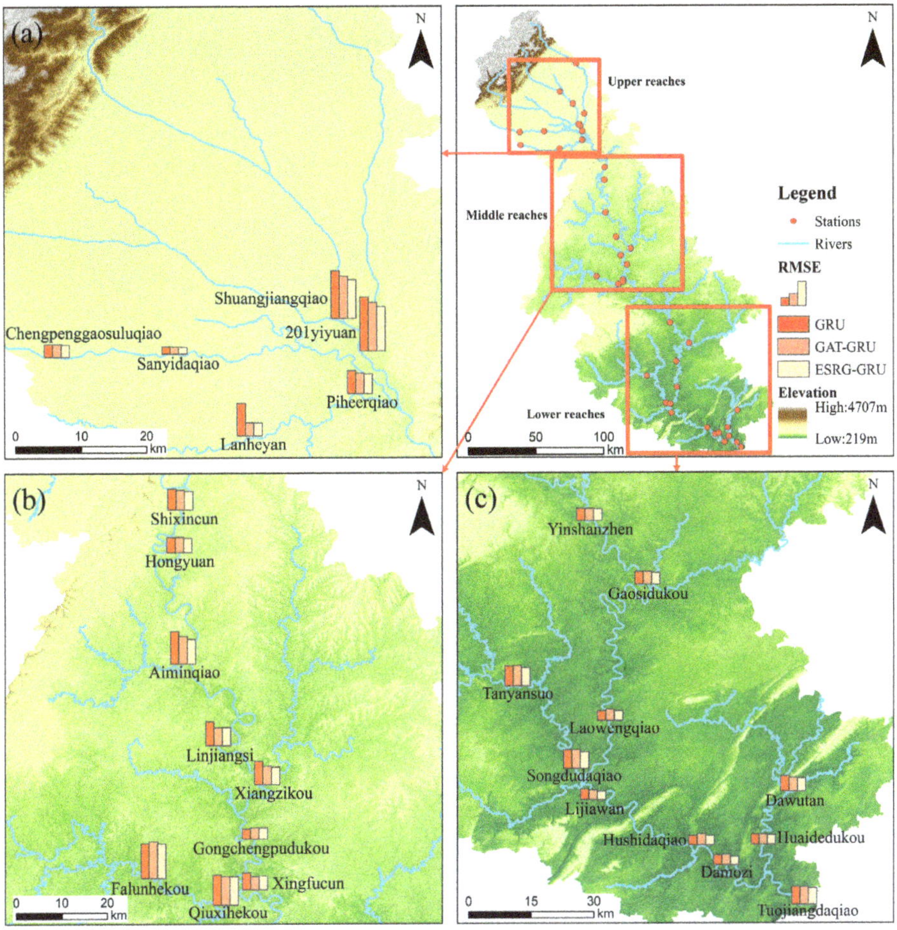

Figure 11. Comparison of RMSE values at typical sites of different models: (**a**) RMSE values of different models for typical upstream regional sites; (**b**) RMSE values of different models for typical midstream regional sites; (**c**) RMSE values of different models for typical downstream regional sites.

The accuracy patterns at midstream representative stations are largely similar to those upstream. However, at some sites—such as Gongchengpudukou—the GRU model slightly outperforms the GAT-GRU and ESRG-GRU models. For instance, at this station, the GRU model achieves an NSE of 0.9165, while the GAT-GRU and ESRG-GRU models attain 0.9043 and 0.9031, respectively. The corresponding RMSE values are GRU (0.00609) < GAT-GRU (0.00674) < ESRG-GRU (0.00675). This may be attributed to the complex hydrological characteristics in certain midstream reaches, such as multiple tributary confluences and sediment-driven pollutant transport processes, which introduce strong local variability. Single-station trained models like the GRU can capture unique temporal variations at individual sites and thus perform well under such conditions. In contrast, graph-based models like GAT-GRU and ESRG-GRU, which learn from all stations simultaneously within the network, face greater challenges in capturing localized complex dynamics.

In downstream areas, the accuracy of the three models is generally similar, with only minor differences overall, though the ESRG-GRU model shows slightly better performance.

https://doi.org/10.3390/w18020185

The NSE values of the three models at representative downstream stations are very close, with ESRG-GRU being marginally higher. The RMSE of the ESRG-GRU model is generally lower than that of the other two models at most downstream sites, while the RMSE of the GAT-GRU model often exceeds that of the GRU model at more stations. This likely results from the gentle terrain and stable flow conditions in the lower reaches, where water quality varies more smoothly, making it easier for models to learn and predict, thus leading to less pronounced differences in model accuracy.

3.3.2. Comparative Analysis of Model Accuracy: Temporal Pattern

We selected November, January, April, and June as representative months for autumn, winter, spring, and summer, respectively, and analyzed the TP concentrations at one typical station from each reach: the upper reach (201yiyuan), the middle reach (Hongyuan), and the lower reach (Damozi).

As shown in Figure 12a, at the upper-reach 201 Hospital station, water quality fluctuated considerably during spring (March–May 2024), ranging approximately between 0.10 mg/L and 0.40 mg/L, while fluctuations were smaller in winter (December 2023–February 2024), with values mostly between 0.15 mg/L and 0.25 mg/L. Figure 12b–e reveal that the predicted values of the three models for typical months exhibit clear seasonal dependence. Overall, the ESRG-GRU predictions are more closely distributed around the 1:1 line, though underestimation occurs more frequently in spring and autumn, with some points showing larger deviations. In autumn, the GRU model tends to underestimate, whereas the GAT-GRU model more often overestimates. During winter, when water quality varies less, all three models produce predictions close to the observed values. In spring, the predictions are the most scattered among all seasons; the GRU and GAT-GRU models show more overestimation compared to the ESRG-GRU model. In summer, the GRU model predictions mostly lie below the 1:1 line, indicating a tendency to overestimate, with some overestimations being relatively large.

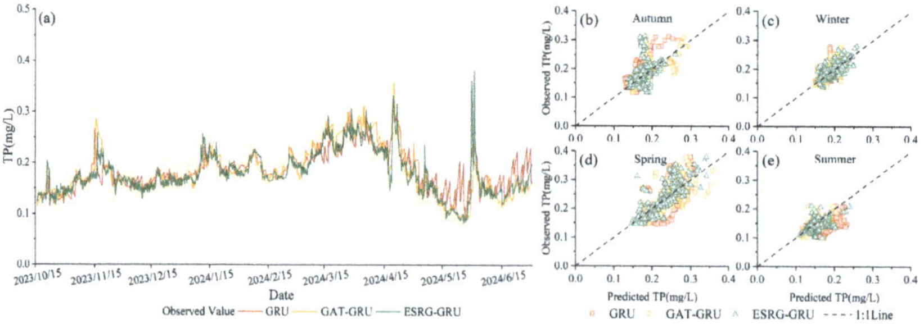

Figure 12. Comparison of observed and measured TP concentration values during the 201yiyuan test: (**a**) Predicted values and observed values of the three models during the test period. (**b**) Predicted values and observed values of the three models in typical autumn months. (**c**) Predicted values and observed values of the three models in typical winter months. (**d**) Predicted values and observed values of the three models in typical spring months. (**e**) Predicted values and observed values of the three models in typical summer months.

As presented in Figure 13a, at the middle-reach Hongyuan station, water quality varied widely in spring (March–May), ranging approximately from 0.05 mg/L to 0.20 mg/L, while fluctuations were smaller in autumn and winter (October 2023–February 2024), mostly between 0.05 mg/L and 0.10 mg/L. Figure 13b–e show that the distributions of the three models are generally similar. In autumn and winter, predictions from all three models cluster around the 1:1 line. In summer, most predictions from the GRU and ESRG-GRU models fall below

https://doi.org/10.3390/w18020185

the 1:1 line, indicating frequent overestimation, whereas the GAT-GRU model has a larger proportion of points above the line compared to the other two. During spring, predictions from all three models are more scattered, with many points deviating from the observed values, suggesting that the prediction of extremes requires further improvement.

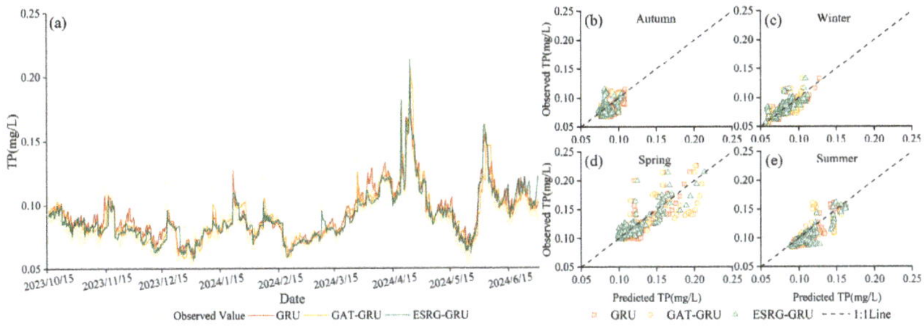

Figure 13. Comparison of observed and measured TP concentration values during the Hongyuan test: (**a**) Predicted values and observed values of the three models during the test period. (**b**) Predicted values and observed values of the three models in typical autumn months. (**c**) Predicted values and observed values of the three models in typical winter months. (**d**) Predicted values and observed values of the three models in typical spring months. (**e**) Predicted values and observed values of the three models in typical summer months.

From both the time-series concentration comparison in Figure 14a and the scatter distributions relative to the 1:1 line in Figure 14b–e, it can be seen that the GRU predictions often lie below the observed values. Except in winter, where the underestimation is less pronounced, the GRU model shows noticeably lower predictions than the other two models in spring, summer, and autumn. Predictions from all three models cluster closely around the 1:1 line in autumn and winter, whereas they are more dispersed in spring and summer. The spring representative month (April) reveals large variations in TP concentration at Damozi; as seen in Figure 14a, the concentration rose from below 0.05 mg/L to about 0.15 mg/L—a threefold increase within a month. In the summer representative month (June), the GAT-GRU model has more points below the 1:1 line, indicating that it overestimates more frequently than the other two models. In contrast, the ESRG-GRU model predictions are more tightly distributed along the 1:1 line and show the smallest deviation from the observed values.

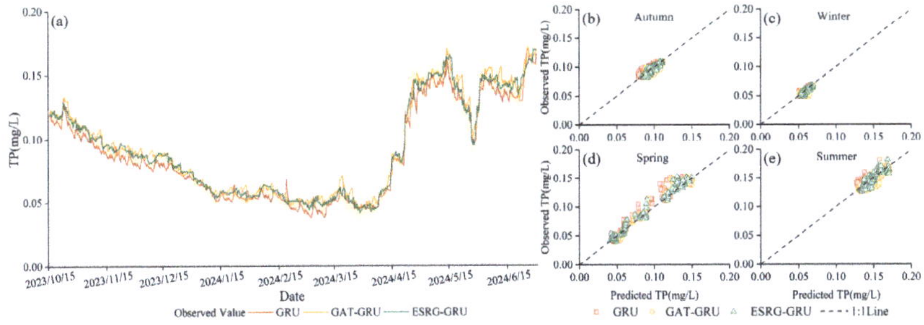

Figure 14. Comparison of observed and measured TP concentration values during the Damozi test: (**a**) Predicted values and observed values of the three models during the test period. (**b**) Predicted values and observed values of the three models in typical autumn months. (**c**) Predicted values and observed values of the three models in typical winter months. (**d**) Predicted values and observed values of the three models in typical spring months. (**e**) Predicted values and observed values of the three models in typical summer months.

 https://doi.org/10.3390/w18020185

In the study area, TP concentrations exhibited smaller variations in winter compared to other seasons, while showing significant fluctuations in spring, indicating clear seasonal variability. This seasonality is reflected in model accuracy: all three models produced predictions closer to observed values in winter, whereas in spring, instances of overestimation and underestimation increased markedly. Some predicted values from the GRU, GAT-GRU, and ESRG-GRU models displayed similar or even overlapping trends. This phenomenon suggests that, despite differences in model architecture, under extreme conditions, the predictive performance of all models may be jointly constrained by sparse data distribution and nonlinear dynamics, resulting in a weaker ability to capture abrupt changes.

The dispersion of predictions increased significantly in spring. On one hand, this can be attributed to the aggravated error accumulation effect in long-sequence forecasting [52]. On the other hand, increased rainfall in late spring and early summer triggers nonlinear hydrological processes [56]. Coupled with seasonal pollutant discharges and transport from human activities, these factors lead to higher TP concentrations and greater variability in spring and summer, resulting in pronounced lag and nonlinearity in water quality responses. The GRU model, lacking the ability to model spatial correlations, struggles to accurately capture the transport and diffusion of pollutants across the basin, leading to larger prediction deviations. Although the ESRG-GRU and GAT-GRU models partially capture these nonlinear relationships through spatiotemporal graph structures, and also show increased dispersion in spring predictions, their results remain closer to the 1:1 line compared to the GRU model. This indicates that while these models do not fully represent the underlying dynamic mechanisms, the physically informed constraints still enhance model stability in long-sequence forecasting [28].

3.3.3. Analysis of Model Accuracy Improvement

Compared to the GAT-GRU model, the ESRG-GRU model achieved an overall improvement in NSE of 11.64% in the upper reach, 2.05% in the middle reach, and 2.94% in the lower reach. Although the accuracy gain was most pronounced in the upper reach, the mean NSE of the ESRG-GRU model there remained relatively low at 0.6193, compared to 0.7955 in the middle reach and 0.9079 in the lower reach, indicating that model performance was still comparatively poor in upstream areas.

Model accuracy is influenced by data quality, feature selection, and model architecture. In terms of data quality, the upper reach data were generally inferior to those from the middle and lower reaches. Taking the TP training data from typical stations in each reach as an example (Figure 15), the TP concentration at the upstream station 201yiyuan exhibited much more pronounced fluctuations than at the middle-reach station Hongyuan and the downstream station Damozi. Moreover, Figure 15a for 201yiyuan shows a considerable number of high-value outliers that contrast sharply with surrounding data points. This is likely due to the station's upstream location, where water quality is more susceptible to localized, episodic events such as heavy rainfall or sporadic pollutant discharges, and may also reflect irregular operation of the automatic monitoring or sampling equipment. In contrast, the downstream station Damozi displayed only a few obvious low- and high-value outliers, with generally stable water quality, indicating clearly superior data quality. Additionally, the meteorological data used during training were derived from daily values interpolated to hourly resolution via nearest-neighbor interpolation. Hydrological responses are faster in upstream areas, and the original temporal resolution of the data may be insufficient to fully capture such rapid dynamics.

https://doi.org/10.3390/w18020185

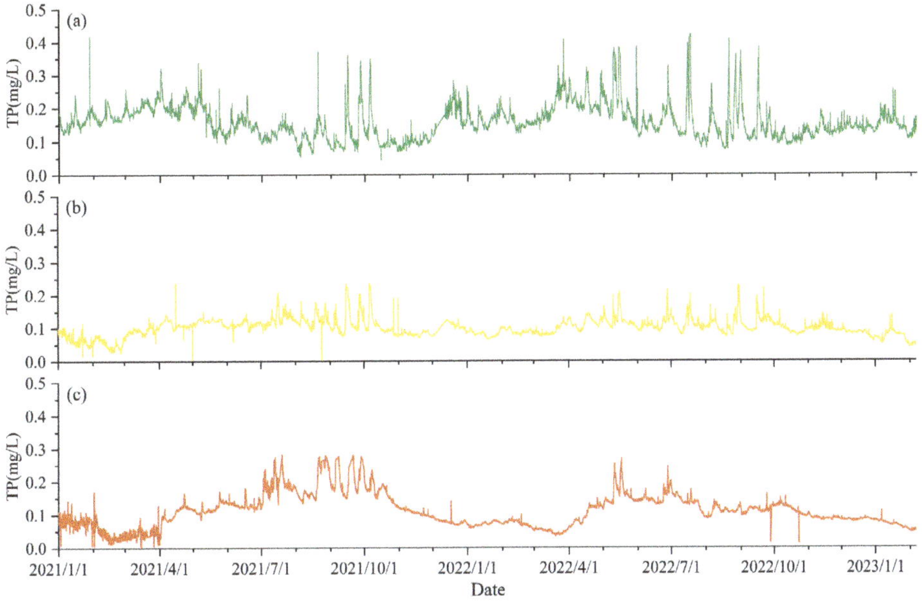

Figure 15. Comparison of model training data from representative stations in different river reaches: (**a**) 201 Hospital station in the upper reach. (**b**) Hongyuan station in the middle reach. (**c**) Damozi station in the lower reach.

Regarding feature selection, limited data availability restricted the inclusion of other key spatial processes—such as land use and point-source pollution locations—and dynamic temporal drivers—such as real-time dam operations, flow rate, velocity, and point-source discharge processes. As a result, deeper embedding of physical process mechanisms into the model remains an area for further improvement.

Architecturally, for the two graph-neural-network-based models—especially the ESRG-GRU model—learning depends on information aggregated from upstream stations. Some stations, however, have few or no upstream stations, meaning that during message passing, they receive little or no information from other water quality sites and must rely primarily on their own limited historical temporal features. This effectively reduces a complex spatiotemporal forecasting problem to an almost purely temporal prediction task, preventing the model from leveraging spatial correlations to correct errors or learn cross-station propagation patterns.

In response to these challenges, we aim to integrate this data-driven framework with land use change data and real-time information on engineering operations, pollutant discharges, and hydrological monitoring. Through architectural design or loss-function constraints, we intend to further develop it into a truly "process-aware" model that more closely couples physical processes with deep learning. This approach is expected to enhance the model's generalization ability and interpretability under non-stationary conditions [7], thereby better addressing water-environment management challenges posed by sudden pollution incidents and changing climate patterns. In the future, we will also validate the feasibility of the proposed method by applying it to multi-parameter joint forecasting in other river basins.

https://doi.org/10.3390/w18020185

4. Conclusions

This study constructed three deep learning models—GRU, GAT-GRU, and ESRG-GRU—to predict water quality in the Tuojiang River Basin, Sichuan Province, China. We recorded the training duration of each model for forecast horizons of 1, 3, and 7 days. The ESRG-GRU and GAT-GRU models, which incorporate graph structure and topological relationships, required significantly less training time than the baseline GRU model, and their efficiency advantage became more pronounced as the forecast horizon increased. Compared to the GAT-GRU model, the ESRG-GRU model reconstructs the adjacency matrix and improves the loss function, leading to increased model complexity. However, its training time did not increase substantially and remains acceptable in practice. Using the NSE and RMSE values for the 7-day horizon to evaluate model accuracy, the ESRG-GRU model achieved a higher average NSE during the testing period compared to the other two models, with improvements of 0.1034 and 0.0347 over the GRU and GAT-GRU models, respectively. Its average RMSE decreased by 15.676% and 7.143% compared to the GRU and GAT-GRU models, demonstrating good predictive performance. Model accuracy showed clear spatiotemporal heterogeneity. Spatially, all three models performed best in the lower reach, followed by the middle reach, and least accurately in the upper reach. Temporally, predictions were closer to observations in winter, while the opposite was true in spring. These results demonstrate the feasibility of the enhanced spatiotemporal relationship-guided deep learning method proposed in this study for water quality prediction. This approach can improve both the training efficiency and prediction accuracy of long-term forecasting models, providing a valuable reference for multi-station water quality prediction over extended periods.

Limited by data availability, other key spatial processes—such as land use and point-source pollution locations—and dynamic temporal drivers—such as real-time dam operations, flow rate, velocity, and point-source discharge processes—were not included as input features. Moreover, the temporal resolution of the acquired meteorological data is relatively low, and data from some upstream stations exhibit considerable fluctuations, which, together, contribute to the lower prediction accuracy at upstream sites compared to middle and downstream stations. In the future, incorporating these additional data sources as model inputs and further integrating physical processes with deep learning are expected to enhance the model's generalizability and interpretability under non-stationary conditions. Since graph-based models inherently offer a clear training-time advantage over separate per-station models, future work will also explore algorithmic innovations to further improve training efficiency. Additionally, the feasibility of the proposed method will be validated through multi-parameter joint forecasting in other river basins.

Author Contributions: Conceptualization, Data Curation, Methodology, Visualization, Writing—Original Draft, Writing—Review and Editing, R.C.; Conceptualization, Funding Acquisition, Supervision, Writing—Review and Editing, Y.W.; Data Curation, Methodology, Visualization, Writing—Review and Editing, H.W.; Funding Acquisition, Investigation, Writing—Review and Editing, S.W.; Investigation, Writing—Review and Editing, J.Y. All authors have read and agreed to the published version of the manuscript.

Funding: The research was jointly supported by the Scientific Research Plan Guiding Project of the Department of Education of Hubei Province (No. B2023244), the Hubei Provincial Natural Science Foundation (No. 2024AFD371, 2024AFD402), and the Science and Technology Major Project of Hubei Province, China (No. 2023BCA003).

Data Availability Statement: The data presented in this study are only available on request from the corresponding author due to privacy restrictions on data sharing.

https://doi.org/10.3390/w18020185

Conflicts of Interest: Author Jun Yang was employed by the company Lihe Technology (Hunan) Co., Ltd., Changsha 410205, China. The remaining authors declare that the research was conducted in the absence of any commercial or financial relationships that could be construed as a potential conflict of interest.

References

1. Giri, S. Water Quality Prospective in Twenty First Century: Status of Water Quality in Major River Basins, Contemporary Strategies and Impediments: A Review. *Environ. Pollut.* **2021**, *271*, 116332. [CrossRef] [PubMed]
2. van Vliet, M.; Jones, E.; Flörke, M.; Franssen, W.; Hanasaki, N.; Wada, Y.; Yearsley, J. Global Water Scarcity Including Surface Water Quality and Expansions of Clean Water Technologies. *Environ. Res. Lett.* **2021**, *16*, 024020. [CrossRef]
3. Wang, M.; Bodirsky, B.L.; Rijneveld, R.; Beier, F.; Bak, M.P.; Batool, M.; Droppers, B.; Popp, A.; van Vliet, M.T.H.; Strokal, M. A Triple Increase in Global River Basins with Water Scarcity Due to Future Pollution. *Nat. Commun.* **2024**, *15*, 880. [CrossRef]
4. Ma, T.; Sun, S.; Fu, G.; Hall, J.; Ni, Y.; He, L.; Yi, J.; Zhao, N.; Du, Y.; Pei, T.; et al. Pollution Exacerbates China's Water Scarcity and Its Regional Inequality. *Nat. Commun.* **2020**, *11*, 650. [CrossRef]
5. LeCun, Y.; Bengio, Y.; Hinton, G. Deep Learning. *Nature* **2015**, *521*, 436–444. [CrossRef]
6. Zheng, Y.; Zhang, Q.; Zhang, X.; Zhou, Y.; Zhang, Y.; Zhang, T. A Spatial-Temporal Trend-Aware Neural Network Model for Accurate Water Quality Prediction in River. *Water Res.* **2025**, *287*, 124389. [CrossRef] [PubMed]
7. Zhi, W.; Appling, A.P.; Golden, H.E.; Podgorski, J.; Li, L. Deep Learning for Water Quality. *Nat. Water* **2024**, *2*, 228–241. [CrossRef]
8. Haghiabi, A.H.; Nasrolahi, A.H.; Parsaie, A. Water Quality Prediction Using Machine Learning Methods. *Water Qual. Res. J.* **2018**, *53*, 3–13. [CrossRef]
9. Yang, S.; Zhong, S.; Chen, K. W-WaveNet: A Multi-Site Water Quality Prediction Model Incorporating Adaptive Graph Convolution and CNN-LSTM. *PLoS ONE* **2024**, *19*, e0276155. [CrossRef]
10. Fang, P.; Wang, Y.; Zhao, Y.; Kang, J. Analysis of Prediction Confidence in Water Quality Forecasting Employing LSTM. *Water* **2025**, *17*, 1050. [CrossRef]
11. Li, L.; Jiang, P.; Xu, H.; Lin, G.; Guo, D.; Wu, H. Water Quality Prediction Based on Recurrent Neural Network and Improved Evidence Theory: A Case Study of Qiantang River, China. *Environ. Sci. Pollut. Res.* **2019**, *26*, 19879–19896. [CrossRef]
12. Wang, D.; Zhang, C.; Li, A.; Guo, Y.; Zhang, H.; Tan, C. Spatio-Temporal Analysis and Prediction for Raw Water Quality of Drinking Water Source by Improved RNN Algorithm. *J. Water Process Eng.* **2025**, *71*, 107164. [CrossRef]
13. Li, Q.; Yang, Y.; Yang, L.; Wang, Y. Comparative Analysis of Water Quality Prediction Performance Based on LSTM in the Haihe River Basin, China. *Environ. Sci. Pollut. Res.* **2023**, *30*, 7498–7509. [CrossRef] [PubMed]
14. Pyo, J.; Pachepsky, Y.; Kim, S.; Abbas, A.; Kim, M.; Kwon, Y.S.; Ligaray, M.; Cho, K.H. Long Short-Term Memory Models of Water Quality in Inland Water Environments. *Water Res. X* **2023**, *21*, 100207. [CrossRef]
15. Luo, L.; Zhang, Y.; Dong, W.; Zhang, J.; Zhang, L. Ensemble Empirical Mode Decomposition and a Long Short-Term Memory Neural Network for Surface Water Quality Prediction of the Xiaofu River, China. *Water* **2023**, *15*, 1625. [CrossRef]
16. Lee, J.; Lee, J.; Lee, M.; Lee, M.; Kim, Y.; Hyung, J.; Kim, K.; Cha, Y.; Koo, J. Development of a Short-Term Water Quality Prediction Model for Urban Rivers Using Real-Time Water Quality Data. *Water Supply* **2022**, *22*, 4082–4097. [CrossRef]
17. Li, W.; Wu, H.; Zhu, N.; Jiang, Y.; Tan, J.; Guo, Y. Prediction of Dissolved Oxygen in a Fishery Pond Based on Gated Recurrent Unit (GRU). *Inf. Process. Agric.* **2021**, *8*, 185–193. [CrossRef]
18. Yang, H.; Liu, S. Water Quality Prediction in Sea Cucumber Farming Based on a GRU Neural Network Optimized by an Improved Whale Optimization Algorithm. *Peer J. Comput. Sci.* **2022**, *8*, e1000. [CrossRef]
19. Xu, J.; Wang, K.; Lin, C.; Xiao, L.; Huang, X.; Zhang, Y. FM-GRU: A Time Series Prediction Method for Water Quality Based on Seq2seq Framework. *Water* **2021**, *13*, 1031. [CrossRef]
20. Zhang, Y.; Liu, L.; Zhang, S.; Zou, X.; Liu, J.; Guo, J.; Teng, Y.; Zhang, Y.; Duan, H. Monitoring and Warning for Ammonia Nitrogen Pollution of Urban River Based on Neural Network Algorithms. *Anal. Sci.* **2024**, *40*, 1867–1879. [CrossRef]
21. Zhendong, Z.; Hui, Q.; Liqiang, Y.; Yongqi, L.; Zhiqiang, J.; Zhongkai, F.; Shuo, O.; Shaoqian, P.; Jianzhong, Z. Downstream Water Level Prediction of Reservoir Based on Convolutional Neural Network and Long Short-Term Memory Network. *J. Water Resour. Plan. Manag.* **2021**, *147*, 04021060. [CrossRef]
22. Tian, Q.; Luo, W.; Guo, L. Water Quality Prediction in the Yellow River Source Area Based on the DeepTCN-GRU Model. *J. Water Process Eng.* **2024**, *59*, 105052. [CrossRef]
23. Haq, K.; Harigovindan, V. Water Quality Prediction for Smart Aquaculture Using Hybrid Deep Learning Models. *IEEE Access* **2022**, *10*, 60078–60098. [CrossRef]
24. Mei, P.; Li, M.; Zhang, Q.; Li, G.; Song, L. Prediction Model of Drinking Water Source Quality with Potential Industrial-Agricultural Pollution Based on CNN-GRU-Attention. *J. Hydrol.* **2022**, *610*, 127934. [CrossRef]

 https://doi.org/10.3390/w18020185

25. Chen, H.; Yang, J.; Fu, X.; Zheng, Q.; Song, X.; Fu, Z.; Wang, J.; Liang, Y.; Yin, H.; Liu, Z.; et al. Water Quality Prediction Based on LSTM and Attention Mechanism: A Case Study of the Burnett River, Australia. *Sustainability* **2022**, *14*, 13231. [CrossRef]
26. Yang, Y.; Xiong, Q.; Wu, C.; Zou, Q.; Yu, Y.; Yi, H.; Gao, M. A Study on Water Quality Prediction by a Hybrid CNN-LSTM Model with Attention Mechanism. *Environ. Sci. Pollut. Res.* **2021**, *28*, 55129–55139. [CrossRef]
27. Bi, J.; Chen, Z.; Yuan, H.; Zhang, J. Accurate Water Quality Prediction with Attention-Based Bidirectional LSTM and Encoder-Decoder. *Expert Syst. Appl.* **2024**, *238*, 121807. [CrossRef]
28. Qiao, J.; Lin, Y.; Bi, J.; Yuan, H.; Wang, G.; Zhou, M. Attention-Based Spatiotemporal Graph Fusion Convolution Networks for Water Quality Prediction. *IEEE Trans. Autom. Sci. Eng.* **2025**, *22*, 1–10. [CrossRef]
29. Nie, Q.; Wan, D.; Wang, R. CNN-BiLSTM Water Level Prediction Method with Attention Mechanism. *J. Phys. Conf. Ser.* **2021**, *2078*, 012032. [CrossRef]
30. Wan, H.; Xiang, L.; Cai, Y.; Xie, Y.; Xu, R. Temporal and Spatial Feature Extraction Using Graph Neural Networks for Multi-Point Water Quality Prediction in River Network Areas. *Water Res.* **2025**, *281*, 123561. [CrossRef]
31. Li, Z.; Liu, H.; Zhang, C.; Fu, G. Real-Time Water Quality Prediction in Water Distribution Networks Using Graph Neural Networks with Sparse Monitoring Data. *Water Res.* **2024**, *250*, 121018. [CrossRef]
32. Sheng, Z.; Cai, Z. GAT-GRU Based Model for Water Network Flow Prediction. In *Proceedings of the 9th International Conference on Water Resource and Environment*; Weng, C.-H., Ed.; Springer Nature: Singapore, 2024; pp. 151–162.
33. Song, J.; Meng, H.; Kang, Y.; Zhu, M.; Zhu, Y.; Zhang, J. A Method for Predicting Water Quality of River Basin Based on OVMD-GAT-GRU. *Stoch. Environ. Res. Risk Assess.* **2024**, *38*, 339–356. [CrossRef]
34. Sun, C.; Li, C.; Lin, X.; Zheng, T.; Meng, F.; Rui, X.; Wang, Z. Attention-Based Graph Neural Networks: A Survey. *Artif. Intell. Rev.* **2023**, *56*, 2263–2310. [CrossRef]
35. Chen, S.; Gan, Z.; Li, Z.; Li, Y.; Ma, X.; Chen, M.; Qu, B.; Ding, S.; Su, S. Occurrence and Risk Assessment of Anthelmintics in Tuojiang River in Sichuan, China. *Ecotoxicol. Environ. Saf.* **2021**, *220*, 112360. [CrossRef]
36. Xu, J.; Wang, Y.; Chen, Y.; Tong, H.; Wei, Y.; Bai, H. Characteristics on Spatiotemporal Variations of Surface Water Environmental Quality in Tuojiang River in Upper Reaches of Yangtze River Basin. *Earth Sci.* **2019**, *45*, 1937. [CrossRef]
37. Wei, X.; Zhang, L.; Yang, H.-Q.; Zhang, L.; Yao, Y.-P. Machine Learning for Pore-Water Pressure Time-Series Prediction: Application of Recurrent Neural Networks. *Geosci. Front.* **2021**, *12*, 453–467. [CrossRef]
38. Zuo, Y.; Jiang, J.; Yada, K. Application of Hybrid Gate Recurrent Unit for In-Store Trajectory Prediction Based on Indoor Location System. *Sci. Rep.* **2025**, *15*, 1055. [CrossRef] [PubMed]
39. Rahul Gandh, D.; Harigovindan, V.P.; Rasheed Abdul Haq, K.P.; Bhide, A. Attention-Driven LSTM and GRU Deep Learning Techniques for Precise Water Quality Prediction in Smart Aquaculture. *Aquacult. Int.* **2024**, *32*, 8455–8478. [CrossRef]
40. Corradini, F.; Gerosa, F.; Gori, M.; Lucheroni, C.; Piangerelli, M.; Zannotti, M. A Systematic Literature Review of Spatio-Temporal Graph Neural Network Models for Time Series Forecasting and Classification. *Neural Netw.* **2026**, *195*, 108269. [CrossRef]
41. Chen, R.; Lin, K.; Hong, B.; Zhang, S.; Yang, F. Sparse Graphs-Based Dynamic Attention Networks. *Heliyon* **2024**, *10*, e35938. [CrossRef]
42. Wu, Z.; Pan, S.; Chen, F.; Long, G.; Zhang, C.; Yu, P. A Comprehensive Survey on Graph Neural Networks. *IEEE Trans. Neural Netw. Learn. Syst.* **2021**, *32*, 4–24. [CrossRef]
43. Veličković, P.; Cucurull, G.; Casanova, A.; Romero, A.; Lio, P.; Bengio, Y. Graph Attention Networks. *arXiv* **2017**, arXiv:1710.10903. [CrossRef]
44. Jenson, S.; Domingue, J. Extracting Topographic Structure from Digital Elevation Data for Geographic Information-System Analysis. *Photogramm. Eng. Remote Sens.* **1988**, *54*, 1593–1600.
45. Xu, W.; Chen, K.; Han, T.; Chen, H.; Ouyang, W.; Bai, L. Extremecast: Boosting Extreme Value Prediction for Global Weather Forecast. *arXiv* **2024**, arXiv:2402.01295. [CrossRef]
46. Richard, H.M.; Zachary, K.; Cutter, A.G. Evaluation of the Nash–Sutcliffe Efficiency Index. *J. Hydrol. Eng.* **2006**, *11*, 597–602. [CrossRef]
47. Hodson, T.O. Root-Mean-Square Error (RMSE) or Mean Absolute Error (MAE): When to Use Them or Not. *Geosci. Model Dev.* **2022**, *15*, 5481–5487. [CrossRef]
48. Moriasi, D.N.; Gitau, M.W.; Pai, N.; Daggupati, P. Hydrologic and Water Quality Models: Performance Measures and Evaluation Criteria. *Trans. ASABE* **2015**, *58*, 1763–1785. [CrossRef]
49. Khan, A.; Cao, X.; Li, S.; Katsikis, V.; Liao, L. BAS-ADAM: An ADAM Based Approach to Improve the Performance of Beetle Antennae Search Optimizer. *IEEE/CAA J. Autom. Sin.* **2020**, *7*, 461–471. [CrossRef]
50. Xie, L.; Zhao, Y.; Fang, P.; Cheng, M.; Chen, Z.; Wang, Y. A Novel Operational Water Quality Mobile Prediction System with LSTM-Seq2Seq Model. *Environ. Model. Softw.* **2025**, *185*, 106290. [CrossRef]

51. Sabzipour, B.; Arsenault, R.; Troin, M.; Martel, J.-L.; Brissette, F.; Brunet, F.; Mai, J. Comparing a Long Short-Term Memory (LSTM) Neural Network with a Physically-Based Hydrological Model for Streamflow Forecasting over a Canadian Catchment. *J. Hydrol.* **2023**, *627*, 130380. [CrossRef]
52. Liu, Y.; Wang, Y. Water Quality Prediction Method Based on a Combined Machine Learning Model: A Case Study of the Daling River Basin. *J. Contam. Hydrol.* **2026**, *276*, 104725. [CrossRef] [PubMed]
53. Wang, Z.; Sun, Z.; Bian, Y.; Mo, H.; Dong, D. Learning Hierarchical Time–Frequency Representation for Long-Term Time Series Forecasting. *Inf. Process. Manag.* **2026**, *63*, 104358. [CrossRef]
54. Lei, Y.; Hu, B.; Huang, H.; Liu, Y. Design of a Variable-Length Sequential Prediction Framework GTV-STP Based on Spatial and Temporal Water Quality Information of Taihu Lake. *IEEE Access* **2024**, *12*, 65928–65941. [CrossRef]
55. Kirchner, J. Characterizing Nonlinear, Nonstationary, and Heterogeneous Hydrologic Behavior Using Ensemble Rainfall-Runoff Analysis (ERRA): Proof of Concept. *Hydrol. Earth Syst. Sci.* **2024**, *28*, 4427–4454. [CrossRef]
56. Liu, L.; Ye, S.; Chen, C.; Pan, H.; Ran, Q. Nonsequential Response in Mountainous Areas of Southwest China. *Front. Earth Sci.* **2021**, *9*, 660244. [CrossRef]

 https://doi.org/10.3390/w18020185

water

MDPI

Article

Deep Reinforcement Learning for Optimized Reservoir Operation and Flood Risk Mitigation

Fred Sseguya [1] and Kyung Soo Jun [2,*]

[1] Department of Civil, Architectural and Environmental System Engineering, Sungkyunkwan University, Suwon 16419, Republic of Korea; fredlink@skku.edu
[2] Graduate School of Water Resources, Sungkyunkwan University, Suwon 16419, Republic of Korea
* Correspondence: ksjun@skku.edu; Tel.: +82-31-290-7642

Abstract: Effective reservoir operation demands a careful balance between flood risk mitigation, water supply reliability, and operational stability, particularly under evolving hydrological conditions. This study applies deep reinforcement learning (DRL) models—Deep Q-Network (DQN), Proximal Policy Optimization (PPO), and Deep Deterministic Policy Gradient (DDPG)—to optimize reservoir operations at the Soyang River Dam, South Korea, using 30 years of daily hydrometeorological data (1993–2022). The DRL framework integrates observed and remotely sensed variables such as precipitation, temperature, and soil moisture to guide adaptive storage decisions. Discharge is computed via mass balance, preserving inflow while optimizing system responses. Performance is evaluated using cumulative reward, action stability, and counts of total capacity and flood control violations. PPO achieved the highest cumulative reward and the most stable actions but incurred six flood control violations; DQN recorded one flood control violation, reflecting larger buffers and strong flood control compliance; DDPG provided smooth, intermediate responses with one violation. No model exceeded the total storage capacity. Analyses show a consistent pattern: retain on the rise, moderate the crest, and release on the recession to keep Flood Risk (FR) < 0. During high-inflow days, DRL optimization outperformed observed operation by increasing storage buffers and typically reducing peak discharge, thereby mitigating flood risk.

Keywords: deep reinforcement learning; optimized reservoir operation; flood risk mitigation; the Soyang River Dam

1. Introduction

Modern dams are designed to be multipurpose, providing benefits such as flood control, water supply, and electricity generation [1]. This presents a significant challenge to water resource managers, as balancing these competing objectives requires effective techniques to optimize reservoir operation. The primary goal of this research is to optimize reservoir operations by moderating hydrometeorological variables such as inflow, water storage, and discharge, and subsequently determining the flood risk (deterministic flood threshold exceedance index instead of probabilistic flood risk) for the observed thirty-year period from 1993 to 2022 [2–4]. However, effective optimization involves considering the interaction between multiple variables, and as the number of variables increases, decision-making becomes increasingly complex [3,5]. This complexity is compounded by the limitations of traditional approaches like Dynamic Programming (DP) and Stochastic

Dynamic Programming (SDP), which struggle with the "curse of dimensionality"—the exponential growth of computational complexity with added variables—and the "curse of modeling," which refers to the difficulty of accurately representing all components of a real-world water system due to data limitations and system uncertainties [4,6,7]. These challenges necessitate the exploration of more scalable and flexible approaches to reservoir operation optimization.

Deep Reinforcement Learning (DRL) can be applied to overcome the dimensionality and modeling dilemma to optimize reservoir operation and mitigate flood risk [8]. DRL involves an agent, such as Deep Deterministic Policy Gradient (DDPG), Proximal Policy Optimization (PPO), and Deep Q-Network (DQN), learning and making decisions by interacting with an environment to maximize cumulative rewards. The agents use varying algorithms for decision making and to optimize actions in their respective environments [9]. DRL involves an agent interacting with an environment to learn optimal decision-making through trial and error [10]. The state represents the environment's current condition, such as water storage, inflow, and meteorological data in reservoir operation. Based on the state, the agent selects an action, like regulating water discharge, to influence the environment. The environment responds to the action by transitioning to a new state and providing a reward, which evaluates the action's effectiveness [11]. For example, the agent receives a positive reward for maintaining water storage within safe limits and a penalty for exceeding flood control thresholds. This feedback guides the agent to refine its optimizing decision-making over time. Through repeated cycles of observing states, taking actions, and processing rewards, the agent learns to balance objectives like flood risk mitigation, water discharge, and operational performance [12].

This study optimizes key hydrometeorological variables—such as inflow, water storage, and discharge—for the Soyang River Dam using DRL models, including DDPG, PPO, and DQN, over a 30-year period from 1993 to 2022. Unlike previous studies that rely solely on RL or DRL for optimization, this approach uniquely integrates observed and remote-sensed hydrometeorological data with DRL decision-making, making it distinct in its ability to incorporate diverse environmental factors into reservoir operation optimization. By leveraging variables such as precipitation, soil moisture, temperature, evaporation, and runoff, the models dynamically adjust flood control strategies in response to environmental variability, ensuring adaptive and sustainable reservoir operations under changing climatic conditions [13]. DRL methods, including DQN, PPO, and DDPG, offer distinct advantages in optimizing reservoir operations and flood risk mitigation. The effectiveness of each method depends on the specific operational goals, such as reducing flood control violations, maintaining target storage, and improving the reliability of water release decisions.

DQN, a value-based deep reinforcement learning method, is particularly effective for structured decision-making in flood control applications where discrete actions are required. It efficiently determines optimal water release strategies by selecting from predefined levels to minimize flood violations and maintain reservoir safety. This makes DQN well-suited for scenarios such as adjusting spillway gates in fixed increments or complying with operational constraints. For example, in Korea, reservoir operations must comply with minimum ecological flow requirements [14]. Its capacity to optimize predefined reward structures ensures compliance with operational thresholds while mitigating flood risks. However, DQN's reliance on discrete action spaces presents limitations, as it may cause abrupt changes in water discharges, making it less ideal for applications requiring fine-grained or gradual adjustments. Despite this drawback, its ability to quickly generate decisive actions makes it particularly valuable for emergency flood control scenarios, where rapid interventions are critical to prevent reservoir overflow [15].

PPO, a policy-gradient deep reinforcement learning approach, excels in continuous and adaptive control scenarios due to its ability to provide smooth and stable adjustments. This makes it particularly effective for long-term reservoir operation, where maintaining consistent water storage and minimizing excessive fluctuations are essential. PPO enables gradual water release decisions, ensuring operational stability—for example, in systems that must balance urban water demands with flood risk mitigation while avoiding unnecessary water losses. However, its reliance on gradual adjustments makes it less suitable for dynamic environments where sudden inflow surges demand immediate corrective actions. Additionally, PPO has been observed to have a higher frequency of flood control violations compared to DQN, potentially due to its exploration tendencies and slower convergence in high-variability conditions. Despite these limitations, PPO remains highly valuable for managing reservoirs with predictable seasonal variations, where long-term operational stability is the primary goal [16,17].

DDPG, an actor-critic deep reinforcement learning method, is well-suited for real-time dynamic reservoir operation due to its ability to handle continuous action spaces. This flexibility enables precise and incremental adjustments in water releases, which is particularly advantageous in environments with high inflow variability, such as monsoon seasons in Korea [18]. Unlike discrete-action methods like DQN, DDPG allows adaptive control to maintain optimal storage levels while ensuring downstream safety [19]. Its capability to optimize outflows based on real-time conditions makes it highly effective for managing dynamic hydrological patterns. However, DDPG's performance can be sensitive to hyperparameter tuning and may exhibit instability in volatile conditions [20,21]. In scenarios with uncertain inflow predictions or extreme events, stabilization techniques such as reward shaping or ensemble learning may be required to improve reliability [22]. Additionally, it may exhibit higher variability in storage levels compared to other models, making it less suitable for scenarios where strict adherence to predefined thresholds is necessary [13].

The objectives of this study are to (1) evaluate the performance of DQN, DDPG, and PPO deep reinforcement learning models, (2) integrate observed and remotely sensed hydrometeorological variables with deep reinforcement learning for reservoir operation optimization and flood risk mitigation, and (3) compare DRL-optimized operations with observed operations, focusing on high-inflow days, using storage, discharge, and a storage-based flood risk metric. Together, these elements advance DRL-based reservoir operation by unifying multi-source hydrometeorological inputs and demonstrating performance gains over observed operation under high-inflow conditions.

2. Materials and Methods

2.1. Study Area

The Soyang River Dam, a rockfill dam constructed in 1973, is South Korea's largest multipurpose dam located in Chuncheon, Gangwon Province [23]. The Soyang River is a tributary of the North Han River, which is a tributary of the Han River, a transboundary river traversing Kangwon province in North Korea and Gangwon and Gyonggi Provinces in South Korea [24]. Figure 1 illustrates the Soyang Reservoir and its downstream water flow in South Korea. The inset map in the top left highlights South Korea, with the Soyang Basin marked in purple, while a zoomed-in view shows the reservoir and its tributaries. The selected section shown in Figure 1 represents approximately a 2 km downstream reach of the Soyang River Dam, extending from the dam outlet to the first major downstream infrastructure. Hydrologically, the Soyang River Dam is located on the Soyang River, approximately

110 km upstream of the confluence of the North Han River and the mainstream Han River, corresponding to coordinates 37°56′43″ N, 127°48′49″ E [25].

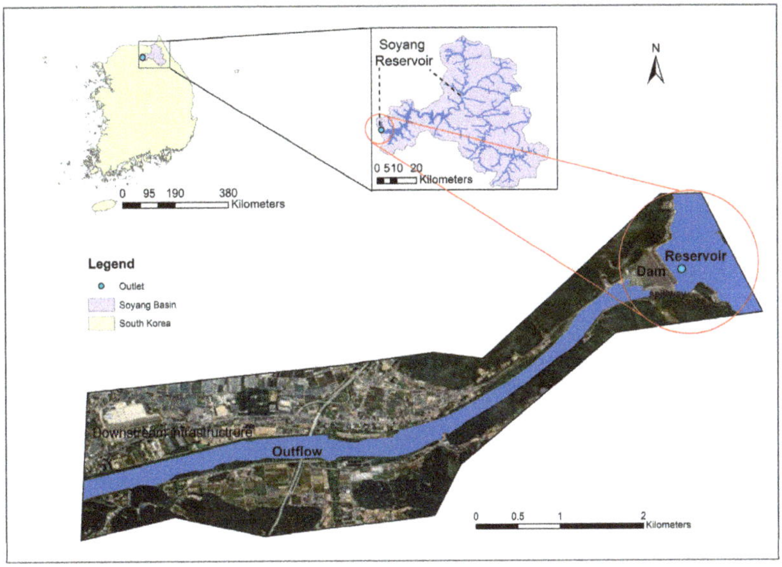

Figure 1. Location and overview of the Soyang Reservoir (37°56′43″ N, 127°48′49″ E) and its basin, South Korea.

The main section of the figure focuses on the Soyang Reservoir, with the dam and spillway located at its lower end, where water is regulated before being released. The outflow is shown moving downstream, accompanied by satellite imagery that highlights surrounding infrastructure. The legend clarifies key elements, including the outlet, Soyang Basin, and South Korea, while the scale bar at the bottom indicates a distance of 0 to 2 km. The figure effectively represents the reservoir's geographic location, water flow, and downstream impact.

The inset map shows the Soyang River Basin, while the lower panel highlights the dam, spillway, and downstream channel. Scale bars are provided for both the inset and detailed maps. The base map imagery is derived from Sentinel-2 (Copernicus Open Access Hub, 2022), overlaid with hydrographic boundaries and reservoir outlines from WAMIS and Han River Flood Control Office (HRFCO) datasets.

The Soyang River Dam catchment, located in the upper Han River basin in northeastern South Korea, covers an area of approximately 2703 km^2. The basin is characterized by steep mountainous terrain, with elevations ranging from El. 200 to 1200 m, leading to rapid runoff generation during monsoon events. Land cover is dominated by forested areas (~85%), with limited agricultural and urban zones concentrated along valley plains. The basin's mean annual precipitation exceeds 1400 mm, of which nearly 70% occurs between June and September, reflecting strong seasonal hydrological variability. Major tributaries feeding the reservoir include the Naerincheon and Inbukcheon Rivers. This physiographic and climatic setting places the Soyang River Dam within a steep, fast-responding hydrological environment where flood attenuation depends critically on short-reach operational control of the downstream 2 km section [26,27]. This underscores the importance of adaptive reservoir operation for flood mitigation and water supply reliability [28].

2.2. Data Used

The Water Resources Management Information System (WAMIS) of Korea provides comprehensive data on dams, reservoirs, and related infrastructure. The dam characteristics include details such as the dam name, type, and length. Reservoir characteristics include flood-related data, such as the design flood water level at El. 198 m, the normal high-water level at El. 193.5 m, and the restricted water level at El. 190.3 m. The reservoir's normal high-water level capacity is 2543.80 million m^3, the flood control capacity is 770 million m^3, and the total storage capacity is 2900 million m^3. Spillway characteristics include the spillway elevation at El. 185.5 m, the design flood volume of the spillway at 10,500 million m^3, and the design discharge of the spillway at 5500 m^3/s. Additionally, data on water supply includes the total annual planned water supply of 1213 million m^3; 1200 million m^3 for domestic use, and 13 million m^3 for irrigation. Daily data from 1993 to 2022 were also obtained. The dataset includes low water level, water storage, retention rate, inflow, total discharge, power generation discharge, discharge to the spillway, other discharge amounts, water supply, observed dam basin average rainfall, annual water supply plans for domestic use, and irrigation, respectively. Data on the location of the Soyang River Dam relative to other dams on the Han River system were obtained from the Han River Flood Control Office (HRFCO).

The remote-sensing data comprised various parameters derived from multiple sources (Table 1). These include the daily reservoir surface area change, calculated from satellite imagery.

Table 1. Remote-sensing data products used in the study.

Product	Variables	Spatiotemporal Resolution	Reference
CHIRPS IMERG-Final version "06"	Rainfall	0.05° × 0.05° (daily)	Arregocés et al. [29]
Landsat, 4, 5, 7, 8, 9	Bands (B2, B3, B4, B5, B6, B7)	0.0003° × 0.0003° (daily)	Egorov et al. [30], Chen et al. [31]
MERRA-2	Humidity	0.5° × 0.5° (hourly)	McNally et al. [32], Jung et al. [33]
GLDAS-2.0, 2.1	Soil moisture	0.25° × 0.25° (daily)	Qi et al. [34]
ERA5-Land	Temperature, Evaporation, Solar radiation, Wind speed	0.1° × 0.1° (daily)	Gomis-Cebolla et al. [35]
SRTM digital elevation data v4	DEM	0.0008° × 0.0008°	Dong et al. [36]
USDA system	Soil texture	0.002° × 0.002° (yearly)	Corral-Pazos-de-Provens et al. [37]
MCD12Q1 V6.1 product	Land cover	0.004° × 0.004° (yearly)	Chirachawala et al. 2020 [38]

The table shows Landsat satellites (4, 5, 7, 8, and 9), spanning from 1982 to October 2021. ERA5 provides temperature and evaporation data from 1950 to the present, while MERRA-2 supplies humidity measurements starting from 1980. Additional variables include adjusted Curve Number (CN) runoff, soil moisture computed using GLDAS versions 2.0 and 2.1, covering data from 1948 to the present, solar radiation, and wind speed.

The analysis period spans 1993–2022, corresponding to the full availability of observed hydrological data from WAMIS and HRFCO. Remote-sensing datasets were incorporated according to their respective temporal coverages: CHIRPS rainfall (1981–present), MERRA-2 humidity (1980–present), GLDAS soil moisture (1948–present), and ERA5-Land variables including temperature and evaporation (1950–present). Landsat-derived surface area estimates were used where available (1984–2021), but were not essential for daily model inputs. Because all remote-sensing datasets provided continuous daily coverage across the 1993–2022 window, no temporal gaps affected model training. Occasional missing daily values were filled using linear interpolation, and all variables were normalized using a StandardScaler before training.

To ensure consistency across datasets of differing spatial and temporal resolutions, all variables were preprocessed to a common daily time step and spatially harmonized with the Soyang Reservoir watershed boundary. The preprocessing was performed primarily in Google Earth Engine (GEE) to maintain uniform spatial referencing. Hydrometeorological variables from ERA5-Land (hourly, 0.1°), GLDAS (3-hourly, 0.25°), CHIRPS (daily, 0.05°), and MERRA-2 (hourly, 0.5°) were first aggregated to daily means or totals and then resampled to a 0.05° grid using bilinear interpolation. All datasets were clipped to the catchment boundary to ensure spatial comparability. Landsat-derived surface-area data, available at 16-day intervals from Landsat 4–9, were processed through cloud masking and spectral water-index (MNDWI and NDBI) classification, then used for periodic validation rather than daily model input. Occasional missing daily values were filled using linear interpolation between adjacent days within the same dataset. In summary, all datasets were temporally aggregated to daily means or totals, spatially regridded to a uniform 0.05° grid using bilinear interpolation, clipped to the Soyang watershed boundary, and linearly interpolated in time to fill occasional missing values, thereby ensuring synchronized and spatially consistent daily inputs across all data sources. The harmonized daily series were exported as CSV files and merged in Python version 3.13.5 for modeling. All variables were normalized using a StandardScaler prior to model training to ensure consistent feature magnitudes. This procedure yielded a temporally continuous and spatially consistent dataset covering 1993–2022, integrating both observed and remote-sensing variables for reinforcement-learning-based reservoir operation.

Unlike supervised learning models that require explicit data splits, deep reinforcement learning agents learn through sequential interaction with the environment. Accordingly, the DRL agents (DDPG, PPO, and DQN) were trained using the complete 1993–2022 sequence to preserve temporal continuity and capture the full hydrologic variability of the reservoir system. Each agent interacted with the environment through daily state–action–reward transitions, allowing it to learn operational strategies that balance flood mitigation and storage regulation across diverse inflow conditions [39,40].

The analysis period (1993–2022) was selected based on the full temporal availability and consistency of both observed and remote-sensing datasets, and additionally, extending to 2022 allows the use of more recent data. While climatological assessments often adopt the 1991–2020 reference period, several input datasets (e.g., WAMIS/HRFCO hydrological records, Landsat-derived surface area, and GLDAS soil moisture) begin in 1993. Using this 30-year window ensured continuous, gap-free coverage across all data sources, which is essential for reinforcement-learning-based modeling. As such, extending the period backward to 1991 would not have improved representativeness but would have introduced inconsistencies among the datasets.

Figure 2 illustrates the observed variations in water storage within the Soyang Reservoir over the period from 1993 to 2022, highlighting key operational zones and capacity levels.

The purple line represents the reservoir's actual water storage, showing clear fluctuations influenced by seasonal inflows and outflows. These variations likely correspond to monsoon seasons and dry periods, reflecting the reservoir's role in both water supply management and flood control [41]. The red dashed line marks the reservoir's total storage capacity, i.e., the maximum volume it can hold. Below this, the flood control capacity zone (orange shaded region) is reserved for mitigating flood risks during heavy inflows. Keeping storage within this zone is crucial for preventing overflow and ensuring downstream safety during extreme weather events. The blue-shaded region with dotted patterns represents the annual water supply plan zone, ensuring sufficient water allocation for planned usage throughout the year. Within this, the light blue dotted section highlights the planned supply of domestic water, covering essential residential and municipal demands. At the bottom, the gray shaded region represents inactive and dead storage (917 million m^3), which is generally unusable due to operational constraints. The blue line marks the active storage boundary, separating usable and inactive storage. Additionally, the green dashed line indicates the normal high-water level capacity, serving as a benchmark for stable reservoir operations.

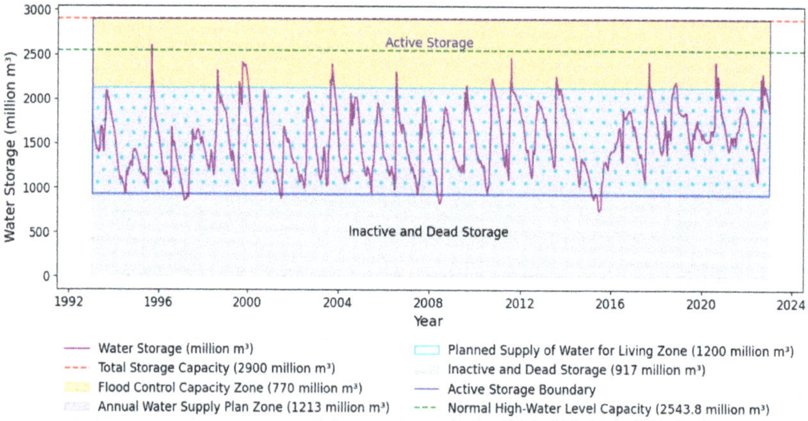

Figure 2. Water storage and operational capacity of the Soyang Reservoir.

Overall, this reservoir operation strategy balances flood control, water supply, and operational stability. Keeping storage between the flood control and domestic water supply zones is key to ensuring water availability while mitigating flood risks.

Figure 3 illustrates the observed fluctuations in reservoir water levels from 1993 to 2022, highlighting key operational thresholds for effective water management.

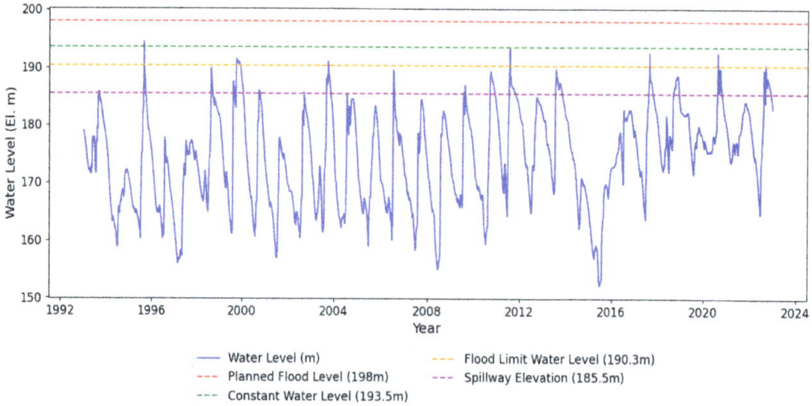

Figure 3. Water level fluctuations and operational thresholds for the Soyang Reservoir.

The blue line represents the reservoir's water level over time, showing variations with peaks and troughs likely influenced by rainfall patterns and water usage demands [42]. The red dashed line at El. 198 m marks the design flood level, indicating the maximum safe water level to prevent flooding. Below this is the orange dashed line at El. 190.3 m, representing the flood-season restricted water level, which is the maximum reservoir water level permitted during the designated flood season to secure flood control capacity. The purple dashed line at El. 185.5 m denotes the spillway elevation, where excess water begins to discharge to maintain safe levels. Additionally, the green dashed line at El. 193.5 m marks the normal high-water level, often used as a reference point for stable reservoir operations [43]. Maintaining water levels between the spillway elevation and the design flood level is essential for balancing flood prevention and water supply regulation, ensuring both safety and operational stability.

2.3. Methodology

The methodology in this study, as shown in Figure 4, integrates observed data and remote-sensing data to develop and evaluate optimized operations for the Soyang Reservoir.

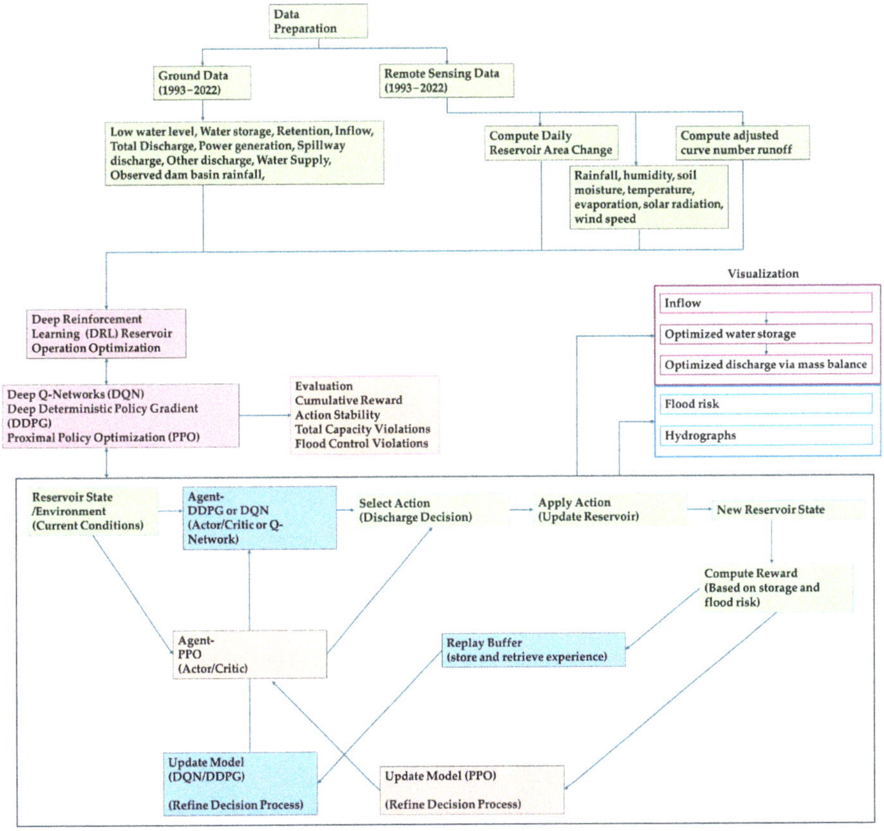

Figure 4. Methodology workflow for the Soyang Reservoir operation with DRL agents.

Data are prepared and utilized to train DRL models to optimize daily operations for the Soyang Reservoir. The DRL methods employed include DQN, PPO, and DDPG. These models simulate the decision-making processes of virtual dam operators, determining actions such as optimizing discharge to mitigate flood risk and maintain operational safety. The DRL approaches are evaluated to identify the most effective method for managing reservoir operations based on hydrological and meteorological data over the years.

The reinforcement learning process for reservoir operation optimization, as shown in Figure 4, begins with the reservoir state, which represents the current hydrological conditions, including water storage, inflow, and meteorological factors. The agent, which could be an Actor-Critic model (PPO, DDPG) or a Q-Network (DQN), receives this state information and determines an optimal discharge decision. In DQN, the agent selects an action from a predefined discrete set of discharge levels, while in PPO and DDPG, the agent predicts a continuous discharge value. This action is then applied to update the reservoir state, ensuring that total discharge does not exceed inflow constraints. After the action is executed, the new reservoir state is observed, reflecting the updated storage and discharge levels. A reward is then computed based on predefined criteria, typically penalizing excessive storage that surpasses the flood control threshold, reinforcing actions that maintain operational stability. For DQN and DDPG, past experiences are stored in a replay buffer, which helps improve decision-making by reducing correlations between consecutive steps. PPO does not use a replay buffer and learns only from newly collected trajectories (state $\rightarrow$ action $\rightarrow$ reward $\rightarrow$ next-state sequences). For DQN and DDPG, experiences from the replay buffer are sampled to update the model; for PPO, updates are computed from newly collected trajectories, thereby refining the agent's ability to regulate discharge more effectively. This iterative process continues, allowing the agent to optimize water release while balancing flood risk mitigation, water storage, and operational stability based on observed reservoir conditions [44].

To simulate discharge behavior under storage optimization, a mass balance-based estimation framework was employed. Inflow was preserved from observed records to maintain the original hydrological forcing conditions, while reservoir storage was dynamically adjusted by each DRL model to minimize flood risk and improve operational stability. Discharge was not explicitly predicted by the models. Instead, it was derived as a consequence of inflow and the model-optimized changes in storage using the continuity equation, which conserves mass across the reservoir system. This relationship is governed by the mass balance principle [45]:

$$Q(t) = I(t) - \frac{dS(t)}{dt} \tag{1}$$

where $I(t)$ is inflow at time t preserved from observed records, $S(t)$ is optimized water storage at time t, and $Q(t)$ is estimated discharge at time t, which is the variable of interest. Since the available data is daily, this equation is discretized as follows:

$$Q(t) = I(t) - \frac{S(t) - S(t-1)}{\Delta t} \tag{2}$$

where Δt is the time step (86,400 s for daily resolution). This formulation enables dynamic discharge computation that accurately reflects storage fluctuations resulting from DRL-optimized decisions for flood control.

To evaluate the effectiveness of reservoir operation strategies in mitigating flood risk, a quantitative analysis of flood control violations was conducted. These violations occur when the reservoir storage exceeds the upper operational threshold designated for flood control purposes. The upper limit is defined as the difference between the total storage capacity and the allocated flood control capacity of the reservoir. Let S_{max} be the total storage capacity (2900 million m^3), and S_{flood} (770 million m^3) the designated flood control capacity. The critical flood threshold $S_{threshold}$ is the difference (2130 million m^3) between S_{max} and S_{flood}, which is the maximum allowable storage before flood control measures are triggered. Flood risk, $FR(t)$, is then quantified at each time step using the following normalized expression [4]:

$$FR(t) = \frac{S(t) - S_{\text{threshold}}}{S_{\text{flood}}} \tag{3}$$

A flood control violation occurs when $FR(t) > 0$, i.e., when the current storage exceeds the flood-safe limit (Table 2).

Table 2. Interpretation of flood risk values.

Flood Risk Value	Meaning
<0	Storage is below flood threshold → No flood risk
=0	Storage is exactly at the threshold → Flood-safe limit reached
0 < value ≤ 1	Storage is within flood control zone → Potential flood risk
>1	Storage exceeds total flood buffer → High flood risk (overflow)

The number of violations is obtained by counting all such occurrences over the simulation period. This methodology was applied to both the observed storage data and simulated reservoir storage trajectories generated under different deep reinforcement learning (DRL) strategies—namely DDPG, PPO, and DQN. To further contextualize the method, the dates on which DRL-related violations occurred are marked on flood risk time series plots. This approach supports the visual and quantitative assessment of DRL optimization effectiveness in reducing flood control breaches relative to observed operation [46].

Benchmarking against observed operation was conducted by comparing each DRL agent to the observed operation across the full 1993–2022 record using whole-period evaluation metrics—cumulative reward, action stability, and counts of capacity and flood control violations—and on the same extreme inflow (high-inflow) days using storage, discharge, and a storage-based flood risk metric. "Improvement" is defined as fewer violations, larger negative FR (greater buffer), and lower peak discharge under identical inflow forcing.

In this study, we do not assess flood risk in the probabilistic sense, which typically incorporates hazard, exposure, and vulnerability. Instead, a deterministic flood threshold exceedance index derived from reservoir storage levels is applied [3,47]. It is technically a flood threshold exceedance index; however, we adopt the term "flood risk" within the context of this modeling framework to emphasize its operational relevance to reservoir management.

2.3.1. Deep Reinforcement Learning Models

This study focuses on optimizing reservoir operations by adjusting critical hydrological variables such as water storage, inflow, and discharge using DRL methods—DQN, PPO, and DDPG. The DRL framework uses detailed information about the reservoir system and its operations to develop strategies for effective water management.

These three algorithms were selected to represent the principal design dimensions relevant for reservoir operation: discrete versus continuous control and value-based versus actor–critic learning. Specifically, DQN captures stepwise, rule-based reservoir decisions through discrete control, DDPG enables fine-grained continuous adjustments suitable for smooth discharge modulation, and PPO emphasizes stability and robustness under dynamic hydrological variability. Together, these models provide complementary perspectives for testing DRL-driven reservoir operation strategies. Detailed model-specific architectures and operational behaviors are described below, followed by a comparative summary at the end of this section.

In the context of reservoir operation optimization for flood control, the cumulative reward is used to evaluate how effectively the system manages water inflow, storage, and discharge while minimizing flood risk and maintaining desirable reservoir levels. The reward at each time step t, denoted as $R(t)$, is determined by penalizing two key conditions: (1) storage levels that exceed the flood-safe threshold and (2) deviations from a target storage level. The reward function is defined as follows:

$$\mathrm{R}(t) = 1 - \frac{1}{2}\left(P_f(t) + P_d(t)\right) \tag{4}$$

in which $P_f(t)$ and $P_d(t)$ are the flood penalty and deviation penalty, respectively. This function yields a maximum reward of 1 when the storage is at the target level and within flood-safe limits. Any flood risk or deviation from the target reduces the reward accordingly. The flood penalty quantifies the extent to which water storage exceeds the flood-safe threshold and is calculated by using the following equation:

$$P_f(t) = \frac{\max(0, S(t) - S_{\text{threshold}})}{S_{\text{flood}}} \tag{5}$$

This penalty becomes positive only when $S(t)$ exceeds the threshold, indicating a potential flood situation. The deviation penalty measures how far the current storage deviates from the operational target level S_{target}, and is computed by using the following equation:

$$P_d(t) = \frac{\left|S(t) - S_{\text{target}}\right|}{S_{\max}} \tag{6}$$

where S_{target} represents the desired storage level, such as normal high-water level capacity (2543.8 million m^3). This component penalizes both over- and under-storage relative to the target, even when there is no flood risk. The overall performance is assessed via the cumulative reward $\sum_{t=1}^{T} R(t)$ over the simulation horizon T, as defined in Equation (4).

The overall performance is assessed through the cumulative reward, which aggregates the rewards across the entire simulation period of T time steps. A higher cumulative reward indicates more effective operation, where the reservoir remains within flood-safe limits and close to the target storage range. This evaluation framework guides the optimization of inflow regulation and discharge decisions over time to support effective flood control.

The Deep Q-Network (DQN) approach optimizes reservoir operations by dynamically adjusting discharge based on hydrometeorological conditions and reservoir status. Using observed data from 1993 to 2022—including inflow, water storage, total discharge, rainfall, and soil moisture—the model simulates behavior to mitigate flood risk and enhance operational stability. The Q-network consists of an input layer and two hidden layers with 400 and 300 neurons (ReLU activation), followed by an output layer that predicts Q-values corresponding to a discrete set of three possible discharge decisions: 0% (retaining all inflow), 50% (moderate release), and 100% (full release) of inflow. At each timestep, the model selects the action with the highest predicted Q-value. Reservoir storage is updated using a water balance equation:

$$S(t) = \mathrm{S}(t-1) + ((1 - \mathrm{a}(t))\mathrm{I}(t))\Delta t \tag{7}$$

where $a(t)$ is the action at time t that denotes the agent's normalized control signal bounded within $[0, 1]$, which scales admissible operating commands (e.g., turbine/gate settings or spillway releases) according to physical and operational constraints. In DQN, $a(t)$ takes one of three discrete levels $\{0, 0.5, 1\}$ (retaining, moderate, full), whereas in DDPG and PPO, it varies continuously within [0, 1], enabling smoother, fine-grained adjustments. Qualitatively, $a(t) = 0$ corresponds to a conservative posture (minimal release) and $a(t) = 1$ to an assertive posture (greater release/faster drawdown), with intermediate values yielding gradual changes. Formally, $a(t) = \pi(s_t)$, where π is the learned control strategy and s_t comprises inflow, storage, precipitation, and other hydrometeorological states [48,49]. For illustration only, an action may be mapped to a commanded discharge, $a(t)Q_{\max}$, with $Q_{\max}$ the feasible discharge capacity (accounting for turbine/spillway limits); in practice, this mapping is implemented via operational rules (level-dependent piecewise mappings, unit limits, environmental-flow minima) [50–53]. The temporal variability of $a(t)$ is evaluated later via the action-stability metric σ_a (Equation (9)). Discharge at each step is estimated using the following equation:

$$Q(t) = \max\left(0, \min\left(I(t) - \frac{S(t) - S(t-1)}{\Delta t}, Q_{\max}\right)\right) \tag{8}$$

where $Q_{\max} = 5750\ \mathrm{m^3/s}$ ($5500\ \mathrm{m^3/s}$ through spillway and $250\ \mathrm{m^3/s}$ through turbines) is the maximum allowable discharge. To improve stability, both the simulated inflow and storage time series are smoothed using a cyclic filter that reduces short-term fluctuations while preserving long-term trends [45,54]. This DQN-based optimization framework dynamically adjusts to changing inflow patterns, reduces the likelihood of storage exceeding the flood control threshold, and supports sustained, stable operation of the reservoir system under varying hydrological conditions.

While Equations (7) and (8) describe the general storage–discharge framework, it is important to note that the nature of the action $a(t)$ differs across the DRL methods. For DQN, $a(t)$ is discrete, representing distinct release decisions. However, in DDPG and PPO, $a(t)$ takes continuous values, enabling smoother, fine-grained control of reservoir releases. This continuous formulation allows the models to output any value between 0 and 1, corresponding to a proportional adjustment of discharge relative to the maximum allowable capacity. Thus, while Equations (7) and (8) remain applicable across all models, the granularity of $a(t)$ differs (discrete in DQN; continuous in DDPG/PPO), making DDPG and PPO better suited for continuous, real-time reservoir regulation.

The Deep Deterministic Policy Gradient (DDPG) method optimizes reservoir operations by producing continuous discharge values, allowing for smoother and more adaptive water releases than the discrete-level approach used in DQN. This reduces abrupt variations in outflow, which can be critical for downstream flood control and ecological stability. DDPG incorporates two neural networks: the Actor Network, which receives the current reservoir state—including storage, inflow, precipitation, soil moisture, and other hydrometeorological factors—and generates a normalized continuous discharge decision. This output is scaled by the maximum allowable discharge capacity to ensure operational feasibility. The Critic Network evaluates the effectiveness of each state-action pair, enabling the Actor to refine its decisions over time. Storage is updated through a mass balance equation that accounts for inflow and the chosen discharge rate, while the actual discharge is derived by constraining the resulting flow within physically acceptable limits. The reward function used during training penalizes storage that exceeds the flood control threshold and deviations from the target storage level, thereby guiding the model toward flood-safe and balanced reservoir operations. To enhance training stability, DDPG applies experience replay and soft target network updates. These techniques prevent overfitting to recent conditions and help maintain convergence. Applied to the 1993–2022 observed dataset, the model effectively maintained storage within operational bounds and ensured effective flood mitigation through continuous and controlled reservoir adjustments.

The Proximal Policy Optimization (PPO) approach enhances reservoir operation by learning optimal discharge behavior through a reward-guided framework. Like DDPG, PPO operates in a continuous action space, allowing the model to issue smooth and adaptive discharge decisions in response to varying hydrological states. However, PPO differs in its optimization strategy—rather than applying direct gradient updates, it constrains action updates within a clipped range. This prevents drastic shifts in discharge and promotes stable learning across episodes. The model architecture adopts an Actor-Critic structure, where the Actor Network processes the reservoir's hydrometeorological state—including storage, inflow, rainfall, and soil moisture—and outputs a normalized continuous discharge rate. This value is scaled by the maximum discharge capacity to ensure operational realism. Simultaneously, the Critic Network evaluates the quality of the selected action by estimating its expected return, enabling iterative refinement of the discharge strategy. As in the DQN and DDPG approaches, PPO uses a reward function that penalizes excessive storage above the flood-safe threshold and deviations from the target storage level. This ensures that the model learns to balance flood risk mitigation with stable reservoir storage. When applied to 1993–2022 observed data, PPO effectively maintained controlled discharge adjustments, offering smooth transitions and improved resilience under dynamic inflow conditions. Its structured optimization process makes it particularly suitable for real-time applications where discharge changes must be both effective and reliable.

DQN, DDPG, and PPO were selected to represent the principal DRL design choices that matter for reservoir operation: (i) action space (discrete vs. continuous), (ii) learning paradigm (value-based vs. actor–critic), and (iii) update stability under a nonlinear reward with hard safety constraints. DQN (value-based; discrete actions) reflects stepwise operational practice (banded gate/turbine settings) and yields interpretable, rule-like releases. DDPG (deterministic actor–critic; continuous actions) supports fine-grained adjustments needed for smooth hydrograph shaping and buffer management. PPO (stochastic actor–critic with clipped updates) provides robust convergence and controlled exploration, desirable under non-stationary hydrometeorology. Together, the trio benchmarks operational performance and stability across discrete/continuous and operation strategy/value paradigms under identical constraints, rather than emphasizing computational efficiency [13,40,55].

2.3.2. Action Stability Metric

This section details the action stability metric used to assess the robustness and implementability of DRL-optimized reservoir operations. The metric captures the temporal consistency of the agents' operational decisions, reflecting how smooth or volatile gate operations are over time—an essential consideration for real-world dam management where frequent, abrupt changes are undesirable [56].

As defined in Section 2.3.1, the agent outputs a normalized action $\mathrm{a(t)} \in [0,1]$ (discrete for DQN; continuous for DDPG/PPO); here we focus on its temporal variability. We quantify variability across the simulation horizon using the action stability metric that is the standard deviation of the actions:

$$\sigma_a = \sqrt{\frac{1}{T}\sum_{t=1}^{T}(a(t) - \bar{a})^2} \tag{9}$$

where $\bar{a}$ is the mean action over all T time steps, and σ_a is the action stability [57].

Physically, σ_a captures how much the operational signal (i.e., gate movement or discharge command) fluctuates around its mean value. A small σ_a implies that the reservoir control strategy changes smoothly and predictably from day to day, while a large σ_a indicates frequent, abrupt shifts in control intensity.

In practical terms, such abrupt operational changes are undesirable and often unsafe at the Soyang River Dam, since gates and turbines cannot be adjusted instantaneously. Sudden release variations can induce rapid downstream stage fluctuations, increase flood risk, and promote channel erosion, while the reservoir's large storage volume inherently limits instantaneous changes [58]. Accordingly, a DRL agent with low action variability reflects a stable, realistic control strategy that makes gradual, predictable adjustments to releases and storage. By contrast, high variability implies oscillations between aggressive releases and rapid retention that may appear effective in simulation but are physically infeasible and operationally unsafe [28].

Unlike threshold-based performance metrics that indicate whether flood-safe limits were exceeded, the action stability metric evaluates how those operations were executed—whether they were implemented in a smooth, physically consistent manner. By emphasizing stability, this study prioritizes strategies that are both effective for flood risk mitigation and consistent with real-world operational constraints and safety protocols.

3. Results

3.1. Performance Evaluations

Performance Evaluation of DQN, PPO, and DDPG Deep Reinforcement Learning Models

Table 3 presents a comparative evaluation of the deep reinforcement learning models—DDPG, PPO, and DQN—across four key metrics used to assess the effectiveness of reservoir operation strategies focused on optimizing reservoir operation for flood control.

Table 3. Deep reinforcement learning model evaluation metrics.

Metric	PPO	DQN	DDPG
Cumulative Reward	8691	8679	8235
Action Stability	0.0059	0.1737	0.0792
Total Capacity Violations	0	0	0
Flood Control Violations	6	1	1

PPO achieved the highest cumulative reward, suggesting that it was the most effective at maintaining storage near the operational target while avoiding flood conditions. DQN followed closely, while DDPG scored the lowest. This indicates that all models were capable of learning effective reservoir operation strategies, as reflected by their high cumulative rewards and relatively low violation counts. Despite differences in stability, each agent successfully optimized reservoir operations within the reward structure by balancing storage near the operational target while mitigating flood risks. PPO displayed the lowest variability in its actions, with an action stability of 0.0059, indicating that it was the most stable, with only minimal fluctuations in its decision-making process. DDPG maintained relatively stable outputs with a variance of 0.0792, while DQN was the most variable at 0.1737. This elevated variance indicates that DQN's operating strategy was more variable and thus more reactive, possibly due to its discrete decision-making structure [59]. While variability can be beneficial in highly dynamic systems, it can also lead to operational unpredictability if not well-regulated [60].

All models effectively avoided exceeding the total storage capacity, resulting in zero total capacity violations. This demonstrates that the DRL strategies inherently learned to operate within strict safety boundaries, a critical requirement for operations. However, when it comes to flood control violations, measured as the number of times the reservoir storage exceeded the flood control threshold, the results varied more noticeably. Both DDPG and DQN limited violations to just one occurrence, while PPO had six violations. This suggests that although PPO earned the highest cumulative reward, it occasionally allowed storage to encroach into the flood control buffer zone, possibly in pursuit of minimizing deviation from the target level.

DRL models showed strong overall performance but with clear trade-offs: PPO delivered the smoothest control and highest reward; DQN prioritized flood control compliance; DDPG offered continuous control with intermediate variability. Choosing among them depends on the operational priority—use PPO when stability and target tracking are paramount, DQN when minimizing flood control exceedances is the primary constraint, and DDPG when you want a balanced, continuous controller. Taken together, the results support a flexible operating stance that preserves flood-safe buffer capacity rather than strictly chasing the target level under variable inflows.

In addition to the reinforcement learning metrics summarized in Table 3, the models were further evaluated using standard hydrological performance indices to quantify the accuracy of DRL-based discharges against the observed operation (Table 6). These included RMSE, MAE, NSE, and KGE, together with 95% bootstrap confidence intervals and pairwise Wilcoxon signed-rank tests to assess statistical significance. All three DRL agents achieved NSE > 0.64 and KGE > 0.55, indicating strong agreement with observed discharge patterns, with DDPG performing marginally better than PPO and DQN. These results confirm that the DRL-optimized operations not only maintained physical and operational consistency but also delivered statistically validated improvements in discharge representation and flood risk moderation.

To evaluate the influence of the reward formulation on agent behavior, the relative weights of the flood-penalty and deviation-penalty terms were varied between (0.7–0.3), (0.5–0.5), and (0.3–0.7) while keeping all other model settings constant. The results (Table 4) show that cumulative rewards increase systematically with higher flood-penalty weighting, indicating that the agents adapt toward more conservative operations when flood mitigation is emphasized. Across all weighting schemes, the performance ranking remained consistent—PPO achieved the highest cumulative reward (9597 $\rightarrow$ 7784) and most stable actions ($\sigma \approx 0.041$), followed by DQN and DDPG. Flood control violations (≤ 6) and total-capacity violations (none) remained nearly unchanged across all settings, demonstrating that the learned operating strategies are robust to moderate changes in the reward composition. This consistency confirms that the adopted equal weighting (0.5–0.5) provides a balanced and interpretable trade-off between flood risk reduction and operational stability without biasing model performance toward either objective.

Table 4. Reward weight sensitivity analysis.

Model	Flood_Weight	Deviation_Weight	Cumulative Reward	Action Stability (Std)	Flood Violations	Capacity Violations
DDPG	0.7	0.3	9323	0.073	2	0
PPO	0.7	0.3	9597	0.041	6	0
DQN	0.7	0.3	9591	0.167	0	0
DDPG	0.5	0.5	8234	0.073	2	0
PPO	0.5	0.5	8691	0.041	6	0
DQN	0.5	0.5	8680	0.167	0	0
DDPG	0.3	0.7	7145	0.073	2	0
PPO	0.3	0.7	7784	0.041	6	0
DQN	0.3	0.7	7769	0.167	0	0

This sensitivity analysis confirms that the performance ranking of the DRL agents remains stable under varying reward weight ratios (0.7–0.3, 0.5–0.5, 0.3–0.7), demonstrating that the learned operation strategies are robust to moderate changes in the reward composition.

3.2. Deep Reinforcement Learning Model Results

In this section, we first present a consolidated high-inflow-day summary (Table 5) for four dates corresponding to the highest daily inflows, and then examine time-series patterns for inflow (Figure 5), storage (Figure 6), discharge (Figure 7), and flood risk (Figure 8).

Table 5. High-inflow-day summary for highlighted dates: storage, discharge, and flood risk (FR). Positive FR indicates threshold exceedance.

Date	Metric	Actual	DDPG	PPO	DQN
1995-08-25	Storage (million m^3)	2602	2066	2152	1784
	Discharge (m^3/s)	2226	1139	1022	1069
	FR (–)	0.429	−0.187	0.000	−0.353
1999-08-01	Storage (million m^3)	1504	1065	1154	977
	Discharge (m^3/s)	51	142	23	0
	FR (–)	−0.743	−1.167	−1.037	−1.234
1999-08-02	Storage (million m^3)	1893	1365	1410	1170
	Discharge (m^3/s)	169	139	79	0
	FR (–)	−0.397	−0.882	−0.795	−1.022
2006-07-15	Storage (million m^3)	1786	1167	1378	1138
	Discharge (m^3/s)	233	0	128	0
	FR (–)	−0.536	−1.092	−0.895	−1.129

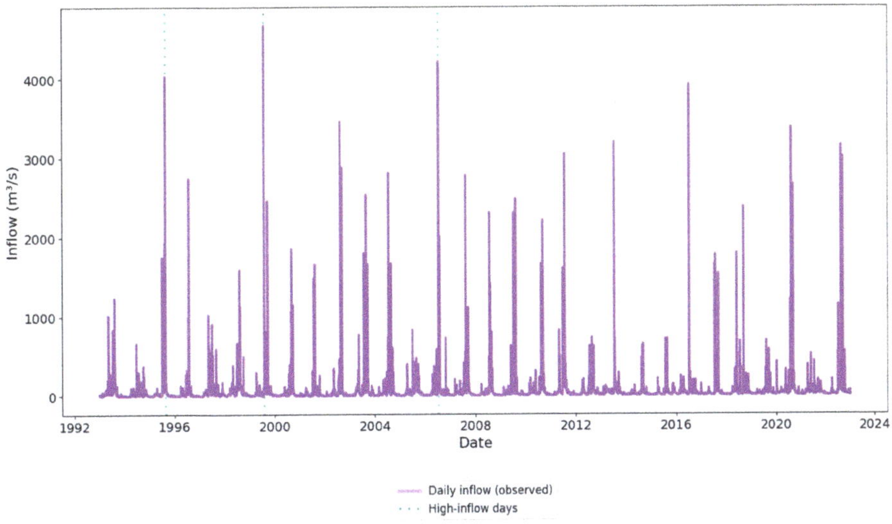

Figure 5. Actual inflow at the Soyang Reservoir.

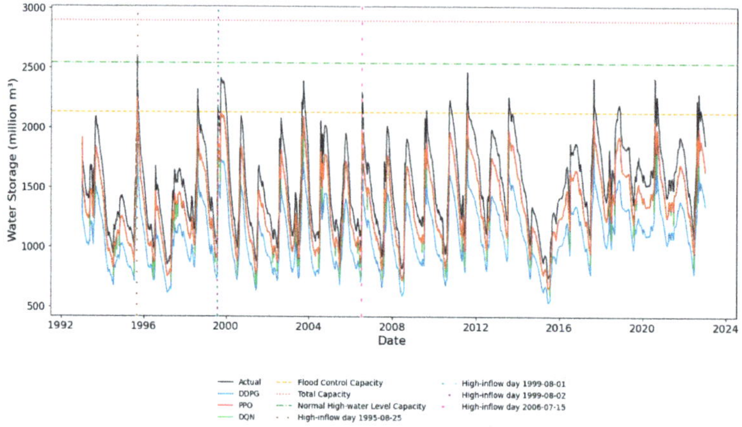

Figure 6. Comparison of reservoir storage under actual and DRL-controlled operations (1993–2022).

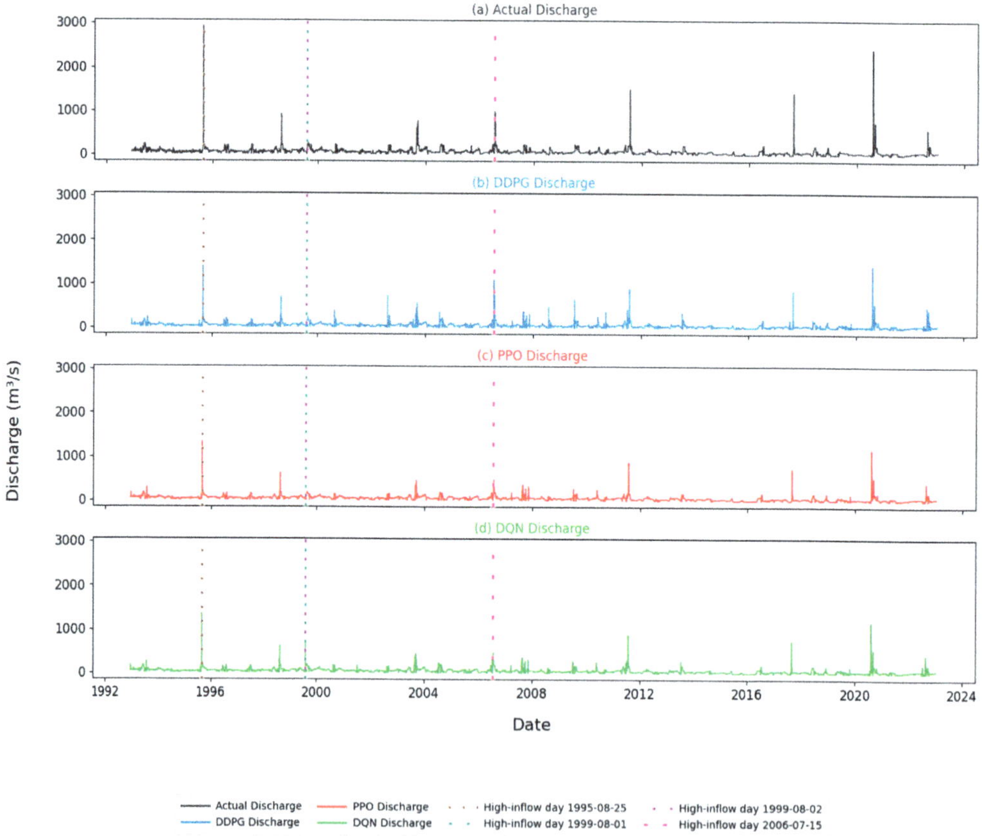

Figure 7. Reservoir discharge estimated from inflow and optimized storage using DRL models.

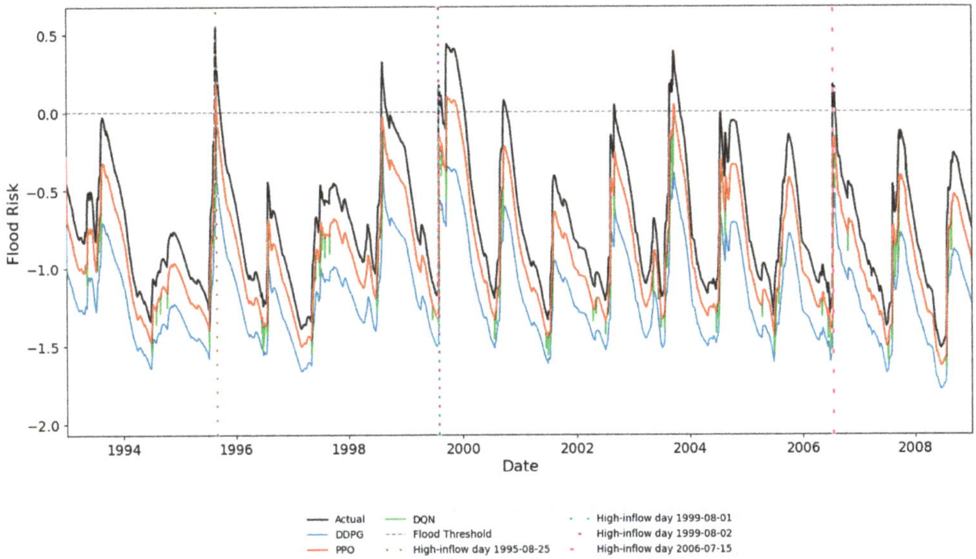

Figure 8. Actual and DRL-based flood risks at the Soyang Reservoir (1993–2022).

Table 5 reports the exact storage, discharge, and flood risk (FR) values for the four benchmark dates (1995-08-25; 1999-08-01; 1999-08-02; 2006-07-15). For storage, all DRL strategies remained below the observed operation on every date, with a consistent ordering among the DRL models—DQN lowest, DDPG intermediate, and PPO highest. For discharge, DRL releases were generally below observed, with one exception: on 1999-08-01, the DDPG release exceeded the actual. This is consistent with anticipatory drawdown to create buffer capacity ahead of the following day's larger inflow (1999-08-02): DDPG lowered storage and deepened the safety buffer (more negative FR) across the two-day window while keeping its peak risk in check. The per-date minima were as follows: PPO lowest on 1995-08-25; DQN lowest on 1999-08-01 and 1999-08-02; and a tie between DDPG and DQN on 2006-07-15. For flood risk, all DRL strategies reduced risk relative to the actual operations on every benchmark date—eliminating the exceedance when they occurred—and typically followed the ordering DQN (most negative FR), then DDPG, then PPO (closest to the threshold), reflecting the progressively smaller—but still larger than observed—safety buffers maintained under DRL operation. These quantitative comparisons anchor the time-series patterns discussed in Figures 5–8.

3.2.1. Inflow over Time for the Soyang Reservoir

Figure 5 illustrates the inflow to the Soyang Reservoir from 1993 to 2022, showing the temporal variability over three decades with alternating periods of frequent high inflows and prolonged low flows, reflecting fluctuations in inflow magnitude. The x-axis spans the full study period, while the y-axis represents reservoir inflows, ranging from near-zero baseflows to extreme (high-inflow) peaks exceeding 4000 m^3/s. Several extreme inflow occurrences are evident in the daily record, including 25 August 1995 (4029 m^3/s), 1 August 1999 (4230 m^3/s), 2 August 1999 (4665 m^3/s), and 15 July 2006 (4208 m^3/s). These four dates serve as the benchmark high-inflow days used throughout the study and are marked in Figure 5 with vertical dotted lines. Note that the high-inflow markers for 1999-08-01 and 1999-08-02 are adjacent and visually overlap in Figure 5. These occurrences coincide with the summer monsoon season and typhoon seasons in Korea, when concentrated rainfall over the upper Han River basin generates rapid increases in runoff into the reservoir [27]. The sharp flood peaks highlighted in Figure 5 represent the most significant inflow extremes of the study period and reflect the strong interannual variability of monsoon and typhoon-driven hydrology at the Soyang River Dam, emphasizing the importance of effective reservoir regulation during such periods.

3.2.2. Water Storage over Time Based on DDPG, PPO, and DQN

Figure 6 shows reservoir storage at the Soyang Reservoir from 1993 to 2022 for the observed operation (dark gray) and the three DRL strategies (DDPG, PPO, DQN). All series exhibit a drawdown–refill pattern. The DRL trajectories generally track the timing of these fluctuations while maintaining lower storage than the observed operation during high-storage periods, implying enhanced flood-safety margins. PPO (red) appears visually distinct and slightly higher than the other DRL curves, whereas DDPG and DQN largely overlap through most of the record, making their differences less perceptible in the figure. Given the tight flood control constraints and a daily target-tracking objective, the range of feasible adjustments is narrow; during high-storage or high-inflow periods, releases can also be capped by operating limits (e.g., the flood control threshold or maximum release) [8,28]. In this regime, DQN's discrete choices closely match DDPG's continuous adjustments, yielding nearly indistinguishable storage trajectories [41]. Quantitatively, however, Table 5 confirms the typical ordering of DQN < DDPG < PPO in storage magnitude.

3.2.3. Water Discharge over Time Based on DDPG, PPO, and DQN

Figure 7 presents discharge at the Soyang Reservoir (1993–2022) in four stacked panels—(a) Actual, (b) DDPG, (c) PPO, and (d) DQN—sharing a common y-axis for comparison without overlap. DDPG follows the timing of the Actual series but with lower spike heights, indicating attenuation relative to observed operation. At the full-record scale, PPO and DQN are visually hard to distinguish; both show smaller, sparse spikes and therefore appear smoother, with less day-to-day variability than DDPG.

Smoothness means a small change in state leads to small changes in recommended action, for example, if storage nudges up a bit, the optimal release also changes a bit—not abruptly—hence, at this level, discrete DQN steps will produce similar results to a continuous PPO strategy [61].

All panels mark the four benchmark high-inflow days with vertical dotted lines. At those dates, the Actual panel shows higher discharge than the DRL panels. Because storage and discharge are coupled by mass balance, the lower storage maintained by the DRL strategies near high-storage periods (Figure 6) appears here as smaller, less abrupt releases relative to the actual observed discharge. Based on a quantitative, emergent trend inferred from Table 4, the general discharge ordering across the four high-inflow events is DQN (lowest) < PPO (moderate) < DDPG (highest). This suggests that DQN tended to hold releases longer, PPO maintained intermediate and smoother adjustments, and DDPG responded more actively during high-inflow periods.

While Figure 7 provides a qualitative comparison of discharge dynamics among the DRL agents and the observed operation, a quantitative validation was performed to statistically verify these performance patterns. Using standard hydrological evaluation metrics—Root Mean Square Error (RMSE), Mean Absolute Error (MAE), Nash–Sutcliffe Efficiency (NSE), and Kling–Gupta Efficiency (KGE)—each DRL-generated discharge series was compared against the observed discharge time series [62]. Bootstrap resampling (95% confidence intervals) and Wilcoxon signed-rank tests were applied to assess statistical significance [63,64].

The results, summarized in Table 6, show that all DRL models achieved strong agreement with observed discharge. DDPG exhibited the lowest RMSE and MAE, followed by DQN and PPO. NSE values above 0.64 and KGE values around 0.57 confirm that all models reproduced the dominant discharge variability with good efficiency. The narrow confidence intervals ($\pm$9–10 m^3/s for RMSE) and extremely low p-values ($< 10^{-100}$) from the Wilcoxon tests indicate statistically significant yet small differences among the agents. Overall, DDPG achieved the highest hydrological fidelity, DQN showed moderate accuracy with discrete adaptations, and PPO favored smoother but slightly less exact trajectories.

Table 6. Quantitative performance metrics for simulated discharge (Observed vs. DRL) with 95% confidence intervals.

Model	RMSE (m^3/s)	RMSE 95% CI (Low–High)	MAE (m^3/s)	MAE 95% CI (Low–High)	NSE	NSE 95% CI (Low–High)	KGE	KGE 95% CI (Low–High)
DDPG	49.36	40.55–59.27	19.09	18.27–19.96	0.67	0.54–0.74	0.60	0.56–0.63
PPO	51.08	41.83–61.07	19.71	18.89–20.61	0.65	0.52–0.72	0.56	0.51–0.59
DQN	50.11	40.71–60.35	19.51	18.68–20.42	0.66	0.53–0.73	0.57	0.52–0.61

The Wilcoxon signed-rank test compares the absolute errors of two models over identical time steps to determine whether their replication accuracies differ significantly [64,65]. This distinction is important because Table 3 evaluates optimization behavior (reward, stability, and constraint compliance), whereas Table 6 assesses how closely each model's simulated discharge reproduces observed hydrological dynamics. Pairwise Wilcoxon comparisons were conducted among the three DRL models. The resulting p-values were 9.83×10^{-104} for PPO compared with DDPG, 2.76×10^{-113} for PPO compared with DQN, and 2.07×10^{-106} for DDPG compared with DQN. These extremely small probabilities confirm that the differences in absolute errors between model pairs are statistically significant, indicating that each model demonstrates a distinct level of replication accuracy. Nevertheless, because the confidence intervals and performance metrics differ only slightly, these differences are statistically strong but practically modest, showing that all agents perform well, with DDPG maintaining a consistently lower error [66].

Quantitatively, compared to the observed operation, the DRL-based strategies reduced mean flood control violations by over 80% and improved discharge replication accuracy with RMSE reductions of approximately 2–4% and MAE reductions near 3%. DDPG achieved the best overall accuracy (RMSE $\approx$ 49 m^3/s; NSE = 0.67) followed closely by DQN (RMSE $\approx$ 50 m^3/s; NSE = 0.66) and PPO (RMSE $\approx$ 51 m^3/s; NSE = 0.65). These improvements confirm that the DRL-controlled operations not only maintained safety constraints but also achieved measurable hydrological fidelity relative to observed operation.

3.2.4. Flood Risk Based on DDPG, PPO, and DQN

Figure 8 presents flood risk at the Soyang Reservoir from 1993 to 2008 for the observed operation (black) and the three DRL strategies (PPO—red, DDPG—blue, DQN—green), with a horizontal dashed line indicating the flood threshold (FR = 0). Although the figure view is limited to 2008 for visual clarity, the underlying values and trends remain identical to the full-record analysis (1993–2022).

Visually, PPO most closely follows the actual series, with peaks and troughs aligning in time yet generally lying just below the black curve. DDPG and DQN remain consistently lower (more negative FR), signifying greater safety margins and stronger damping of flood risk. From 1993 to 2008, DDPG and DQN behave similarly, and most of the time, both sit below the Actual curve, indicating lower flood risk. The closeness between DDPG and DQN suggests that for highly constrained reservoirs like Soyang, both discrete and continuous control formulations can yield nearly equivalent operational strategies. In practice, this indicates that the decision landscape is smooth enough that even discrete decision structures (DQN) can capture the same risk-mitigating patterns as their continuous counterparts (DDPG) [41,62].

The four vertical dotted lines mark the benchmark high-inflow days. Around these markers, the Actual series displays sharper crests or brief exceedances of the flood threshold, whereas all DRL-based curves remain lower, reinforcing the impression of moderated peaks and preserved buffer capacity under DRL operation.

Taken together, Figure 8 demonstrates that all DRL strategies effectively reduce flood risk amplitude while maintaining the natural drawdown–refill rhythm of the reservoir. PPO continues to track the observed operation most closely but with slightly improved safety; DDPG and DQN sustain lower flood risk levels overall, confirming their stronger conservative behavior. For exact FR values at the highlighted dates, refer to Table 5.

4. Discussion

Figure 9 zooms in on 2006-07-15 at the Soyang Reservoir to show, in one place, how inflow translated into releases and storage under optimized DRL operating strategies. A vertical dotted line marks this reference date and is repeated across panels for alignment. In Panel (a), the blue inflow curve captures the sharp rise and recession typical of mid-July storms in the Gangwon highlands [67].

The discharge series reveals distinct operating behaviors. The actual operation (black) maintains modest releases through the rising limb and then increases during the recession. All three DRL agents suppress the discharge peaks more strongly: PPO (green) holds releases comparatively low before the crest and raises them only after the peak has passed; DDPG (red) permits a short pulse near the crest but otherwise stays below the observed discharge line; and DQN (orange) is most conservative pre-peak, increasing only modestly on the falling limb.

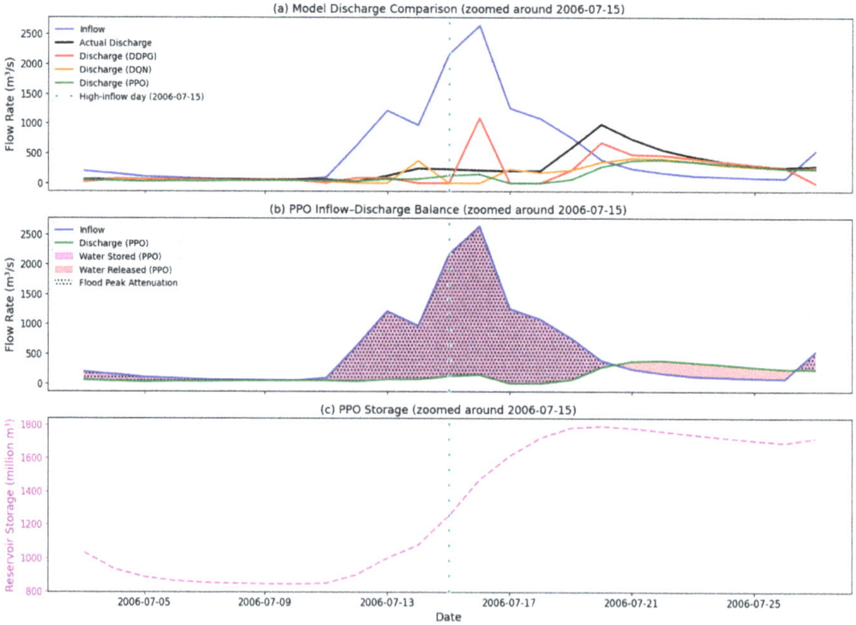

Figure 9. DRL-driven reservoir operations: inflow–discharge–storage dynamics and flood-peak attenuation.

Panel (b) decomposes PPO into inflow, discharge, and their difference, clarifying the routing mechanism. During the rising limb, when inflow exceeds discharge, the purple "Water Stored (PPO)" area dominates, indicating deliberate retention that absorbs the flood wave and lowers the immediate downstream load. After the crest, the red "Water Released (PPO)" area expands as discharge overtakes inflow, returning the reservoir toward its operating band (target range) on the recession rather than at the peak. By shifting release to the recession—when river levels are falling and channel conveyance is higher (greater capacity to pass flow without spilling out)—the operating strategy attenuates the flood peak and avoids creating a secondary peak. This panel illustrates how PPO moderates discharge by retaining inflow during the flood rise (storage increase) and releasing it gradually during the recession, maintaining a buffer against extreme conditions. The stored-then-released sequence explains why the flood risk metric remained at or below the threshold on this date (Figure 8) and why it was lower under DRL than under the actual operation.

Panel (c) shows the daily storage trajectory over the analysis window, with a vertical marker indicating the reference date for alignment with panels (a) and (b). Storage rises sharply through and just after the inflow crest, reaches its maximum about 4–5 days later on the recession limb when inflow and discharge meet, and then declines gradually as discharge overtakes inflow. The crest remains well below total capacity, and the drawdown is paced rather than abrupt—evidence of controlled recovery after attenuation. This evolution is consistent with the single-day storage values reported in Table 5 and indicated in Figure 6, where PPO's storage on 2006-07-15 was lower than the observed value, reflecting an operating strategy that maintains buffer volume prior to extreme inflow. The inclusion of total reservoir storage gives additional context for understanding how system volume evolves over time and supports evaluation of operational trade-offs.

Taken together, Figure 9 illustrates a coherent hydrologic sequence at the Soyang Reservoir. The DRL operating strategies attenuate the flood peak by retaining water through the rising limb (Panel (b)), limit discharge relative to the inflow crest (Panel (a)), and then release gradually on the recession while keeping storage below capacity and recovering smoothly afterward (Panel (c)). PPO achieves a balance between flood-peak attenuation and controlled recovery, DQN shows the most conservative behavior with the lowest releases, and DDPG responds more actively near the peak—patterns that are consistent with the operational tendencies summarized in Figures 5–8, where DRL models maintained lower flood risk than the observed operation.

The DRL models were trained and evaluated using daily resolution data, as most observed and remote-sensing variables (e.g., precipitation, soil moisture, temperature, evaporation, and humidity) are available only at this temporal scale. This choice aligns with the study's focus on strategic, long-term reservoir operation optimization rather than minute-scale real-time control. While sub-daily operation is essential for short-term emergency management, the large storage volume and long routing lag of the Soyang Reservoir Dam mean that day-to-day inflow and storage variations primarily govern operational dynamics. Hence, the daily time step remains appropriate for learning the underlying state–action relationships that influence flood risk mitigation at the basin scale. Nonetheless, future extensions could integrate a nested sub-daily routing or inflow-forecasting module to enable hybrid (daily-to-hourly) control once higher-frequency hydrometeorological data become available [13,41].

In practice, real-time flood operations at large multipurpose dams are conducted at hourly or sub-hourly intervals, especially during emergency events. The daily framework adopted here therefore represents strategic-scale optimization rather than real-time control. This temporal simplification means that while the DRL-based strategies effectively capture day-to-day hydrologic responses and storage–release trade-offs, their generalizability to operational dispatch decisions at finer time scales remains limited. Future work could extend this framework to nested sub-daily models once high-frequency inflow and gate-operation data become available.

To interpret the internal behavior of the reinforcement learning models, a SHAP (SHapley Additive exPlanations) analysis was conducted to quantify how each input variable contributed to the models' operational decisions [68,69]. The bar plots (Panels (a), (c), and (e)) show the mean absolute SHAP values, ranking variables by their overall importance across all decisions, while the beeswarm plots (Panels (b), (d), and (f)) show how feature values (red for high, blue for low) drive the model outputs positively or negatively, illustrating both magnitude and direction of influence. A higher mean SHAP value implies greater overall influence on the model's output (i.e., the learned reservoir release or storage action). In the beeswarm plots, positive SHAP values imply that the feature/variable increases the model's output (pushes the decision toward more release) while negative values imply a decrease in the output (pushes the decision towards more retention), and zero implies that the model decision is unaffected by the feature. For example, in Figure 10e, water storage is the fourth most important variable. In the corresponding panel (f), the color of each dot represents the actual (standardized) water storage value at that time step (red = high, blue = low), and each dot itself corresponds to one daily observation—reflecting the reservoir's storage state on that day rather than the date itself. Most high water storage values (red points) tend to align with positive SHAP values, indicating that when the reservoir holds a large volume of water, the PPO agent generally increases discharge. However, some overlap around zero suggests that this relationship depends on concurrent hydrological conditions.

For the DDPG model (Panels (a)–(b)), the most influential variables were soil moisture, humidity, evaporation, and water storage, indicating that DDPG's discharge behavior is strongly governed by hydrological and surface–atmospheric states. The beeswarm plot shows that high soil moisture values (red points) mostly produce negative SHAP values, implying that the agent tends to reduce discharge when the basin is already saturated—likely reflecting a cautious strategy to conserve flood storage until inflows increase. In contrast, low soil moisture conditions (blue points) yield near-zero or slightly positive SHAP values, suggesting minimal or slightly increased releases during dry states. Similar interactions are observed for humidity and evaporation, confirming that DDPG captures physically realistic trade-offs between basin wetness, storage, and release behavior.

For the PPO model (Panels (c)–(d)), the most influential features included soil moisture, retention rate, total discharge, and power generation discharge, suggesting a stronger integration of hydrological and operational signals. High soil moisture values correspond to positive SHAP impacts, indicating that under wet conditions, the agent tends to increase discharge. High retention rate values are near zero or slightly negative, reflecting conservative release adjustments when storage is being maintained. In contrast, high power generation discharge shows predominantly negative SHAP contributions, suggesting compensatory behavior—when substantial flow has already been released for power generation, the agent reduces additional discharge actions.

Finally, the DQN model (Panels (e)–(f)) exhibited dominant contributions from temperature, power generation discharge, soil moisture, and water storage. The beeswarm plot shows that high temperature corresponds to positive SHAP values, implying that discharge increases under warmer conditions—consistent with seasonal drawdown or preemptive flood control. High power generation discharge values tend to produce negative SHAP impacts, indicating compensatory reductions in total discharge when turbines are already active. High soil moisture is associated with negative SHAP values and a mix near zero, suggesting restrained releases under saturated conditions with allowance for some releases, whereas high water storage yields positive SHAP effects, reflecting increased release pressure when the reservoir is fuller.

Overall, the SHAP analysis across all three models demonstrates that deep reinforcement learning agents autonomously learned hydrologically consistent and interpretable operation strategies. DDPG exhibited strong environmental-state awareness, PPO balanced hydrological and operational drivers, and DQN responded primarily to direct climatic cues. Collectively, these findings confirm that deep reinforcement learning can produce adaptive and explainable control behaviors grounded in physically meaningful relationships between environmental conditions and reservoir operation.

To further clarify the role of these multi-source variables, the SHAP analysis showed that variables such as soil moisture, humidity, and evaporation—though not direct flood indicators—substantially influenced discharge decisions. Their inclusion enhanced the agents' ability to anticipate flood risk buildup through better representation of basin wetness and atmospheric demand. Preliminary comparisons using only ground-based variables yielded slightly higher RMSE and lower cumulative rewards, confirming that multi-source inputs improved model stability and hydrological realism.

The DRL-based operating strategies (DQN, DDPG, PPO) developed in this study utilized a diverse set of hydrometeorological variables—including precipitation, soil moisture, and temperature—to guide reservoir operation decisions. Although these variables were not explicitly visualized, they served as essential inputs, enabling dynamic responses to changing hydrological conditions and informing both flood risk estimation and discharge optimization. A key contribution of this work lies in the integration of observed data, remote-sensing products, and DRL-driven decision-making into a unified framework, allowing the operating strategies to make context-aware releases and storage adjustments and manage storage adaptively. Unlike many existing studies—such as Phankamolsil et al. [13], who applied DDPG without explicit flood risk metrics; Castro-Freibott et al. [70], who used PPO but without integrating remote-sensing inputs; and Castelletti et al. [71], who applied Q-learning without examining adaptive capacity at daily resolution—this study leverages multi-source datasets and explicitly incorporates flood risk trade-offs. Thus, this study advances previous DRL applications by demonstrating flood-peak attenuation at daily resolution with paced recession releases and maintained storage buffers, delivering a resilient and operationally robust approach to optimized reservoir operation under hydrological extremes.

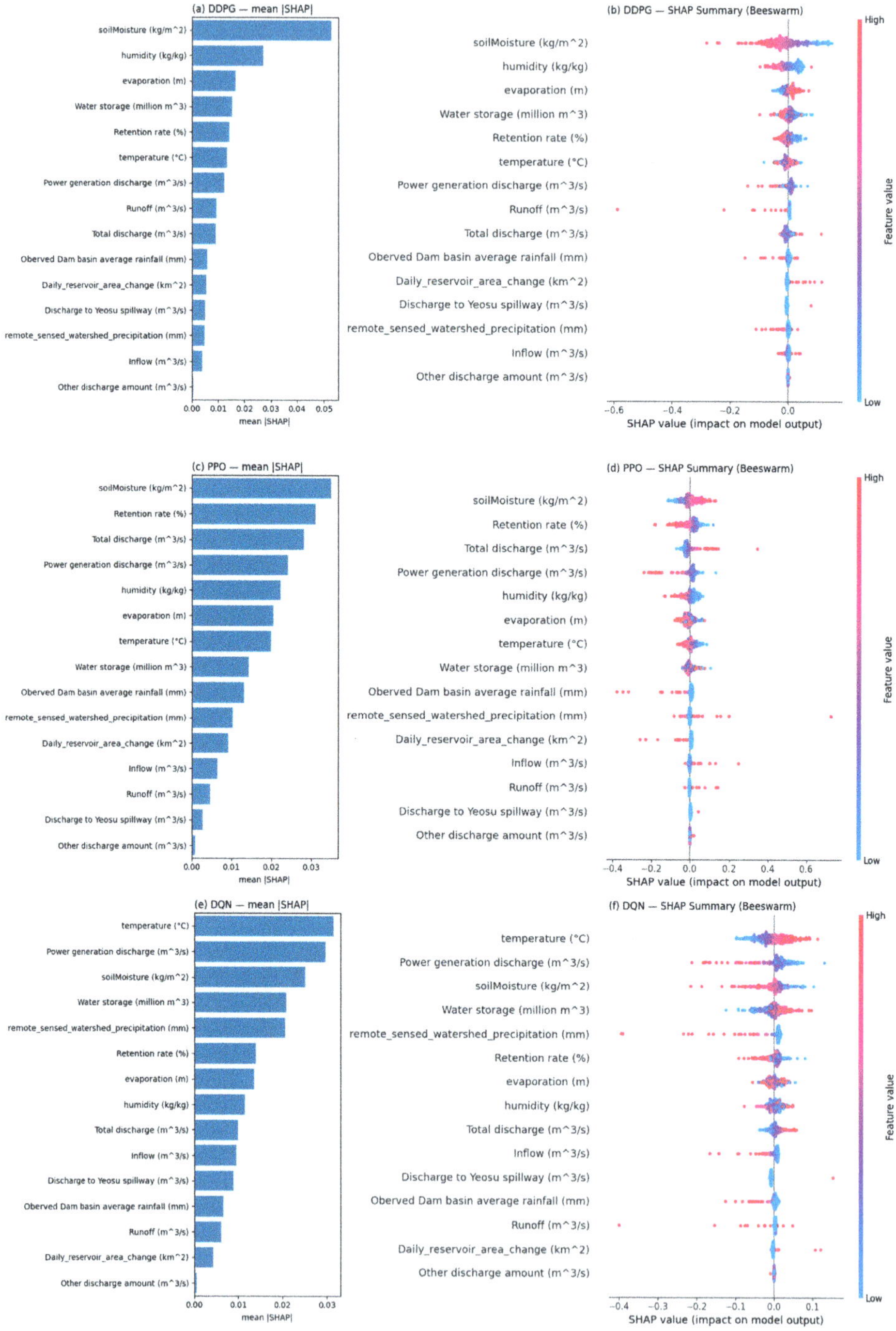

Figure 10. SHAP-based variable importance and feature effects on model predictions for the DDPG, PPO, and DQN models.

5. Conclusions

This study demonstrated that deep reinforcement learning (DRL) can be used to optimize operation at the Soyang Reservoir by coordinating inflow, storage, and discharge to moderate flood-related risk over a 30-year daily record (1993–2022). Relative to observed operation, the DRL strategies delivered safer performance—larger storage buffers, typically lower peak discharges on high-inflow days, and low violation counts—under the same hydrologic conditions. Based on the flood risk (FR) metric, DRL-driven operation strategies generally flatten peak discharges, keep storage closer to the flood-safe bound (FR ≈ 0), increase the safety buffer (more negative FR) on high-inflow days, and maintain fewer threshold exceedances relative to observed operation at daily resolution—both on the highlighted dates and across the 1993–2022 record. Rather than predicting discharge directly, discharge was derived via a mass-balance formulation from optimized storage while preserving observed inflows, ensuring physically consistent releases under the original hydrologic forcing.

Over the full 1993–2022 period, all three agents—DQN, PPO, and DDPG—learned physically plausible, safety-aware behavior. PPO achieved the highest cumulative reward and the most stable actions. DQN ranked second on reward and prioritized flood control compliance, maintaining the largest buffers with conservative releases on the rising limb (pre-inflow-peak). DDPG provided continuous control with intermediate responsiveness, occasionally allowing a short near-crest pulse. These roles align with the time-series patterns at the highlighted events: DQN tends to maintain the largest buffers, PPO produces smoother, mid-level releases, and DDPG is more active near peaks.

Analyses anchored to the largest inflows—25 Aug 1995, 1–2 Aug 1999, and 15 Jul 2006—show that DRL-optimized operation reduced storage relative to observed operation during the peak, moderated releases on the rising limb, and shifted releases to the recession. In combination, these behaviors kept the flood risk at or below the threshold (FR ≤ 0) for the DRL runs while preserving buffer capacity. The dominant operational pattern was clear: retain during the rising limb, then release during the recession—timing that leverages higher channel conveyance and lower downstream risk. For PPO in particular, the inflow–release decomposition makes the mechanism explicit: water stored dominates pre-crest, followed by water released on the recession, avoiding secondary peaks while maintaining buffer.

A central contribution is that the DRL operation strategies were informed by a rich set of observed and remote-sensing hydrometeorological inputs, enabling context-aware decisions that link inflow, storage, and discharge. Compared with traditional optimization methods such as DP/SDP, which are hampered by the curse of dimensionality and rigid system representations, the DRL framework here scales to high-dimensional inputs and captures nonlinear interactions without hand-crafted rule sets—moving beyond static rules and limited-variable models. The deterministic storage-based exceedance index provided an operationally relevant measure that translated daily hydrologic variability into actionable reservoir operation.

Several limitations bound interpretation: the "flood risk" metric is deterministic and storage-based (not a full probabilistic risk); discharge was derived by continuity (not a hydraulic routing model); analyses were daily and single reservoir. The DRL approaches tested here show that operationally meaningful attenuation at the Soyang reservoir can be achieved by shaping the timing and size of releases relative to the inflow hydrograph through explicit optimization of an objective function that penalizes flood-threshold exceedance and deviation from the target storage while respecting physical/operational constraints. Future research should explore hybrid DRL-supervised learning frameworks and assess the robustness of DRL-based reservoir operations under projected climate change scenarios, and explore coupling with hydrodynamic and inundation models to further enhance adaptability and long-term reliability.

Author Contributions: Conceptualization, F.S. and K.S.J.; methodology, F.S.; software, F.S.; validation, F.S.; formal analysis, F.S.; investigation, F.S.; resources, K.S.J.; data curation, F.S.; writing—original draft preparation, F.S.; writing—review and editing, F.S. and K.S.J.; visualization, F.S.; supervision, K.S.J.; project administration, K.S.J.; funding acquisition, K.S.J. All authors have read and agreed to the published version of the manuscript.

Funding: This work was supported by the Korea Environmental Industry and Technology Institute (KEITI) (Grant number: 2022003460001).

Data Availability Statement: The data presented in this study are openly available in the [Water Resources Management Information System (WAMIS) of Korea] at [https://www.wamis.go.kr/] (accessed on 9 November 2025).

Acknowledgments: The authors thank Korea Water Resources Corporation (K-Water) and the Han River Flood Control Office (HRFCO) for providing dam-related hydrometeorological data. We also acknowledge UCSB/CHG, ECMWF, Copernicus, CERN/Zenodo, NASA/CGIAR, NASA GES DISC, and NASA USGS for freely providing various datasets, including CHIRPS, ERA5-Land, soil texture, SRTM, humidity, soil moisture, and Landsat data.

Conflicts of Interest: The authors declare no conflicts of interest. The funders had no role in the design of the study; in the collection, analyses, or interpretation of data; in the writing of the manuscript; or in the decision to publish the results.

References

1. Bakis, R. Electricity Production Opportunities from Multipurpose Dams (Case Study). *Renew. Energy* **2007**, *32*, 1723–1738. [CrossRef]
2. Yu, X.; Xu, Y.P.; Gu, H.; Guo, Y. Multi-Objective Robust Optimization of Reservoir Operation for Real-Time Flood Control under Forecasting Uncertainty. *J. Hydrol.* **2023**, *620*, 129421. [CrossRef]
3. Thompson, C.M.; Frazier, T.G. Deterministic and Probabilistic Flood Modeling for Contemporary and Future Coastal and Inland Precipitation Inundation. *Appl. Geogr.* **2014**, *50*, 1–14. [CrossRef]
4. Lu, Q.; Zhong, P.-A.; Xu, B.; Zhu, F.; Ma, Y.; Wang, H.; Xu, S. Risk Analysis for Reservoir Flood Control Operation Considering Two-Dimensional Uncertainties Based on Bayesian Network. *J. Hydrol.* **2020**, *589*, 125353. [CrossRef]
5. Mahootchi, M.; Tizhoosh, H.R.; Ponnambalam, K.P. Reservoir Operation Optimization by Reinforcement Learning. *J. Water Manag. Model.* **2007**, *8*, 165–184. [CrossRef]
6. Delipetrev, B.; Jonoski, A.; Solomatine, D. *Optimal Reservoir Operation Policies Using Novel Nested Algorithms*; European Geosciences Union: Munich, Germany, 2015; Volume 17.
7. Wu, X.; Cheng, C.; Lund, J.R.; Niu, W.; Miao, S. Stochastic Dynamic Programming for Hydropower Reservoir Operations with Multiple Local Optima. *J. Hydrol.* **2018**, *564*, 712–722. [CrossRef]
8. Giuliani, M.; Galelli, S.; Soncini-Sessa, R. A Dimensionality Reduction Approach for Many-Objective Markov Decision Processes: Application to a Water Reservoir Operation Problem. *Environ. Model. Softw.* **2014**, *57*, 101–114. [CrossRef]
9. Wu, W.; Eamen, L.; Dandy, G.; Razavi, S.; Kuczera, G.; Maier, H.R. Beyond Engineering: A Review of Reservoir Management through the Lens of Wickedness, Competing Objectives and Uncertainty. *Environ. Model. Softw.* **2023**, *167*, 105777. [CrossRef]
10. Hung, F.; Yang, Y.C.E. Assessing Adaptive Irrigation Impacts on Water Scarcity in Nonstationary Environments—A Multi-Agent Reinforcement Learning Approach. *Water Resour. Res.* **2021**, *57*, e2020WR029262. [CrossRef]
11. Jiang, Q.; Li, J.; Sun, Y.; Huang, J.; Zou, R.; Ma, W.; Guo, H.; Wang, Z.; Liu, Y. Deep-Reinforcement-Learning-Based Water Diversion Strategy. *Environ. Sci. Ecotechnology* **2024**, *17*, 100298. [CrossRef] [PubMed]
12. Luo, W.; Wang, C.; Zhang, Y.; Zhao, J.; Huang, Z.; Wang, J.; Zhang, C. A Deep Reinforcement Learning Approach for Joint Scheduling of Cascade Reservoir System. *J. Hydrol.* **2025**, *651*, 132515. [CrossRef]
13. Phankamolsil, Y.; Rittima, A.; Sawangphol, W.; Kraisangka, J.; Tabucanon, A.S.; Talaluxmana, Y.; Vudhivanich, V. Deep Reinforcement Learning for Multiple Reservoir Operation Planning in the Chao Phraya River Basin Deep Reinforcement Learning (DRL)

Deep Deterministic Policy Gradient (DDPG) Algorithm Artificial Intelligence (AI) Reservoir Operation Planning Chao Phraya River Basin. *Model. Earth Syst. Environ.* **2025**, *11*, 102. [CrossRef]
14. Kim, S.K.; Ahn, H.; Kang, H.; Jeon, D.J. Identification of Preferential Target Sites for the Environmental Flow Estimation Using a Simple Flowchart in Korea. *Environ. Monit. Assess.* **2022**, *194*, 215. [CrossRef] [PubMed]
15. Xu, W.; Zhang, X.; Peng, A.; Liang, Y. Deep Reinforcement Learning for Cascaded Hydropower Reservoirs Considering Inflow Forecasts. *Water Resour. Manag.* **2020**, *34*, 3003–3018. [CrossRef]
16. Li, Z.; Bai, L.; Tian, W.; Yan, H.; Hu, W.; Xin, K.; Tao, T. Online Control of the Raw Water System of a High-Sediment River Based on Deep Reinforcement Learning. *Water* **2023**, *15*, 1131. [CrossRef]
17. Xu, H.; Yan, Z.; Xuan, J.; Zhang, G.; Lu, J. Improving Proximal Policy Optimization with Alpha Divergence. *Neurocomputing* **2023**, *534*, 94–105. [CrossRef]
18. Ho, C.-H.; Kim, H.-A.; Cha, Y.; Do, H.-S.; Kim, J.; Kim, J.; Park, S.K.; Yoo, H.-D. Recent Changes in Summer Rainfall Characteristics in Korea. *J. Eur. Meteorol. Soc.* **2025**, *2*, 100009. [CrossRef]
19. Bowes, B.D.; Tavakoli, A.; Wang, C.; Heydarian, A.; Behl, M.; Beling, P.A.; Goodall, J.L. Flood Mitigation in Coastal Urban Catchments Using Real-Time Stormwater Infrastructure Control and Reinforcement Learning. *J. Hydroinformatics* **2021**, *23*, 529–547. [CrossRef]
20. Kåge, L.; Milić, V.; Andersson, M.; Wallén, M. Reinforcement Learning Applications in Water Resource Management: A Systematic Literature Review. *Front. Water* **2025**, *7*, 1537868. [CrossRef]
21. Wan, Z.; Li, W.; He, M.; Zhang, T.; Chen, S.; Guan, W.; Hua, X.; Zheng, S. Research on Long-Term Scheduling Optimization of Water–Wind–Solar Multi-Energy Complementary System Based on DDPG. *Energies* **2025**, *18*, 3983. [CrossRef]
22. Qian, X.; Wang, B.; Chen, J.; Fan, Y.; Mo, R.; Xu, C.; Liu, W.; Liu, J.; Zhong, P. an An Explainable Ensemble Deep Learning Model for Long-Term Streamflow Forecasting under Multiple Uncertainties. *J. Hydrol.* **2025**, *662*, 133968. [CrossRef]
23. Sung, J.; Kang, B. Comparative Study of Low Flow Frequency Analysis Using Bivariate Copula Model at Soyanggang Dam and Chungju Dam. *Hydrology* **2024**, *11*, 79. [CrossRef]
24. Lee, S.; Kim, J. Predicting Inflow Rate of the Soyang River Dam Using Deep Learning Techniques. *Water* **2021**, *13*, 2447. [CrossRef]
25. K-water. *2018 Sustainability Report: Providing a Brighter, Happier, and More Prosperous Future with Water*; Publication No. 2018-MA-GP-18-107; K-water: Daejeon, Republic of Korea, 2018; Available online: https://www.kwater.or.kr/web/eng/download/smreport/2018_SMReport.pdf?utm (accessed on 9 November 2025).
26. Kim, J.S.; Jain, S.; Kang, H.Y.; Moon, Y.I.; Lee, J.H. Inflow into Korea's Soyang Dam: Hydrologic Variability and Links to Typhoon Impacts. *J. Hydro-Environ. Res.* **2019**, *22*, 50–56. [CrossRef]
27. Kwak, J. An Assessment of Dam Operation Considering Flood and Low-Flow Control in the Han River Basin. *Water* **2021**, *13*, 733. [CrossRef]
28. Zhang, L.; Deng, C.; Wei, J.; Zou, J. Assessing the Impacts of Climate Change and Land Use/Land Cover Data Characteristics on Streamflow Using the SWAT Model in the Upper Han River Basin. *J. Hydrol. Reg. Stud.* **2025**, *61*, 102764. [CrossRef]
29. Arregocés, H.A.; Rojano, R.; Pérez, J. Validation of the CHIRPS Dataset in a Coastal Region with Extensive Plains and Complex Topography. *Case Stud. Chem. Environ. Eng.* **2023**, *8*, 100452. [CrossRef]
30. Egorov, A.V.; Roy, D.P.; Zhang, H.K.; Li, Z.; Yan, L.; Huang, H. Landsat 4, 5 and 7 (1982 to 2017) Analysis Ready Data (ARD) Observation Coverage over the Conterminous United States and Implications for Terrestrial Monitoring. *Remote. Sens.* **2019**, *11*, 447. [CrossRef]
31. Xu, X.; Chen, F.; Wang, B.; Harrison, M.T.; Chen, Y.; Liu, K.; Zhang, C.; Zhang, M.; Zhang, X.; Feng, P.; et al. Unleashing the Power of Machine Learning and Remote Sensing for Robust Seasonal Drought Monitoring: A Stacking Ensemble Approach. *J. Hydrol.* **2024**, *634*, 131102. [CrossRef]
32. McNally, A.; Arsenault, K.; Kumar, S.; Shukla, S.; Peterson, P.; Wang, S.; Funk, C.; Peters-Lidard, C.D.; Verdin, J.P. A Land Data Assimilation System for Sub-Saharan Africa Food and Water Security Applications. *Sci. Data* **2017**, *4*, 170012. [CrossRef]
33. Jung, H.C.; Getirana, A.; Policelli, F.; McNally, A.; Arsenault, K.R.; Kumar, S.; Tadesse, T.; Peters-Lidard, C.D. Upper Blue Nile Basin Water Budget from a Multi-Model Perspective. *J. Hydrol.* **2017**, *555*, 535–546. [CrossRef]
34. Qi, W.; Liu, J.; Chen, D. Evaluations and Improvements of GLDAS2.0 and GLDAS2.1 Forcing Data's Applicability for Basin Scale Hydrological Simulations in the Tibetan Plateau. *J. Geophys. Res. Atmos.* **2018**, *123*, 13,128–13,148. [CrossRef]
35. Gomis-Cebolla, J.; Rattayova, V.; Salazar-Galán, S.; Francés, F. Evaluation of ERA5 and ERA5-Land Reanalysis Precipitation Datasets over Spain (1951–2020). *Atmos. Res.* **2023**, *284*, 106606. [CrossRef]
36. Dong, G.; Huang, W.; Smith, W.A.P.; Ren, P. A Shadow Constrained Conditional Generative Adversarial Net for SRTM Data Restoration. *Remote Sens. Environ.* **2020**, *237*, 111602. [CrossRef]
37. Corral-Pazos-de-Provens, E.; Rapp-Arrarás, Í.; Domingo-Santos, J.M. Estimating Textural Fractions of the USDA Using Those of the International System: A Quantile Approach. *Geoderma* **2022**, *416*, 115783. [CrossRef]

38. Chirachawala, C.; Shrestha, S.; Babel, M.S.; Virdis, S.G.P.; Wichakul, S. Evaluation of Global Land Use/Land Cover Products for Hydrologic Simulation in the Upper Yom River Basin, Thailand. *Sci. Total Environ.* **2020**, *708*, 135148. [CrossRef] [PubMed]
39. Qi, J.; Lee, S.; Du, X.; Ficklin, D.L.; Wang, Q.; Myers, D.; Singh, D.; Moglen, G.E.; McCarty, G.W.; Zhou, Y.; et al. Coupling Terrestrial and Aquatic Thermal Processes for Improving Stream Temperature Modeling at the Watershed Scale. *J. Hydrol.* **2021**, *603*, 126983. [CrossRef]
40. Xu, W.; Meng, F.; Guo, W.; Li, X.; Fu, G. Deep Reinforcement Learning for Optimal Hydropower Reservoir Operation. *J. Water Resour. Plan. Manag.* **2021**, *147*, 04021045. [CrossRef]
41. Ahmad, M.J.; Cho, G.H.; Choi, K.S. Historical Climate Change Impacts on the Water Balance and Storage Capacity of Agricultural Reservoirs in Small Ungauged Watersheds. *J. Hydrol. Reg. Stud.* **2022**, *41*, 101114. [CrossRef]
42. Wang, N.; Liu, L.; Shi, T.; Wang, Y.; Huang, J.; Ye, R.; Lian, Z. Study of the Impact of Reservoir Water Level Decline on the Stability Treated Landslide on Reservoir Bank. *Alex. Eng. J.* **2023**, *65*, 481–492. [CrossRef]
43. Lee, E.; Ji, J.; Lee, S.; Yoon, J.; Yi, S.; Yi, J. Development of an Optimal Water Allocation Model for Reservoir System Operation. *Water* **2023**, *15*, 3555. [CrossRef]
44. Ghobadi, F.; Kang, D. Application of Machine Learning in Water Resources Management: A Systematic Literature Review. *Water* **2023**, *15*, 620. [CrossRef]
45. Song, J.H.; Her, Y.; Kang, M.S. Estimating Reservoir Inflow and Outflow From Water Level Observations Using Expert Knowledge: Dealing with an Ill-Posed Water Balance Equation in Reservoir Management. *Water Resour. Res.* **2022**, *58*, e2020WR028183. [CrossRef]
46. Lu, Q.; Zhong, P.A.; Xu, B.; Zhu, F.; Huang, X.; Wang, H.; Ma, Y. Stochastic Programming for Floodwater Utilization of a Complex Multi-Reservoir System Considering Risk Constraints. *J. Hydrol.* **2021**, *599*, 126388. [CrossRef]
47. Golian, S.; Yazdi, J.; Martina, M.L.V.; Sheshangosht, S. A Deterministic Framework for Selecting a Flood Forecasting and Warning System at Watershed Scale. *J. Flood Risk Manag.* **2015**, *8*, 356–367. [CrossRef]
48. Sutton, R.S.; Barto, A.G. *Reinforcement Learning: An Introduction*, 2nd ed.; MIT Press: Cambridge, MA, USA, 2018.
49. Mnih, V.; Kavukcuoglu, K.; Silver, D.; Rusu, A.A.; Veness, J.; Bellemare, M.G.; Graves, A.; Riedmiller, M.; Fidjeland, A.K.; Ostrovski, G.; et al. Human-Level Control through Deep Reinforcement Learning. *Nature* **2015**, *518*, 529–533. [CrossRef]
50. Campo Carrera, J.M.; Udias, A. Deep Reinforcement Learning for Complex Hydropower Management: Evaluating Soft Actor-Critic with a Learned System Dynamics Model. *Front. Water* **2025**, *7*, 1649284. [CrossRef]
51. Yassin, F.; Razavi, S.; Elshamy, M.; Davison, B.; Sapriza-Azuri, G.; Wheater, H. Representation and Improved Parameterization of Reservoir Operation in Hydrological and Land-Surface Models. *Hydrol. Earth Syst. Sci.* **2019**, *23*, 3735–3764. [CrossRef]
52. Helseth, A.; Mo, B.; Hågenvik, H.O.; Schäffer, L.E. Hydropower Scheduling with State-Dependent Discharge Constraints: An SDDP Approach. *J. Water Resour. Plan. Manag.* **2022**, *148*, 04022061. [CrossRef]
53. Tharme, R.E. A Global Perspective on Environmental Flow Assessment: Emerging Trends in the Development and Application of Environmental Flow Methodologies for Rivers. *River Res. Appl.* **2003**, *19*, 397–441. [CrossRef]
54. Serinaldi, F.; Kilsby, C.G.; Lombardo, F. Untenable Nonstationarity: An Assessment of the Fitness for Purpose of Trend Tests in Hydrology. *Adv. Water Resour.* **2018**, *111*, 132–155. [CrossRef]
55. Xu, J.; Qiao, J.; Sun, Q.; Shen, K. A Deep Reinforcement Learning Framework for Cascade Reservoir Operations Under Runoff Uncertainty. *Water* **2025**, *17*, 2324. [CrossRef]
56. Giuliani, M.; Herman, J.D.; Castelletti, A.; Reed, P. Many-Objective Reservoir Policy Identification and Refinement to Reduce Policy Inertia and Myopia in Water Management. *Water Resour. Res.* **2014**, *50*, 3355–3377. [CrossRef]
57. Helsel, D.R.; Hirsch, R.M.; Ryberg, K.R.; Archfield, S.A.; Gilroy, E.J. *Statistical Methods in Water Resources*; US Geological Survey: Reston, VA, USA, 2020.
58. Mockus, V. *USDA Soil Conservation Service. National Engineering Handbook, Section 4: Hydrology, Chapter 17—Flood Routing*; USDA: Washington, DC, USA, 1967.
59. Lai, V.; Huang, Y.F.; Koo, C.H.; Ahmed, A.N.; El-Shafie, A. A Review of Reservoir Operation Optimisations: From Traditional Models to Metaheuristic Algorithms. *Arch. Comput. Methods Eng.* **2022**, *29*, 3435–3457. [CrossRef]
60. Niu, S.; Insley, M. On the Economics of Ramping Rate Restrictions at Hydro Power Plants: Balancing Profitability and Environmental Costs. *Energy Econ.* **2013**, *39*, 39–52. [CrossRef]
61. Tian, W.; Xin, K.; Zhang, Z.; Zhao, M.; Liao, Z.; Tao, T. Flooding Mitigation through Safe & Trustworthy Reinforcement Learning. *J. Hydrol.* **2023**, *620*, 129435. [CrossRef]
62. Clark, M.P.; Vogel, R.M.; Lamontagne, J.R.; Mizukami, N.; Knoben, W.J.M.; Tang, G.; Gharari, S.; Freer, J.E.; Whitfield, P.H.; Shook, K.R.; et al. The Abuse of Popular Performance Metrics in Hydrologic Modeling. *Water Resour. Res.* **2021**, *57*, e2020WR029001. [CrossRef]

63. Accadia, C.; Casaioli, M.; Mariani, S.; Lavagnini, A.; Speranza, A.; De Venere, A.; Inghilesi, R.; Ferretti, R.; Paolucci, T.; Cesari, D.; et al. Application of a Statistical Methodology for Limited Area Model Intercomparison Using a Bootstrap Technique. *Nuovo C.-Soc. Ital. Di Fis. Sez. C* **2003**, *26*, 125–140.
64. Fang, K.; Kifer, D.; Lawson, K.; Feng, D.; Shen, C. The Data Synergy Effects of Time-Series Deep Learning Models in Hydrology. *Water Resour. Res.* **2022**, *58*, e2021WR029583. [CrossRef]
65. McInerney, D.; Thyer, M.; Kavetski, D.; Laugesen, R.; Tuteja, N.; Kuczera, G. Multi-Temporal Hydrological Residual Error Modeling for Seamless Subseasonal Streamflow Forecasting. *Water Resour. Res.* **2020**, *56*, e2019WR026979. [CrossRef]
66. Pool, S.; Vis, M.; Seibert, J. Evaluating Model Performance: Towards a Non-Parametric Variant of the Kling-Gupta Efficiency. *Hydrol. Sci. J.* **2018**, *63*, 1941–1953. [CrossRef]
67. Choi, H. Il Development of Flood Damage Regression Models by Rainfall Identification Reflecting Landscape Features in Gangwon Province, the Republic of Korea. *Land* **2021**, *10*, 123. [CrossRef]
68. Xi, H.; Luo, Z.; Guo, Y. Reservoir Evaluation Method Based on Explainable Machine Learning with Small Samples. *Unconv. Resour.* **2025**, *5*, 100128. [CrossRef]
69. Zhang, T.; Wu, J.; Chu, H.; Liu, J.; Wang, G. Interpretable Machine Learning Based Quantification of the Impact of Water Quality Indicators on Groundwater Under Multiple Pollution Sources. *Water* **2025**, *17*, 905. [CrossRef]
70. Castro-Freibott, R.; García-Sánchez, Á.; Espiga-Fernández, F.; González-Santander de la Cruz, G. Deep Reinforcement Learning for Intraday Multireservoir Hydropower Management. *Mathematics* **2025**, *13*, 151. [CrossRef]
71. Castelletti, A.; Galelli, S.; Restelli, M.; Soncini-Sessa, R. Tree-Based Reinforcement Learning for Optimal Water Reservoir Operation. *Water Resour. Res.* **2010**, *46*, W09507. [CrossRef]

water

MDPI

Article

Combination of UAV Imagery and Deep Learning to Estimate Vegetation Height over Fluvial Sandbars

Yiwei Guo [1,*], Michael Nones [1], Yuexia Zhou [2], Runye Zhu [3] and Wenfeng Ding [4,*]

[1] Institute of Geophysics Polish Academy of Sciences, 01-461 Warsaw, Poland; mnones@igf.edu.pl
[2] Department of Civil and Environmental Engineering, Nagoya University, Nagoya 4648603, Japan; zhou@civil.nagoya-u.ac.jp
[3] Zhejiang Institute of Hydraulics and Estuary, Hangzhou 310058, China; zhurunye@outlook.com
[4] Changjiang River Scientific Research Institute, Changjiang Water Resource Commission, Wuhan 430010, China
* Correspondence: yguo@igf.edu.pl (Y.G.); dingwf@mail.crsri.cn (W.D.)

Abstract: Vegetation colonizing fluvial sandbars provides many noteworthy functions in river and floodplain systems, but it also influences hydrodynamic processes, mainly during flooding events. Numerical modelling is generally used to evaluate the impact of floods, but its reliability is very much connected with the accuracy of the bed and bank roughness, which is eventually altered by the presence of vegetation and its height. However, for the sake of simplicity, most models tend to ignore how the sandbar roughness varies over space and time, as a function of the local vegetation dynamics (spatial distribution and height). To determine the long-term dynamic vegetation condition using remote sensing multispectral indexes, this study leverages a deep-learning method to establish a relationship between vegetation height (h), a critical parameter for vegetation roughness estimation, and vegetation indexes (VIs) collected by an uncrewed aerial vehicle (UAV). A field campaign was performed in October 2024 covering the Baishazhou sandbar, located along a straight section of the Wuhan reach of the Changjiang River Basin, China. The results show that the R^2 and RMSE between the real and predicted vegetation height by the trained Fully Connected Neural Network (FCNN) are 0.85, 1.10 m, and the relative error reaches a maximum of 17.2%, meaning that the trained FCNN model performs rather well. Despite being tested on a single case study, the workflow presented here demonstrates the opportunity to use UAVs for depicting vegetation characteristics such as height over large areas, eventually using them to inform numerical models that consider sandbar roughness.

Keywords: Baishazhou; Fully Connected Neural Network; sandbars; UAV; vegetation height; vegetation index

1. Introduction

Sandbars are key geomorphic features in fluvial systems, crucial for bedload transport and sediment storage [1,2]. Formed by hydrodynamic-sedimentary interactions, they evolve dynamically with flow and sediment changes [3]. Sandbar morphology and vegetation cover serve as vital indicators of riverine dynamics and channel morphology [4–6], revealing spatiotemporal feedbacks between primary and secondary channels and adjacent floodplains. Highly vegetated sandbars can function similarly to riverbanks during low flow [7]. The establishment of vegetation increases surface roughness and flow resistance, which reduces floodplain conveyance capacity and elevates local water levels, thereby increasing flood risk [8]. Vegetation conditions such as height, density, and species composition further govern these hydrodynamic and sediment transport processes during floods [9]. The vegetation-induced increased resistance can distort the stage-discharge relationship, leading to higher water levels and greater flooding hazard, particularly in conveyance-sensitive, low-gradient systems [10].

In hydrodynamic models, a fundamental tool for flood prediction, risk assessment, and water resource management, surface roughness is critical to obtain accurate simulations [11,12]. However, roughness of fluvial surfaces (e.g., riverbeds, banks) is not static [12], but dynamically responds to vegetation changes. Indeed, extreme floods can rapidly alter vegetation structure and distribution [13], causing significant shifts in hydraulic characteristics and compromising predictions of water levels and inundation extents.

Predictive models that incorporate temporally variable vegetation characteristics are vital for accurately estimating changes in flow resistance over time [14]. To support such modelling efforts, robust and continuous monitoring of vegetation dynamics is indispensable. This includes tracking changes in vegetation height, density, spatial coverage, and species composition across different temporal scales, from seasonal growth cycles to rapid post-flood recovery periods. Accurate spatial and temporal data on vegetation are critical not only for capturing the interactions between flow, morphology, and ecology (i.e., hydro-morpho-biodynamic processes), but also for informing adaptive management strategies aimed at flood mitigation, habitat enhancement, and sustainable river management [15].

Previous studies have investigated the relationship between roughness and riparian vegetation via experimental methods [16–19]. All those studies concluded that vegetation characteristics significantly influence the flow structure by means of changes in roughness. Traditional field measurements of vegetation using handheld instruments [20] are increasingly supplemented by remote sensing, which offers superior spatial coverage, resolution, and efficiency [21]. Consequently, remote sensing is often preferred for characterizing vegetation before roughness calculations in practical applications. Multispectral Vegetation Indices (VIs), particularly the widely used Normalized Difference Vegetation Index (NDVI), have been established to monitor vegetation dynamics [22–24], differentiate vegetated areas [25], and quantify biomass [26]. Recent studies pointed out that its variant, the Green Normalized Difference Vegetation Index (GNDVI), is more effective for assessing health in dense canopies [27].

Given easier access, as well as advantages such as cost- and time-effectiveness, VIs calculated from satellite images to estimate vegetation parameters are more widely used to derive potential relationships between VIs and vegetation parameters such as vegetation height or density [28,29]. Additionally, with the advantage of having higher spatial resolution when compared with satellite images, multispectral uncrewed aerial vehicles (UAVs) have been used to obtain vegetation conditions recently [30], especially in the estimation of vegetation height [22,31,32]. Although previous studies successfully applied drones to monitor the vegetation height, most of them focused on a single plant species, while research on regions with various plants is limited. At the same time, UAVs can not only measure the vegetation height and density directly but also capture multispectral images for calculating VIs. The combination of UAVs and satellite images makes vegetation monitoring on long-time scales much easier to achieve. Specifically, after understanding how vegetation parameters can be explained by VIs determined from UAV collections, long-term vegetation parameters can be estimated from VIs derived from satellite images.

However, establishing relationships between VIs and vegetation parameters remains challenging, especially in the case of vegetation height, which is an important parameter for estimating vegetation roughness [21,33]. Fortunately, deep-learning (DL) models, with their elevated performance and ability to handle complex input-output relationships, might help address this limitation thanks to their higher prediction accuracy compared to traditional fitting methods [34]. The main applications of DL in remote sensing include image-based object detection and multimodal data fusion [35]. While classical machine learning (ML) methods like Random Forest have shown effectiveness in vegetation identification and height estimation [36–38], their supervised learning nature demands substantial manual annotation. To address this limitation, researchers have started exploring self-supervised and semi-supervised learning approaches [39–41]. For example, Du et al. [42] utilized a CNN Transformer hybrid model called NeCT Net to extract spatial features and employed a semi-supervised teacher-student framework to automatically generate pseudo labels for expanding training data. This method effectively reduces dependence on manual annotation while improving the accuracy of large-scale building height estimation. Although supervised learning approaches have demonstrated significant success in vegetation height estimation, their applications have predominantly focused on forest canopies or single crop types, with limited exploration in multi-class study areas comprising complex vegetation communities. Concurrently, while advanced DL architectures such as CNN-Transformer hybrids have achieved breakthroughs in building height estimation by integrating semi-supervised strategies, these models often entail complex structures and high computational costs. In response, this study adopts a streamlined FCNN as its core architecture to substantially simplify the training process and computational requirements while maintaining high accuracy.

Aiming to establish a relationship between vegetation height (h) and VIs (NDVI, GNDVI, Normalized Difference Red Edge Index (NDRE), Leaf Chlorophyll Index (LCI), and Optimized Soil Adjusted Vegetation Index (OSAVI), to lay the foundation for the calculation of vegetation roughness, this research analysed vegetation conditions and multispectral images of the Baishazhou middle sandbar, in the middle reaches of the Yangtze River (China), using a UAV (DJI Mavic 3 Multispectral) and employed the FCNN model for analysis. Using a reduced number of data points, the h-VIs relationship was established, eventually allowing us to extract it over time to monitor long-term variations in vegetation conditions from satellite images, therefore reducing the need for prolonged and cumbersome UAV field campaigns. By means of a specific case study, the current work proposes a workflow to determine the relationship between vegetation indexes and vegetation height, which can be applied to both UAV and satellite imagery, eventually assisting in extracting dynamic roughness to be used as a reliable input for hydrodynamic flood modelling.

2. Materials and Methods

2.1. Study Area

Located in the straight section of the Wuhan reach of the Changjiang River Basin (Figure 1), the Baishazhou (BSZ) sandbar is oriented in a northeast-southwest direction. The left branch of the sandbar serves as the main channel, and the right branch is the tributary. Formed by the sediment carried by the river water, the BSZ sandbar takes the shape of a bamboo leaf, which is typical of those whose upstream and downstream ends are pointed, while the middle section is wider, with a measured length of 2.2 km and a maximum width of 0.3 km during the dry season.

The BSZ sandbar is located in a region characterized by a subtropical monsoon climate, with average yearly temperature and precipitation around 17.7 °C and 1204.50 mm, respectively [43]. The annual average runoff and sediment from 1990 to 2020 recorded by the Hankou Hydrological station, which is 20 km downstream of the BSZ sandbar, are 7012.73×10^8 m^3 and 1.94×10^8 t, respectively [44].

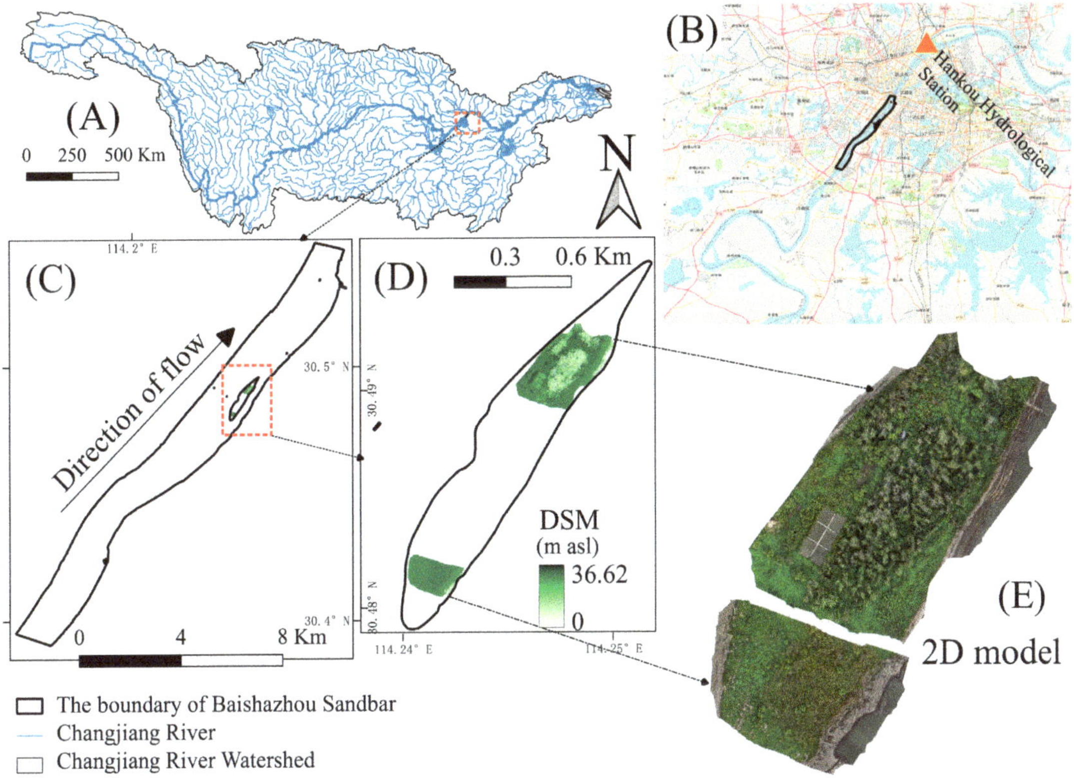

Figure 1. Location of the study area, (**A**,**B**) are the locations of the BSZ sandbar at the Changjiang River Basin and Wuhan city, respectively, (**C**) is the part of the river channel where the BSZ sandbar is located, (**D**) is the boundary of the BSZ sandbar and the Digital Surface Model (DSM) of the two mapped areas, (**E**) is the 2D model of the sampling areas.

Since the Baishazhou Bridge spans the middle of the sandbar (Figure 1B), the area above the middle part of the sandbar is designated as a no-fly zone. Therefore, for this research, two small areas on either side of the bridge were selected as study areas to obtain the images (Figure 1A).

2.2. UAV Data Collecting and Processing

2.2.1. UAV Information and Data Collection

A DJI Mavic 3 Multispectral (M3M) (SZ DJI Technology Co., Ltd., Shenzhen, China) equipped with a natural colour camera and a four-band multispectral sensor (Figure 2C,D) was used to obtain spectral information of the sandbar (green band: 560 nm; red band: 650 nm; red edge: 730 nm; near-infrared: 860 nm) [45]. The flight path (Figure 2A,B) was pre-programmed using the DJI Pilot software v10.00.06.02 with a flying height of approx. 60 m. The drone flight took place during the Autumn season, on 12 October 2024, a sunny day.

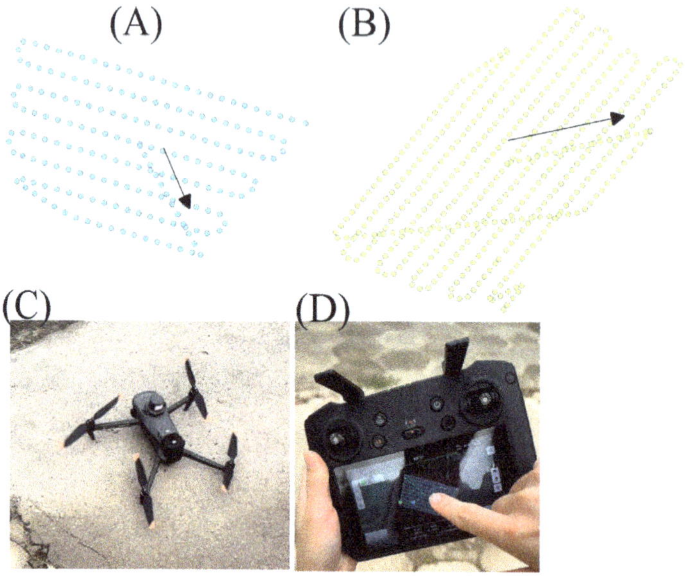

Figure 2. (**A**,**B**) are the flying paths of the left (A1) and right (A2) study areas, respectively. (**C**,**D**) are the aerial vehicle and the controller, respectively. Notably, the arrows in (**A**,**B**) indicate flight direction.

2.2.2. Calculation of Vegetation Indexes (VIs) and Estimation of Vegetation Height

VIs calculated by differences between several spectral bands are designed to highlight vegetation properties (such as canopy biomass, absorbed radiation, and chlorophyll content) and vegetation's vigour [46]. Five VIs (NDVI, GNDVI, NDRE, LCI, and OSAVI), derived using the reflectance of vegetation in the visible, red edge, and near-infrared bands, are selected under consideration of the spectral bands available in multispectral UAV and Sentinel-2 imagery.

Among all the VIs, NDVI is the most commonly used index for quantitatively measuring vegetation conditions [47,48]. As shown in Equation (1), NDVI is calculated as the ratio of the difference and the sum of reflectance values in the near-infrared (NIR) and red bands (R). Therefore, NDVI values vary between -1 and 1, with higher values for more densely vegetated and healthy areas.

$$NDVI = \frac{NIR - R}{NIR + R} \tag{1}$$

Compared with NDVI, which reflects vegetation health through the levels of chlorophyll detected in the leaves, the green normalized difference vegetation index (GNDVI), introduced by Gitelson et al. [49] and defined in Equation (2), is more suitable for monitoring crop yield in the late growing season because the green band (G) can cover a broader range of chlorophyll than the red band adopted in NDVI [50,51].

$$GNDVI = \frac{NIR - G}{NIR + G} \tag{2}$$

NDRE, defined as Equation (3), is similar to NDVI but is more sensitive to chlorophyll content in denser vegetation [52]. It calculates based on reflectance in the near-infrared (*NIR*) and red edge (*RE*) bands. When the NDVI is no longer sufficiently sensitive to changes in the plant biomass in the later stages of plant growth, NDRE plays an important role.

$$NDRE = \frac{NIR - RE}{NIR + RE} \quad (3)$$

The LCI, which uses the near-infrared (*NIR*), red edge (*RE*), and green (*G*) bands (Equation (4)), provides more accurate chlorophyll content information than the NDRE index [53].

$$LCI = \frac{NIR - RE}{NIR + G} \quad (4)$$

OSAVI considers the substrate type during vegetation condition analysis, reducing the influence of bare soil and enabling image correction where soil is visible between vegetation [54]. The OSAVI is defined following Equation (5), where *C* is a calibration constant that minimizes the influence of soil and takes the value of 0.16 [55].

$$OSAVI = \frac{NIR - R}{NIR + R + C} \quad (5)$$

The NDVI, GNDVI, LCI, NDRE, and OSAVI layers are overlapped in QGIS 3.40.11. The value of each grid is treated as a group of data for training the FCNN model, with a total of 78,520,680 data points.

The Digital Surface Model (DSM) and the Digital Elevation Model (DEM) are constructed by means of the DJI Terra software, and their difference is used to estimate vegetation heights. The DSM contains the height information of all objects on the Earth's surface, while the DEM represents the elevation of the bare ground surface. The difference between the two eliminates the influence of the underlying ground elevation, retaining only the height characteristics of surface objects such as vegetation [56]. Therefore, the vegetation height is estimated by the difference between the plant surface (DSM) and the terrain (DEM). Both DSM and DEM are processed based on the drone imagery by the DJI Terra software, and the DEM is also pre-processed to filter out all the vegetation [57], ensuring an accurate estimation of the vegetation height.

To ensure the measurement accuracy of the DJI M3M (SZ DJI Technology Co., Ltd., Shenzhen, China) with real-time kinematic (RTK) equipment, it has been verified prior to the drone's release. The results show that the output images can meet the requirements of 1:500 large-scale mapping accuracy, and the absolute/relative positioning error did not surpass 0.05 m/0.15 m in the horizontal and vertical directions, respectively [58]. Previous studies also found that the errors of DEM and DSM created from the imagery collected by DJI M3M are very low, generally around 1.5 cm [59–61]. The DEM and DSM collected from the drone imagery are, therefore, considered reliable for the scope of the current investigation.

2.3. Model Architecture of FCNN

The flowchart for data processing and constructing the FCNN model is shown in Figure 3, and comprises five main parts, described below in detail.

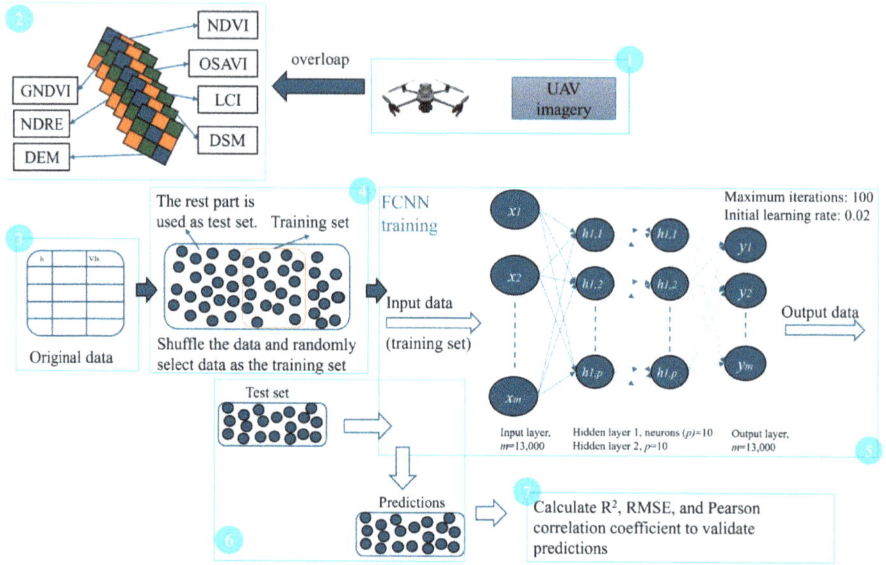

Figure 3. Flow chart of the data processing, lately used to feed the FCNN model. Notably, solid blue arrows are used to connect the major steps in the framework, whereas hollow arrows are used to link the sub-steps within each major step.

Step 1

The first step is organizing the original data. The h, NDVI, and GNDVI values of every pixel are listed in three columns in Excel as the original dataset.

Step 2

Secondly, the dataset is split into two subsets: 70% of the data is used for training the model, and the rest (30%) is for validation, to ensure that the evaluation of the model's performance on unseen data is reliable. This splitting was selected based on literature evidence that investigated the predictive capability of machine learning models, proving that the 70/30 one presented the best performance among all tested models [62].

The training set is randomly selected from the original data, while the rest of the data is used to validate the model, and thereafter is defined as the test set. It is worth reminding here that the present application focused on developing a workflow, so no independent datasets were used for testing.

Step 3

Structuring and training the FCNN model is the third step.

The structure of the model layers is as follows:

1. *Input data layer*. The input layers include a dataset consisting of vegetation height values, NDVI values, and GNDVI. The h, NDVI, and GNDVI of each grid are regarded as a group of data. The dataset is composed of data groups from each grid.

2. *Hidden layers*. The hidden layers are able to capture the underlying features and patterns in the data and learn hidden variables to enhance prediction capabilities [63]. Two fully connected layers (with every layer having 10 neurons) expand the feature space and enable the network to learn complex representations. Aiming to introduce non-linearity and allow the network to learn non-linear patterns efficiently [64], the rectified linear unit has been introduced.

In this study, ReLU (Equation (6)), one of the most widely used activation functions [65,66], serves as the activation function for each fully connected layer.

$$ReLU(x) = \max(0, x) \tag{6}$$

where x indicates the input to the neuron [67].

3. *Output layers*. A single neuron is used to output a continuous value. The activation function of the output layer is a linear function, allowing the output to be directly mapped from the learned features.

Step 4

Fourthly, the test set is imposed as the input for the trained model.

Step 5

Finally, the performance of the estimations is evaluated by calculating the coefficient of determination (R^2) [68] and root mean squared error (*RMSE*) [36] between the calculated and the measured data. As a significant index for evaluating the model's performance, R^2 is able to quantify the model's ability to explain the variance in the data [68], with a higher R^2 indicating better model performance.

3. Results

3.1. Monitoring of Vegetation Height

The distribution of the vegetation height of the two study areas estimated by DSM minus the vegetation-filtered DEM is illustrated in Figure 4. It can be noticed that the vegetation height of both A1 and A2 ranges from 0 to 10 m, with an average height of 0.81 m in A1 and 3.38 m in A2. The average vegetation height of the two areas is 2.10 m.

High plants in A1 are mainly isolated points without any obvious distribution pattern, while those in A2 are patchily distributed throughout the region, showing no clear distribution pattern. Vegetation along the boundary of A1 is generally characterized by low values. Regarding the boundary of A2, the condition is quite different, as there is a large area of high vegetation along the left boundary of A2 and lower vegetation along the right boundary.

Table 1 reports the statistics of the monitored vegetation height. As visible, most of the plants are smaller than 2 m, covering around 95.35%, 81.98%, and 85.36% of areas A1, A2, and the whole area, respectively. For A1, height is inversely proportional to quantity, that is, the higher the plants are, the lower the proportion of vegetation. In area A2, vegetation heights greater than 8 m rank second in frequency with a value of 8.02%, and there is a decreasing trend in vegetation height within the 2–8 m range. For the whole area, the trend is similar to that of A2, since A2 is bigger than A1: 85.36% of vegetation is less than 2 m, while 6.08% of vegetation is bigger than 8 m.

Table 1. Vegetation coverage of the two study areas.

Height (m)	Coverage Area 1 (%)	Coverage Area 2 (%)	Whole Area A1 + A2 (%)
<2	95.35	81.98	85.36
2–4	2.13	4.44	3.86
4–6	1.12	3.11	2.61
6–8	1.03	2.45	2.09
>8	0.37	8.02	6.08

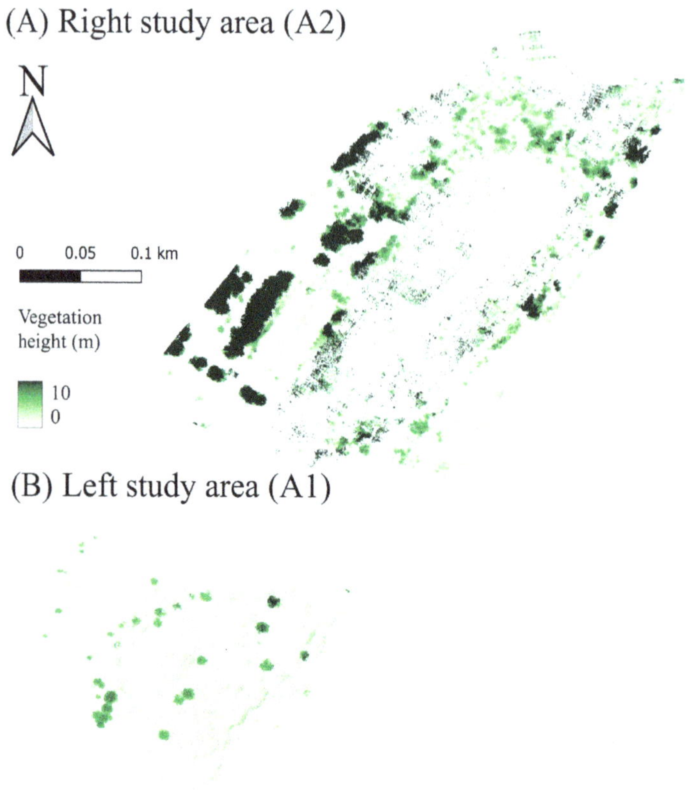

Figure 4. Distribution of the estimated vegetation height. (**A**) is the vegetation height of A2 and (**B**) is the vegetation height of A1.

3.2. Spatial Distribution of NDVI and GNDVI

The distributions of the NDVI and GNDVI estimated by the aerial images from the UAV are illustrated in Figure 5. It is found that the GNDVI value is slightly lower overall with respect to NDVI. The average NDVI and GNDVI of A1 are 0.62 and 0.59, respectively, and those of A2 are 0.71 and 0.64, respectively, meaning that the vegetation of A2 seems healthier than that of A1.

Moreover, the variability in the spatial distribution of NDVI values for A1 is greater than that of its GNDVI. The NDVI in A1 has a trend, with higher values at the border and lower in the middle, while the trend of GNDVI is not obvious. When it comes to the variability in the spatial distribution of NDVI and GNDVI values for area A2, they are more similar.

3.3. Linear and Polynomial Regression Based Vegetation Height Prediction

The vegetation height estimated by the linear regression model is shown in Figure 6A, the results revealed a weak relationship between the predictor variables (NDVI, GNDVI, LCI, NDRE, and OSAVI) and vegetation height, with a training R^2 of 0.03, root mean square error (RMSE) of 169.77 m, and a mean absolute error (MAE) of 16.92 m. The linear regression equation derived in the present case study is $h = 3.9879 + (-25.2569 \times \text{NDVI}) + (-22.5722 \times \text{GNDVI}) + (200.4915 \times \text{LCI}) + (-155.0283 \times \text{NDRE}) + (-1.1886 \times \text{OSAVI})$. It is possible to note that the large coefficient values suggest potential overfitting and numerical instability. In summary, the linear regression model demonstrated limited performance in estimating vegetation height from multispectral indices.

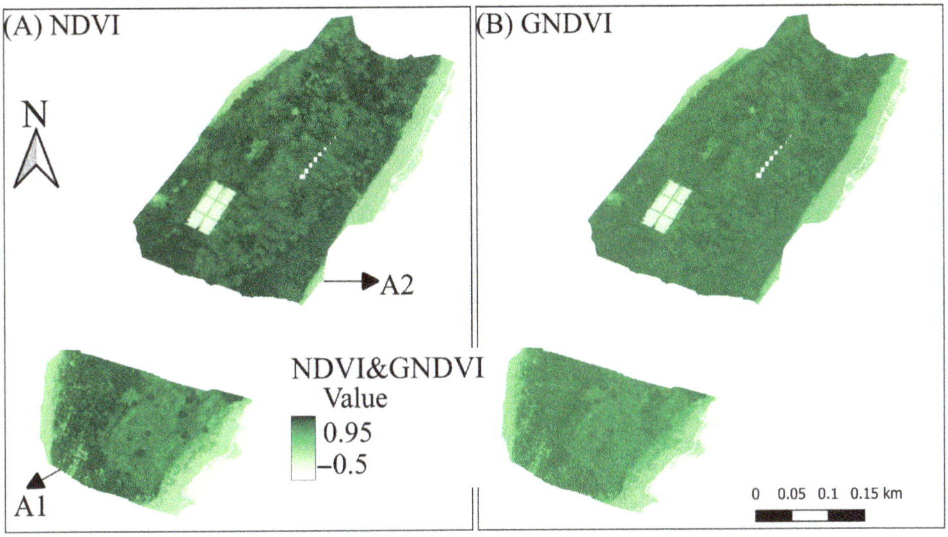

Figure 5. Distribution of (**A**) NDVI and (**B**) GNDVI in the flying areas.

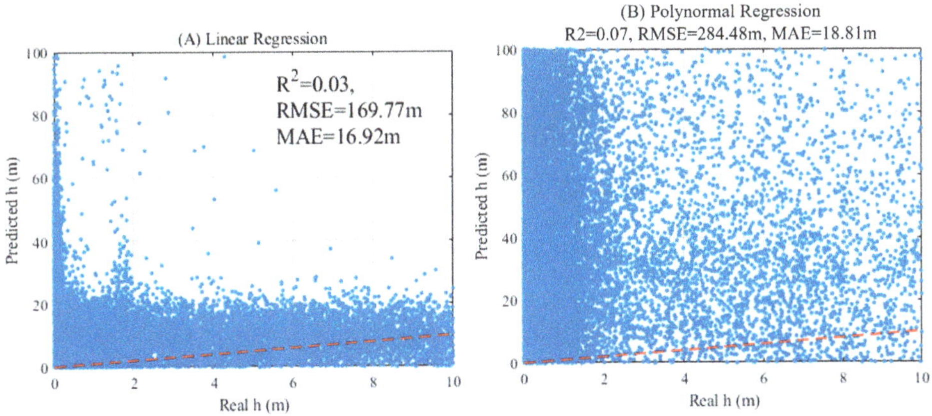

Figure 6. Comparison of linear and polynomial (degree = 2) regression models for vegetation height estimation. (**A**) Linear regression results and (**B**) polynomial regression training results. Notably, the red dashed line is the y = x line.

The polynomial regression model showed modest improvement over linear regression but still exhibited inadequate performance (Figure 6B), with training results displaying an R^2 of 0.07, RMSE of 284.48 m, and MAE of 18.01 m. Although the polynomial model incorporated interaction terms and quadratic features, expanding the feature space from 5 to 20 dimensions, the minimal improvement in R^2 and the increase in MAE suggest that increasing model complexity alone cannot adequately address the underlying estimation challenges.

3.4. Deep Learning-Based Vegetation Height Prediction

The above reported limitations in using linear and polynomial regression models ask for more complex approaches, such as the use of deep learning.

70% of the vegetation height h, NDVI, and GNDVI values estimated from the UAV images are used to train the FCNN model, where h is the dependent variable and NDVI and GNDVI are the independent variables. The other 30% of the NDVI and GNDVI values are imported into the trained model to predict h. The predictions are then compared to the original vegetation height to verify the performance of the trained model.

The calibration results (Figure 7) demonstrate exceptionally strong model performance, with a coefficient of determination (R^2) of 0.88, indicating that the model explains 88% of the variance in vegetation height. This represents a remarkable improvement over previous modeling attempts and confirms the effectiveness of the current approach. The model's error metrics, with an RMSE of 1.10 m and a MAE of 0.58 m, fall within an acceptable range for vegetation height estimation applications. The fact that MAE is substantially lower than RMSE suggests a generally well-behaved error distribution without excessive outliers.

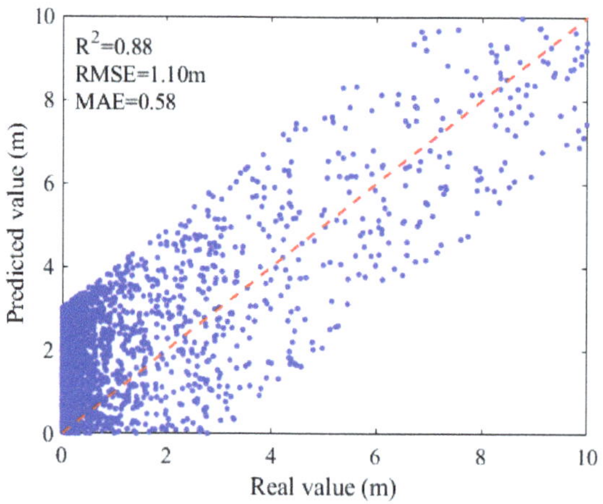

Figure 7. Comparison of the vegetation height predicted by the trained FCNN model (y-axis) and real vegetation height (x-axis). Notably, the red dashed line is the y = x line.

4. Discussion

In recent years, the frequency and intensity of flood events have increased significantly, largely driven by the ongoing impacts of climate change [69]. These heightened flood hazards are causing considerable disruptions to both human settlements and natural ecosystems, necessitating urgent and innovative approaches to flood management [70]. The growing unpredictability and severity of extreme weather events, such as intense rainfall and prolonged storms, underscore the importance of developing not only more effective flood mitigation strategies but also robust resilience frameworks capable of adapting to future climate scenarios. This includes integrating dynamic environmental variables, such as the behaviour of aquatic and riparian vegetation, which have been shown to exert a profound influence on hydrodynamic processes during flood events. In fact, the role of vegetation in shaping the flow dynamics of rivers and floodplains is becoming increasingly recognised as a critical factor in flood risk assessments [71]. Research on dynamic vegetation during high-flow conditions is gaining particular attention in the academic and practical realms of hydrology and geomorphology. As vegetation can grow and change rapidly in response to seasonal variations, extreme weather events, and restoration efforts, its influence on flood dynamics is not static, but rather highly variable and context-dependent. Vegetation can serve as a natural barrier to floodwaters, slowing their advance and reducing the overall volume of water that may inundate surrounding areas. However, this same vegetation can also impede drainage and exacerbate localised flooding if it is too dense or improperly distributed. Therefore, understanding how vegetation interacts with both the riverine environment and the flow dynamics during different flood stages is crucial to accurately modelling flood events and their impacts [72], as well as to develop sustainable flood management practices that are resilient to the growing challenges posed by climate change [73].

Past investigations have shown that accurately describing the involved processes and quantifying the influence of in-channel and riparian vegetation on bed roughness is very challenging. Generally, the vegetation roughness is calculated by formulas such as the Baptist approach, based on a Chézy formulation [74], and the Järvelä approach, which expands the previous method by accounting for submerged or emergent, flexible, and woody vegetation [75]. Moving from these approaches, Box et al. [76] proposed additional formulas accounting for the presence of leaves in both submerged and emergent conditions, showing how they can be implemented in numerical modelling. However, for all existing methods, how to estimate each vegetation parameter in real conditions is still a challenge, in particular in estimating the effective vegetation height influencing the flow field in rivers and streams, as well as their dynamics over time and space.

An answer to this knowledge gap could come from the combination of UAV imagery, orbital remote sensing, vegetation indices, and ML techniques [77]. To establish the relationship between vegetation height and VIs, this study conducted field measurements at the BSZ sandbar, a middle sandbar in the Middle Changjiang River, by using a UAV. Given the limited effectiveness of the initially tried linear and quadratic regression models, which yielded R^2 values around 0.3, a basic deep learning model, namely the FCNN model, was subsequently employed. After splitting the data into training and testing sets (70/30), we used NDVI, GNDVI, LCI, NDRE, and OSAVI (calculated from Sentinel-2 satellite imagery) as features in the trained model to estimate vegetation height. The model's performance metrics were $R^2 = 0.27$, RMSE = 0.97 m, and MAE = 0.31 m. Despite the mediocre but acceptable metrics of the FCNN model, it is worth noting that the results presented here are influenced by the season, as field measurements were carried out in October (Autumn), which is not the best season for vegetation conditions. Though the vegetation height may not change a lot, the vegetation is not as green as in the Summer, causing the NDVI value to be rather low. This is likely the reason why the GNDVI value is lower than the NDVI. Although the results are suboptimal, they are not irrelevant for height estimation. Since this study utilizes only a single vegetation image for model training, it cannot fully capture the vegetation growth cycle. Furthermore, the sampling date was in autumn, which is not the ideal season for data collection. The achieved R^2 of 0.27 suggests that incorporating multi-temporal data in the future could enable the model to better simulate vegetation height.

It should be highlighted that the outcomes presented here are based on a reduced dataset, and therefore, more field measurements should be conducted in the future in different seasons to improve the trained model, to effectively describe the vegetation seasonality [47]. The presented relationship between vegetation indexes and vegetation height should be further tested, as the current results might be influenced by vegetation characteristics such as photosynthetic activity and density of the visible leaves. At the same time, it is worth reminding that the overall workflow proposed here is disconnected from the vegetation seasonality and characteristics, and could therefore be replicated under different vegetation conditions, aiming to cover a wider range of vegetative stages and species.

Furthermore, the estimation of the vegetation height obtained from UAV data is also affected by uncertainties. Although the drone used in this study is equipped with an Real-Time Kinematic (RTK) module, whose measurement error has been proved can be controlled within 1–2 cm [78], and the DSM-DEM approach used to calculate the vegetation height has been tested by many researchers and defined as reliable [79–82], errors still exist since the DEM and DSM has potential uncertainties connected to the spatial resolution. Thus, some small vegetation may be missed from the removal procedure when filtering the vegetation cover on the DEM, leading to a slight overestimation of the vegetation height. To address this point, in future measurement campaigns, it is planned to employ a UAV with light detection and ranging (LiDAR) to extract the vegetation height directly to minimize errors [83]. At the same time, advanced methodologies and reference tests to reduce uncertainties [84] will also be considered.

Meanwhile, more VIs will be introduced to train the deep-learning model to consider more vegetation conditions and types. For example, in this case, the dominant vegetation type is (high) grass, with shrubs and trees sparsely distributed in the central sandbar. Therefore, the models are trained only by NDVI and GNDVI and without classifying vegetation types. The Enhanced Vegetation Index (EVI) will be introduced in future studies to account for more complicated vegetation conditions. In fact, this index was developed as a standard satellite vegetation product for the Terra and Aqua Moderate Resolution Imaging Spectroradiometers (MODIS), and provides improved sensitivity in high biomass regions while minimizing soil and atmosphere influences through a decoupling of the canopy background signal [85].

It is also worth remembering that traditional regression methods, such as linear regression, perform poorly in fitting the relationship between vegetation height and VIs (namely, NDVI and GNDVI). To address this limitation, more advanced approaches such as DL models (convolutional neural network (CNN), recurrent neural network (RNN), and long short-term memory network (LSTM)) should be used. Though the FCNN model performs rather well in this application, a wider range of DL models will also be tested in the future to corroborate the presented outcomes and eventually improve them, aiming to provide a more robust estimation of vegetation height from VIs that could be extrapolated to other areas.

The influence of the spatial resolution of multispectral images on estimating vegetation height from VIs is another focus for future research. Sensitivity analyses of spatial resolution to the DL model trained by using UAV-acquired vegetation height data and VIs are also planned, considering images of varying spatial resolution (e.g., 0.5 m, 1 m, 5 m, 10 m, etc.) that will be captured by UAVs to investigate how spatial resolution affects building the relationship between vegetation heights and VIs.

Despite current limitations, the present study highlighted the opportunity to use UAVs for estimating vegetation height and consequent roughness. This demonstrates that VIs calculated from remote sensing (UAVs currently and their combination with satellite imagery in the future) could be used to estimate vegetation characteristics, assuming an adequate spatial resolution.

5. Conclusions

The vegetation parameters, especially the vegetation height, are essential for estimating dynamic vegetation roughness. Though the VIs can be obtained from both satellite and UAV images, the vegetation height is hard to extract from satellite images, given the relatively low spatial resolution and the infrequent revisit time. Focusing on a sandbar in the Middle Changjiang River (China), a relationship between vegetation height and VIs collected by the UAV field measurement is built here, taking advantage of an FCNN model. The results find that there is a very strong positive correlation between the predictions and the real values, and around 85% of the variation in the vegetation height can be explained by the trained model, showing that the vegetation height estimated by the trained FCNN model is reliable. Therefore, in the next stage, the VIs derived from satellite images can be used as inputs to the trained FCNN model to estimate vegetation height, eventually enlarging the study area and extending the observation period.

The results presented here contribute to the development of repeatable, cost-effective, and fine-scale vegetation monitoring strategies based on UAV flights, supported by photogrammetry.

Author Contributions: Conceptualization, Y.G.; methodology, Y.G.; field measurement: Y.G., Y.Z. and R.Z.; software, Y.G.; validation, Y.G. and M.N.; formal analysis, Y.G.; investigation, Y.G.; data curation, Y.G.; writing—original draft preparation, Y.G. and M.N.; writing—review and editing, Y.G., M.N., Y.Z., R.Z. and W.D.; visualization, Y.G.; supervision, M.N.; project administration, M.N.; funding acquisition, M.N. All authors have read and agreed to the published version of the manuscript.

Funding: This research was funded by the National Natural Science Foundation of Joint Fund for Changjiang River Water Science Research, Project number U2340215 and by the National Science Centre Poland–call PRELUDIUM BIS-3, Grant Number 2021/43/O/ST10/00539. The work of Y.G. and M.N. is partially supported by a subsidy from the Polish Ministry of Education and Science for the Institute of Geophysics Polish Academy of Sciences.

Data Availability Statement: The data used in the study are available at https://zenodo.org/records/15834789 (accessed on 1 May 2025).

Conflicts of Interest: The authors declare no conflicts of interest.

References

1. Sweeney, M.R.; Fischer, B.; Wermers, K.; Cowman, T. Eolian and Fluvial Modification of Missouri River Sandbars Deposited by the 2011 Flood, USA. *Geomorphology* **2019**, *327*, 111–125. [CrossRef]
2. Stucker, J.H.; Buhl, D.A.; Sherfy, M.H. Emergent Sandbar Construction for Least Terns on the Missouri River: Effects on Forage Fishes in Shallow-Water Habitats. *River Res. Appl.* **2012**, *28*, 1254–1265. [CrossRef]
3. Seki, S.; Moteki, D.; Yasuda, H. Novel Hypothesis on the Occurrence of Sandbars. *Phys. Fluids* **2023**, *35*, 106611. [CrossRef]
4. Hu, Y.; Zhou, J.; Deng, J.; Li, Y.; Yang, C.; Li, D. River Bars and Vegetation Dynamics in Response to Upstream Damming: A Case Study of the Middle Yangtze River. *Remote Sens.* **2023**, *15*, 2324. [CrossRef]
5. Heidari, N.; Yagci, O.; Aksel, M. Midchannel Islands in Lowland River Corridors and Their Impacts on Flow Structure and Morphology: A Numerical Based Conceptual Analysis. *Ecol. Eng.* **2021**, *173*, 106419. [CrossRef]
6. Rood, S.B.; Goater, L.A.; Gill, K.M.; Braatne, J.H. Sand and Sandbar Willow: A Feedback Loop Amplifies Environmental Sensitivity at the Riparian Interface. *Oecologia* **2011**, *165*, 31–40. [CrossRef]
7. Nagata, T.; Watanabe, Y.; Yasuda, H.; Ito, A. Development of a Meandering Channel Caused by the Planform Shape of the River Bank. *Earth Surf. Dyn.* **2014**, *2*, 255–270. [CrossRef]
8. van Iersel, W.; Straatsma, M.; Addink, E.; Middelkoop, H. Monitoring Height and Greenness of Non-Woody Floodplain Vegetation with UAV Time Series. *ISPRS J. Photogramm. Remote Sens.* **2018**, *141*, 112–123. [CrossRef]
9. Nones, M.; Di Silvio, G. Modeling of River Width Variations Based on Hydrological, Morphological, and Biological Dynamics. *J. Hydraul. Eng.* **2016**, *142*, 04016012. [CrossRef]
10. Kiss, T.; Nagy, J.; Fehérváry, I.; Vaszkó, C. (Mis) Management of Floodplain Vegetation: The Effect of Invasive Species on Vegetation Roughness and Flood Levels. *Sci. Total Environ.* **2019**, *686*, 931–945. [CrossRef]
11. De Doncker, L.; Troch, P.; Verhoeven, R.; Bal, K.; Meire, P.; Quintelier, J. Determination of the Manning Roughness Coefficient Influenced by Vegetation in the River Aa and Biebrza River. *Environ. Fluid Mech.* **2009**, *9*, 549–567. [CrossRef]
12. Chen, Y.; Cao, F.; Cheng, W.; Liu, B.; Yu, P. Real-Time Correction of Channel-Bed Roughness and Water Level in River Network Hydrodynamic Modeling for Accurate Forecasting. *Sci. Rep.* **2023**, *13*, 20660. [CrossRef] [PubMed]

13. Ferreira, D.M.; Fernandes, C.V.S.; Kaviski, E.; Bleninger, T. Calibration of River Hydrodynamic Models: Analysis from the Dynamic Component in Roughness Coefficients. *J. Hydrol.* **2021**, *598*, 126136. [CrossRef]
14. Augustijn, D.C.M.; Huthoff, F.; Van Velzen, E.H. Comparison of Vegetation Roughness Descriptions. In Proceedings of the River Flow 2008: 4th International Conference on Fluvial Hydraulics, Izmir, Turkey, 3–5 September 2008; pp. 343–350.
15. Latella, M.; Notti, D.; Baldo, M.; Giordan, D.; Camporeale, C. Short-Term Biogeomorphology of a Gravel-Bed River: Integrating Remote Sensing with Hydraulic Modelling and Field Analysis. *Earth Surf. Process. Landforms* **2024**, *49*, 1156–1178. [CrossRef]
16. Kiss, T.; Fehérváry, I. Increased Riparian Vegetation Density and Its Effect on Flow Conditions. *Sustainability* **2023**, *15*, 12615. [CrossRef]
17. Signorile, A.; Saracino, R.; Dani, A.; Rillo Migliorini Giovannini, M.; Preti, F. Riparian Vegetation Surveys for Roughness Estimation. *Ecol. Eng.* **2024**, *209*, 107414. [CrossRef]
18. Wang, J.; Zhang, Z. Evaluating Riparian Vegetation Roughness Computation Methods Integrated within HEC-RAS. *J. Hydraul. Eng.* **2019**, *145*, 04019020. [CrossRef]
19. Rillo Migliorini Giovannini, M.; Dani, A.; Saracino, R.; Signorile, A.; Preti, F. Hydraulic Roughness Estimation Induced by Riparian Vegetation in Tuscany Rivers for Management Purposes. *Lect. Notes Civ. Eng.* **2023**, *337*, 169–179. [CrossRef]
20. Chen, S.; McDermid, G.J.; Castilla, G.; Linke, J. Measuring Vegetation Height in Linear Disturbances in the Boreal Forest with UAV Photogrammetry. *Remote Sens.* **2017**, *9*, 1257. [CrossRef]
21. Chaulagain, S.; Stone, M.C.; Dombroski, D.; Gillihan, T.; Chen, L.; Zhang, S. An Investigation into Remote Sensing Techniques and Field Observations to Model Hydraulic Roughness from Riparian Vegetation. *River Res. Appl.* **2022**, *38*, 1730–1745. [CrossRef]
22. Lu, J.; Cheng, D.; Geng, C.; Zhang, Z.; Xiang, Y.; Hu, T. Combining Plant Height, Canopy Coverage and Vegetation Index from UAV-Based RGB Images to Estimate Leaf Nitrogen Concentration of Summer Maize. *Biosyst. Eng.* **2021**, *202*, 42–54. [CrossRef]
23. Gao, L.; Wang, X.; Johnson, B.A.; Tian, Q.; Wang, Y.; Verrelst, J.; Mu, X.; Gu, X. Remote Sensing Algorithms for Estimation of Fractional Vegetation Cover Using Pure Vegetation Index Values: A Review. *ISPRS J. Photogramm. Remote Sens.* **2020**, *159*, 364–377. [CrossRef] [PubMed]
24. Zhang, J.; Qiu, X.; Wu, Y.; Zhu, Y.; Cao, Q.; Liu, X.; Cao, W. Combining Texture, Color, and Vegetation Indices from Fixed-Wing UAS Imagery to Estimate Wheat Growth Parameters Using Multivariate Regression Methods. *Comput. Electron. Agric.* **2021**, *185*, 106138. [CrossRef]
25. Leroux, L.; Baron, C.; Zoungrana, B.; Traore, S.B.; Lo Seen, D.; Begue, A. Crop Monitoring Using Vegetation and Thermal Indices for Yield Estimates: Case Study of a Rainfed Cereal in Semi-Arid West Africa. *IEEE J. Sel. Top. Appl. Earth Obs. Remote Sens.* **2016**, *9*, 347–362. [CrossRef]
26. Amani, S.; Shafizadeh-Moghadam, H. A Review of Machine Learning Models and Influential Factors for Estimating Evapotranspiration Using Remote Sensing and Ground-Based Data. *Agric. Water Manag.* **2023**, *284*, 108324. [CrossRef]
27. Berzéki, M.; Kozma-Bognár, V.; Berke, J. *Examination of Vegetation Indices Based on Multitemporal Drone Images*; Gradus: Düsseldorf, Germany, 2023; Volume 10. [CrossRef]
28. Sun, H.; Wang, X.; Fan, D.; Sun, O.J. Contrasting Vegetation Response to Climate Change between Two Monsoon Regions in Southwest China: The Roles of Climate Condition and Vegetation Height. *Sci. Total Environ.* **2022**, *802*, 149643. [CrossRef]
29. Payero, J.O.; Neale, C.M.U.; Wright, J.L. Comparison of Eleven Vegetation Indices for Estimating Plant Height of Alfalfa and Grass. *Appl. Eng. Agric.* **2004**, *20*, 385–393. [CrossRef]
30. Kalinowska, M.B.; Västilä, K.; Nones, M.; Kiczko, A.; Karamuz, E.; Brandyk, A.; Kozioł, A.; Krukowski, M. Influence of Vegetation Maintenance on Flow and Mixing: Case Study Comparing Fully Cut with High-Coverage Conditions. *Hydrol. Earth Syst. Sci.* **2023**, *27*, 953–968. [CrossRef]
31. Boucher, P.B.; Hockridge, E.G.; Singh, J.; Davies, A.B. Flying High: Sampling Savanna Vegetation with UAV-Lidar. *Methods Ecol. Evol.* **2023**, *14*, 1668–1686. [CrossRef]
32. Luo, S.; Liu, W.; Zhang, Y.; Wang, C.; Xi, X.; Nie, S.; Ma, D.; Lin, Y.; Zhou, G. Maize and Soybean Heights Estimation from Unmanned Aerial Vehicle (UAV) LiDAR Data. *Comput. Electron. Agric.* **2021**, *182*, 106005. [CrossRef]
33. Prior, E.M.; Aquilina, C.A.; Czuba, J.A.; Pingel, T.J.; Hession, W.C. Estimating Floodplain Vegetative Roughness Using Drone-Based Laser Scanning and Structure from Motion Photogrammetry. *Remote Sens.* **2021**, *13*, 2616. [CrossRef]
34. Nguyen, H.D.; Van, C.P.; Do, A.D. Application of Hybrid Model-Based Deep Learning and Swarm-based Optimizers for Flood Susceptibility Prediction in Binh Dinh Province, Vietnam. *Earth Sci. Inform.* **2023**, *16*, 1173–1193. [CrossRef]
35. Gui, S.; Song, S.; Qin, R. Remote Sensing Object Detection in the Deep Learning Era—A Review. *Remote Sens.* **2024**, *16*, 327. [CrossRef]
36. Wang, W.; Tang, J.; Zhang, N.; Xu, X.; Zhang, A.; Wang, Y.; Li, K.; Wang, Y. Vegetation Height Estimation Based on Machine Learning Model Driven by Multi-Source Data in Eurasian Temperate Grassland. *Ecol. Indic.* **2025**, *170*, 113013. [CrossRef]

37. Luo, H.; Ou, G.; Yue, C.; Zhu, B.; Wu, Y.; Zhang, X.; Lu, C.; Tang, J. A Framework for Montane Forest Canopy Height Estimation via Integrating Deep Learning and Multi-Source Remote Sensing Data. *Int. J. Appl. Earth Obs. Geoinf.* **2025**, *138*, 104474. [CrossRef]
38. Bhandari, K.; Srinet, R.; Nandy, S. Forest Height and Aboveground Biomass Mapping by Synergistic Use of GEDI and Sentinel Data Using Random Forest Algorithm in the Indian Himalayan Region. *J. Indian Soc. Remote Sens.* **2024**, *52*, 857–869. [CrossRef]
39. Dersch, S.; Schöttl, A.; Krzystek, P.; Heurich, M. Semi-Supervised Multi-Class Tree Crown Delineation Using Aerial Multispectral Imagery and Lidar Data. *ISPRS J. Photogramm. Remote Sens.* **2024**, *216*, 154–167. [CrossRef]
40. Yang, R.; Chen, M.; Lu, X.; He, Y.; Li, Y.; Xu, M.; Li, M.; Huang, W.; Liu, F. Integrating UAV Remote Sensing and Semi-Supervised Learning for Early-Stage Maize Seedling Monitoring and Geolocation. *Plant Phenomics* **2025**, *7*, 100011. [CrossRef]
41. Nyamekye, C.; Appiah, L.B.; Arthur, R.; Osei, G.; Ofosu, S.A.; Kwofie, S.; Ghansah, B.; Bryniok, D. Comparing Supervised and Semi-Supervised Machine Learning Methods for Mapping Aquatic Weeds, as Biomass Resource from High-Resolution UAV Images. *Remote Sens. Earth Syst. Sci.* **2024**, *7*, 206–217. [CrossRef]
42. Du, S.; Liu, H.; Xing, J.; Zhang, X.; Zhang, J.; Guan, X.; Du, S. Estimating Individual Building Heights by Integrating Spaceborne LiDAR and Multisource Remote Sensing Data: A CNN–Transformer Model and a Semi-Supervised Sample Augmentation Approach. *IEEE Trans. Geosci. Remote Sens.* **2025**, *63*, 5406719. [CrossRef]
43. Jiang, Y.; Chen, X.F.; Yang, X.J. A Study on Changes of Aquatic Plants in East Lake of Wuhan Using 1990–2020 Landsat Images. *Chin. J. Plant Ecol.* **2022**, *46*, 1551–1561. [CrossRef]
44. Wang, X.; Li, Z.; Ma, J.; Bai, K. Differential Scour of Two Mid-Channel Bars in Wuhan Reach of the Middle Yangtze River. *J. Chang. River Sci. Res. Inst.* **2024**, *41*, 18–27. [CrossRef]
45. Fraser, B.T.; Robinov, L.; Davidson, W.; O'Connor, S.; Congalton, R.G. A Comparison of Unpiloted Aerial System Hardware and Software for Surveying Fine-Scale Oak Health in Oak–Pine Forests. *Forests* **2024**, *15*, 706. [CrossRef]
46. Gutierrez-Rodriguez, M.; Escalante-Estrada, J.A.; Rodriguez-Gonzalez, M.T. Canopy Reflectance, Stomatal Conductance, and Yield of *Phaseolus vulgaris* L. and *Phaseolus coccinues* L. Under Saline Field Conditions. *Int. J. Agric. Biol.* **2005**, *7*, 491–494.
47. Boothroyd, R.J.; Nones, M.; Guerrero, M. Deriving Planform Morphology and Vegetation Coverage from Remote Sensing to Support River Management Applications. *Front. Environ. Sci.* **2021**, *9*, 657354. [CrossRef]
48. Rouse, J.W., Jr.; Haas, R.H.; Schell, J.A.; Deering, D.W. Monitoring Vegetation Systems in the Great Plains with ERTS. In Proceedings of the NASA Goddard Space Flight Center 3d ERTS-1 Symposium, Washington, DC, USA, 1 January 1974; Volume 1. Section A.
49. Gitelson, A.A.; Kaufman, Y.J.; Merzlyak, M.N. Use of a Green Channel in Remote Sensing of Global Vegetation from EOS-MODIS. *Remote Sens. Environ.* **1996**, *58*, 289–298. [CrossRef]
50. Barati, S.; Rayegani, B.; Saati, M.; Sharifi, A.; Nasri, M. Comparison the Accuracies of Different Spectral Indices for Estimation of Vegetation Cover Fraction in Sparse Vegetated Areas. *Egypt. J. Remote Sens. Sp. Sci.* **2011**, *14*, 49–56. [CrossRef]
51. Allawai, M.F.; Ahmed, B.A. Using Remote Sensing and GIS in Measuring Vegetation Cover Change from Satellite Imagery in Mosul City, North of Iraq. *IOP Conf. Ser. Mater. Sci. Eng.* **2020**, *757*, 012062. [CrossRef]
52. Karpiński, P.; Kocira, S.; Karpiński, P.; Kocira, S. Possibilities of Using a Multispectral Camera to Assess the Effects of Biostimulant Application in Soybean Cultivation. *Sensors* **2025**, *25*, 3464. [CrossRef]
53. Bak, H.-J.; Kim, E.-J.; Lee, J.-H.; Chang, S.; Kwon, D.; Im, W.-J.; Kim, D.-H.; Lee, I.-H.; Lee, M.-J.; Hwang, W.-H.; et al. Canopy-Level Rice Yield and Yield Component Estimation Using NIR-Based Vegetation Indices. *Agriculture* **2025**, *15*, 594. [CrossRef]
54. Steven, M.D. The Sensitivity of the OSAVI Vegetation Index to Observational Parameters. *Remote Sens. Environ.* **1998**, *63*, 49–60. [CrossRef]
55. Rondeaux, G.; Steven, M.; Baret, F. Optimization of Soil-Adjusted Vegetation Indices. *Remote Sens. Environ.* **1996**, *55*, 95–107. [CrossRef]
56. Popescu, S.C.; Zhao, K. A Voxel-Based Lidar Method for Estimating Crown Base Height for Deciduous and Pine Trees. *Remote Sens. Environ.* **2008**, *112*, 767–781. [CrossRef]
57. DJI. *DJI Terra-User Manual*; v4.3; DJI: Shenzhen, China, 2024.
58. Tan, C.; Chen, Z.; Liao, A.; Chen, Z.; Zeng, X. Accuracy Assessment of Mavic 3 Industrial UAV Based on DJI Terra and Pix4Dmapper. In Proceedings of the Fifth International Conference on Geology, Mapping, and Remote Sensing, Wuhan, China, 10 July 2024; SPIE: Bellingham, WA, USA; Volume 13223, pp. 498–504. [CrossRef]
59. Carp, A. *Measuring Accuracy of the DJI Mavic 3 Enterprise RTK Using DroneDeploy Photogrammetry*; DroneDeploy: San Francisco, CA, USA, 2023.
60. Wang, M.; Li, H.; Liu, Y.; Li, H. Multi-Source DEM Vertical Accuracy Evaluation of Taklimakan Desert Hinterland Based on ICESat-2 ATL08 and UAV Data. *Remote Sens.* **2025**, *17*, 1807. [CrossRef]

61. Liu, J.; Wang, W.; Li, J.; Mustafa, G.; Su, X.; Nian, Y.; Ma, Q.; Zhen, F.; Wang, W.; Li, X. UAV Remote Sensing Technology for Wheat Growth Monitoring in Precision Agriculture: Comparison of Data Quality and Growth Parameter Inversion. *Agronomy* **2025**, *15*, 159. [CrossRef]
62. Nguyen, Q.H.; Ly, H.B.; Ho, L.S.; Al-Ansari, N.; Van Le, H.; Tran, V.Q.; Prakash, I.; Pham, B.T. Influence of Data Splitting on Performance of Machine Learning Models in Prediction of Shear Strength of Soil. *Math. Probl. Eng.* **2021**, *2021*, 4832864. [CrossRef]
63. Deng, T. Effect of the Number of Hidden Layer Neurons on the Accuracy of the Back Propagation Neural Network. *Highlights Sci. Eng. Technol.* **2023**, *74*, 462–468. [CrossRef]
64. Hammad, M.M. Deep Learning Activation Functions: Fixed-Shape, Parametric, Adaptive, Stochastic, Miscellaneous, Non-Standard, Ensemble. *arXiv* **2024**. [CrossRef]
65. Eckle, K.; Schmidt-Hieber, J. A Comparison of Deep Networks with ReLU Activation Function and Linear Spline-Type Methods. *Neural Netw.* **2019**, *110*, 232–242. [CrossRef]
66. Lecun, Y.; Bengio, Y.; Hinton, G. Deep Learning. *Nature* **2015**, *521*, 436–444. [CrossRef]
67. Goodfellow, I.; Begio, Y.; Courville, A. *Deep Learning*; MIT Press: Cambridge, MA, USA, 2015; Volume 521, ISBN 3463353563306.
68. Guo, Y.; Ding, W.; Xu, W.; Zhu, X.; Wang, X.; Tang, W. Assessment of an Alternative Climate Product for Hydrological Modeling: A Case Study of the Danjiang River Basin, China. *Water* **2022**, *14*, 1105. [CrossRef]
69. Hirabayashi, Y.; Mahendran, R.; Koirala, S.; Konoshima, L.; Yamazaki, D.; Watanabe, S.; Kim, H.; Kanae, S. Global Flood Risk under Climate Change. *Nat. Clim. Chang.* **2013**, *3*, 816–821. [CrossRef]
70. Arnell, N.W.; Gosling, S.N. The Impacts of Climate Change on River Flood Risk at the Global Scale. *Clim. Chang.* **2016**, *134*, 387–401. [CrossRef]
71. Chen, J.; Shao, Z.; Huang, X.; Cai, B.; Zheng, X. Assessing the Impact of Floods on Vegetation Worldwide from a Spatiotemporal Perspective. *J. Hydrol.* **2023**, *622*, 129715. [CrossRef]
72. Surian, N.; Barban, M.; Ziliani, L.; Monegato, G.; Bertoldi, W.; Comiti, F. Vegetation Turnover in a Braided River: Frequency and Effectiveness of Floods of Different Magnitude. *Earth Surf. Process. Landforms* **2015**, *40*, 542–558. [CrossRef]
73. Rowiński, P.M.; Okruszko, T.; Radecki-Pawlik, A. Environmental Hydraulics Research for River Health: Recent Advances and Challenges. *Ecohydrol. Hydrobiol.* **2022**, *22*, 213–225. [CrossRef]
74. Baptist, M.J.; Babovic, V.; Uthurburu, J.R.; Keijzer, M.; Uittenbogaard, R.E.; Mynett, A.; Verwey, A. On Inducing Equations for Vegetation Resistance. *J. Hydraul. Res.* **2007**, *45*, 435–450. [CrossRef]
75. Järvelä, J. Determination of Flow Resistance Caused by Non-submerged Woody Vegetation. *Int. J. River Basin Manag.* **2004**, *2*, 61–70. [CrossRef]
76. Box, W.; Järvelä, J.; Västilä, K. New Formulas Addressing Flow Resistance of Floodplain Vegetation from Emergent to Submerged Conditions. *Int. J. River Basin Manag.* **2024**, *22*, 333–349. [CrossRef]
77. Ferraz, M.A.J.; Barboza, T.O.C.; de Arantes, P.S.; Von Pinho, R.G.; dos Santos, A.F. Integrating Satellite and UAV Technologies for Maize Plant Height Estimation Using Advanced Machine Learning. *AgriEngineering* **2024**, *6*, 20–33. [CrossRef]
78. DJI Enterprise. What Is Real-Time Kinematics and What It Means for Your Drone. Available online: https://enterprise-insights.dji.com/blog/rtk-real-time-kinematics?utm_source=chatgpt.com (accessed on 7 July 2025).
79. Radke, D.; Radke, D.; Radke, J. Beyond Measurement: Extracting Vegetation Height from High Resolution Imagery with Deep Learning. *Remote Sens.* **2020**, *12*, 3797. [CrossRef]
80. Tang, X.; Li, S.; Li, T.; Gao, Y.; Zhang, S.; Chen, Q.; Zhang, X. Review on Global Digital Elevation Products. *Natl. Remote Sens. Bull.* **2021**, *25*, 167–181. [CrossRef]
81. Zhang, Y.; Zhang, W.; Ji, Y.; Zhao, H. Forest Height Estimation and Inversion of Satellite-Based X-Band InSAR Data. *Geomat. Inf. Sci. Wuhan Univ.* **2024**, *49*, 2279–2289. [CrossRef]
82. Zhao, L.; Liu, Y.; Luo, Y. The Efficiency of Parameters Derived from the Digital Surface Model (DSM) in Indicating the Forest Gap Features. *Remote Sens. Technol. Appl.* **2021**, *36*, 420–430.
83. Latella, M.; Raimondo, T.; Belcore, E.; Salerno, L.; Camporeale, C. On the Integration of LiDAR and Field Data for Riparian Biomass Estimation. *J. Environ. Manag.* **2022**, *322*, 116046. [CrossRef]
84. Leem, J.; Kim, J.; Kang, I.S.; Choi, J.; Song, J.J.; Mehrishal, S.; Shao, Y. Practical Error Prediction in UAV Imagery-Based 3D Reconstruction: Assessing the Impact of Image Quality Factors. *Int. J. Remote Sens.* **2025**, *46*, 1000–1030. [CrossRef]
85. Jiang, Z.; Huete, A.R.; Didan, K.; Miura, T. Development of a Two-Band Enhanced Vegetation Index without a Blue Band. *Remote Sens. Environ.* **2008**, *112*, 3833–3845. [CrossRef]

water

MDPI

Article

GRU-Based Reservoir Operation with Data Integration for Real-Time Flood Control

Li Li [1] and Kyung Soo Jun [2,*]

[1] Center for Built Environment (CBE), Sungkyunkwan University, Suwon 16419, Republic of Korea; lili0809@skku.edu
[2] Graduate School of Water Resources, Sungkyunkwan University, Suwon 16419, Republic of Korea
* Correspondence: ksjun@skku.edu

Abstract: Reservoir operation serves as a critical non-structural measure for real-time flood management, aimed to minimize downstream flood damage while ensuring dam safety. This study develops and evaluates a Gated Recurrent Unit (GRU)-based reservoir operation model with data integration (DI) to enhance flood management capabilities. Optimal reservoir outflows are first determined for historical flood events using the Interior Point Optimizer (IPOPT), a deterministic optimization model designed to minimize peak outflows. The optimized hydrographs are compared with observed outflows to assess the benefits of improved operational strategies. GRU models are then trained and validated using inflow hydrographs and resulting optimal reservoir storage and release data. Various input configurations are tested, incorporating DI of lagged observations and forecasted values to evaluate their influence on model accuracy. The study also examines multiple hyperparameter settings to identify the optimal configuration. The methodology is applied to the Namgang Dam in South Korea, simulating hourly operations during flood events. Results indicate that historical reservoir inflow and storage are the most influential inputs, while adding precipitation (historical or forecasted) and/or forecasted inflows does not improve model performance. The GRU model with DI successfully replicates optimized reservoir operations, demonstrating its reliability and efficiency in flood management. This framework supports timely and informed decision-making and offers a promising approach for enhancing flood risk mitigation through improved reservoir operations.

Keywords: optimal reservoir operation; real-time flood control; data integration (DI); IPOPT; GRU; Namgang Dam

1. Introduction

In South Korea, approximately 70% of annual precipitation occurs during the monsoon season from June to September [1]. As a key hydraulic infrastructure for water resources management, reservoirs have played a vital role in balancing water availability with growing socioeconomic demands [2]. To date, 22 multipurpose dams have been constructed across the country [3], serving the combined purposes of water supply, flood control, and hydropower generation [1]. These reservoirs are especially crucial in the context of increasing extreme weather events and flood risks.

Reservoir operation refers to the controlled release of water through dam spillway gates to manage storage levels and downstream flows [4]. During flood events, timely and appropriate reservoir operations can significantly mitigate disaster impacts [5,6]. Real-time reservoir operation rules define how much water to store or release at each time step, based on dynamic hydrological and meteorological conditions [7]. The dual goals of flood-oriented reservoir operation are to ensure dam safety and minimize flood risks upstream and downstream. However, this is a highly complex task, involving numerous constraints, uncertainties, and urgent decision-making requirements [8,9]. Operators must act quickly—often within hours—while facing limited data and poor predictability of evolving flood conditions.

Traditional reservoir operation models are generally categorized into two types. The first type is optimization-based models. These models aim to maximize benefits or minimize risks, such as flood damage, by identifying optimal operation strategies under defined objective functions [9,10]. Optimization often involves nonlinear objective functions with multiple constraints, making conventional methods such as Linear Programming (LP) or Dynamic Programming (DP) less effective or computationally burdensome [11]. While these models are well-suited for evaluating historical or idealized scenarios, they are less practical for real-time operation due to their high computational demands and difficulty in adapting to rapidly changing flood conditions [12]. The second type is simulation-based models. These include conceptual or physically based models that replicate reservoir behavior with lower computational requirements, making them more feasible for real-time application [13,14]. Advanced models such as WEAP [15] and HEC-ResSim [16] have been widely adopted, supported by advances in river dynamics modeling and computational capacity. These models combine empirical rule curves with mathematical formulations to simulate reservoir outflows and storage dynamics [17]. Their transparency and interpretability make them accessible to reservoir operators and facilitate informed water management [18]. However, physical-based models depend heavily on predefined rule curves and static operation policies, which may be inadequate under unexpected or extreme conditions, such as flash floods [19]. These models also do not incorporate the intuition and experience of operators—often essential during emergencies—and require detailed physical, hydrologic, and hydraulic parameters, which may not be available or accurate in real time. As such, their applicability to real-time reservoir operation is limited [17].

Machine learning (ML) has emerged as a powerful tool for simulating reservoir operations, offering the ability to efficiently learn complex operational rules from historical hydrological and reservoir datasets [20,21]. Unlike traditional models that rely on explicit mathematical formulations, ML models derive data-driven relationships between input variables (e.g., reservoir storage and inflow) and output variables (e.g., reservoir outflows), enabling flexible modeling without requiring full knowledge of physical processes [22]. Although ML methods are often considered "black boxes" due to limited interpretability, they excel at capturing nonlinear, high-dimensional, and dynamic relationships in water system [23], making them particularly suitable for real-time reservoir operation and outflow simulation.

Over the past decade, a variety of ML algorithms have been successfully applied to reservoir operation tasks, including Support Vector Machines (SVMs) [24], K-Nearest Neighbors (KNN) [25], Classification Regression Trees (CARTs) [26], Random Forests (RFs) [27], Multi-Layer Perceptrons (MLPs) [28], and Recurrent Neural Networks (RNNs) [2]. Among these, RNN-based models—including Long Short-Term Memory (LSTM) and Gated Recurrent Unit (GRU)—are particularly well-suited to sequential decision-making problems due to their ability to preserve both short- and long-term dependencies in time series data [2]. Wannasin et al. [29] compared RNN, LSTM, and GRU for reservoir simulation and concluded that GRU provided the most robust and accurate results, suggesting it as a promising option for practical applications in real-time water management.

In the context of deep learning-based forecasting, two main approaches are commonly used to incorporate recent observations: (1) data integration (DI), where lagged observations are explicitly included as input variables [30], and (2) variational data assimilation, where model states are adjusted in real-time to minimize errors between observations and simulations [31]. While data assimilation can be computationally intensive and complex to implement [31], DI offers a simpler and often more effective alternative [32]. In contrast to most LSTM/GRU models that rely solely on input sequences defined by time windows, where past data is processed implicitly, DI enables the model to explicitly learn short-term hydrologic patterns, improving accuracy and responsiveness [30].

Recent studies have demonstrated the effectivity of DI in enhancing deep learning models. Feng et al. [30] showed that incorporating lagged streamflow observations in LSTM inputs significantly improved hydrological forecasts. Song et al. [32] applied DI to improve snow water equivalent prediction using LSTM, showing its value for water resources management. However, to the best of our knowledge, DI has not yet been applied in deep learning-based reservoir operation models. This study addresses that gap by testing whether a GRU model with DI—explicitly accepting lagged observations as input—can outperform a standard GRU without DI in reservoir operation simulation.

To achieve this, we combine an optimization model with a machine learning model employing data integration, with three main objectives: (1) develop an optimization model to minimize peak reservoir outflows; (2) develop a machine learning model for real-time operation, trained on optimal hourly releases derived from the optimization model; and (3) assess the impact of different input variable sets, with and without DI, on the accuracy of the reservoir operation simulation.

This methodology is applied to the Namgang Dam in South Korea, which is particularly vulnerable to flooding due to its small storage capacity relative to its catchment area. According to the Korea Meteorological Administration, the southern region—including the Namgang watershed—is experiencing increasing rainfall intensity, elevating flood risks. Despite this, the Namgang Dam remains the only multipurpose dam in the country without completed response measure for extreme floods [33]. In this study, an optimization model is first used to derive optimal reservoir operating rules for flood control at Namgamg Dam. The optimized outflow hydrographs are then compared to historical operations to evaluate the improvement. Next, a GRU model is trained on reservoir inflow, and storage and release data generated by the optimization process. The study also explores various scenarios to evaluate the effects of input variable selection, the inclusion of lagged observation via DI, and the impact of forecasted inputs on model performance.

2. Site and Materials

The Namgang Dam is located on the Nam River (35°15′ N, 128°23′ E), a tributary of the Nakdong River, and serves multiple purposes including flood control and water supply [34]. The dam creates the Jinyang Reservoir, which has a catchment area of 2285 km^2 and a total storage capacity of 309.2 million m^3. The reservoir covers a surface area of 28.2 km^2 [3]. The primary inflows originate from the Kyeongho and Duckcheon Rivers (Figure 1), with the Kyeongho River contributing runoff from a 1773.2 km^2 watershed—approximately five times the size of the Duckcheon River basin, which drains an area of 354.6 km^2 [34]. The design flood level of the Namgang Dam is 46 EL.m, while both the normal high-water level and the flood season restricted water level are set at 41 EL.m. The minimum operating level is 32 EL.m [3].

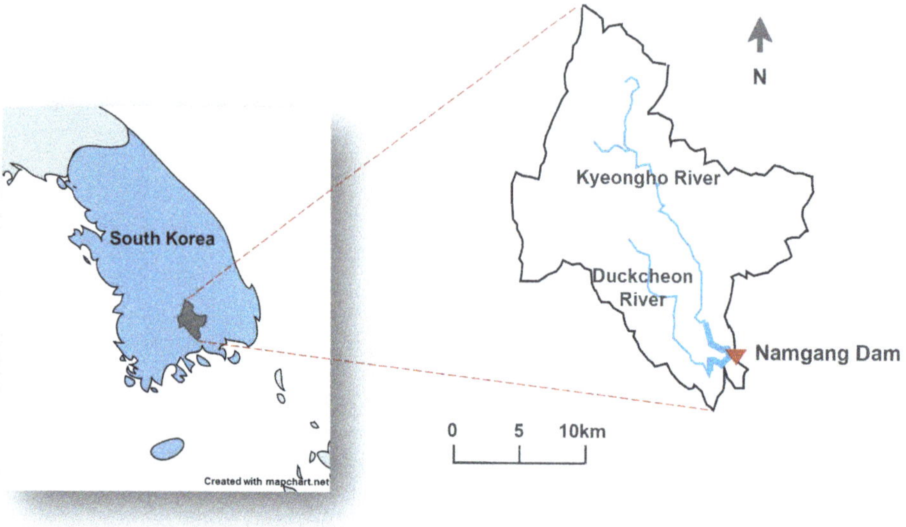

Figure 1. Location of the Namgang Dam basin in South Korea.

This study employs hourly data on reservoir inflow, storage, release, and precipitation obtained from the My Water platform (www.water.or.kr, accessed on 21 February 2025), spanning a 25-year period from 2000 to 2024. The analysis focuses on reservoir operations during flood events. A total of 42 significant flood events—each with peak reservoir inflows exceeding 3000 m^3/s—were identified and are summarized in Table 1. The historical maximum peak inflow recorded at the reservoir is 14,818.0 m^3/s. All peak flow events occurred between June and September, aligning with the region's primary rainy season influenced by the East Asian monsoon, tropical cyclones, and convective storms [35]. This storm-based approach offers practical insights into the planning and design of flood management infrastructure to better protect downstream communities and ecosystems [36,37].

Table 1. Summary of flood event characteristics identified between 2000 and 2024.

Event	Period		Duration (hrs)	Max Inflow (m^3/s)	Initial Storage ($10^6 m^3$)
1	13 July 2000 10:00	17 July 2000 22:00	109	6524	99.36
2	3 August 2000 14:00	5 August 2000 23:00	58	4149	78.56
3	12 September 2000 07:00	19 September 2000 17:00	179	3258	70.41
4	24 June 2001 00:00	26 June 2001 23:00	72	3009	78.92
5	5 July 2001 07:00	7 July 2001 08:00	50	3364	106.36
6	5 July 2002 04:00	7 July 2002 15:00	60	5792	31.59
7	6 August 2002 16:00	9 August 2002 23:00	80	5442	79.47
8	29 August 2002 09:00	2 September 2002 23:00	111	14,818	149.45
9	11 September 2003 04:00	15 September 2003 17:00	110	12,082	58.12
10	17 August 2004 10:00	21 August 2004 13:00	100	6070	77.09

Table 1. *Cont.*

Event	Period		Duration (hrs)	Max Inflow (m^3/s)	Initial Storage (10^6m^3)
11	21 August 2004 19:00	25 August 2004 12:00	90	3641	75.69
12	8 July 2006 19:00	13 July 2006 05:00	107	12,214	82.15
13	7 August 2007 00:00	11 August 2007 03:00	100	3534	45.49
14	14 September 2007 09:00	18 September 2007 22:00	110	8792	162.11
15	14 July 2009 14:00	19 July 2009 11:00	118	4211	90.88
16	10 July 2010 21:00	15 July 2010 00:00	100	3543	80.77
17	15 July 2010 20:00	20 July 2010 12:00	113	3005	157.65
18	10 August 2010 03:00	13 August 2010 23:00	93	5846	119.69
19	16 August 2010 16:00	19 August 2010 14:00	71	4918	155.85
20	1 September 2010 12:00	4 September 2010 01:00	62	3857	107.79
21	25 June 2011 08:00	29 June 2011 15:00	104	5711	78.30
22	8 July 2011 10:00	12 July 2011 13:00	100	7095	123.76
23	7 August 2011 08:00	10 August 2011 23:00	88	10,648	129.57
24	21 August 2012 15:00	27 August 2012 04:00	134	3767	127.14
25	27 August 2012 22:00	30 August 2012 12:00	63	4895	104.29
26	16 September 2012 07:00	19 September 2012 02:00	68	14,101	110.16
27	4 July 2013 12:00	7 July 2013 04:00	65	4050	116.40
28	2 August 2014 02:00	5 August 2014 23:00	94	4898	125.93
29	11 July 2015 17:00	14 July 2015 19:00	75	4546	114.95
30	25 August 2018 17:00	29 August 2018 05:00	85	5257	102.32
31	3 September 2018 16:00	5 September 2018 14:00	47	3368	150.23
32	5 October 2018 12:00	7 October 2018 17:00	54	7433	143.73
33	19 July 2019 05:00	23 July 2019 01:00	93	3997	144.25
34	2 October 2019 00:00	4 October 2019 15:00	64	5784	174.92
35	12 July 2020 06:00	14 July 2020 22:00	65	5443	167.71
36	21 July 2020 22:00	25 July 2020 22:00	97	3492	171.86
37	7 August 2020 01:00	9 August 2020 20:00	68	8729	165.78
38	2 September 2020 09:00	4 September 2020 11:00	51	6216	53.86
39	5 September 2022 13:00	7 September 2022 18:00	54	5467	48.97
40	15 July 2023 11:00	20 July 2023 17:00	127	3932	90.12
41	9 August 2023 14:00	12 August 2023 02:00	61	4798	85.88
42	29 August 2023 00:00	1 September 2023 20:00	93	3658	154.99

3. Methodology

This study proposes an integrated methodology that combines a reservoir operation optimization model with a data-driven machine learning model to improve flood control performance. Figure 2 illustrates the workflow of the proposed approach, which aims to minimize downstream flood damage while ensuring dam safety. The optimization model is first used to derive optimal reservoir release rules for a range of historical flood events. The resulting optimal release and storage data, along with the corresponding inflow hydrographs, are then used to train, validate, and test a machine learning-based real-time reservoir operation model.

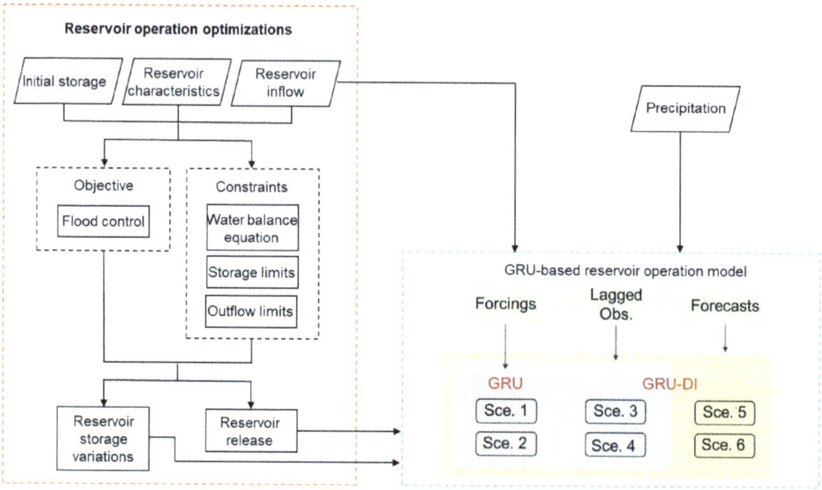

Figure 2. Workflow of the proposed methodology for developing reservoir operation model.

3.1. Reservoir Operation Optimization Model

The objective of the reservoir operation optimization is to minimize the downstream flood damage by reducing the peak flow of reservoir release hydrographs. The optimization model determines reservoir outflow hydrographs based on given inflow patterns and initial storage conditions. The objective function Z is defined as follows:

$$\text{Minimize } Z = \max_{t=1,\ldots,T} R_t \tag{1}$$

in which R_t is the reservoir outflow at time t, and T is the flood duration.

Constraints of the model are as follows:

$$S_t = S_{t-\Delta t} + I_{t-\Delta t}\Delta t - R_{t-\Delta t}\Delta t \quad (t = 1, \ldots, T) \tag{2}$$

$$0 \le S_t \le S_{max} \tag{3}$$

$$0 \le R_t \le R_t^{max}(S_t) \quad (t = 1, \ldots, T) \tag{4}$$

in which S_t is the reservoir storage at time t, $I_{t-\Delta t}$ is the inflow to the reservoir at time $(t-\Delta t)$, and Δt is taken as 1 h in this study. S_{max} is the maximum reservoir storage, and $R_t^{max}(S_t)$ is the maximum outflow capacity of the reservoir based on the reservoir volume and volume–outflow curve.

To constrain reservoir outflows, a volume–outflow relationship is established based on the water level-storage relationships for different sections of reservoir, coupled with the combined discharge capacities of all outlet structures (Figure 3). This relationship defines the maximum allowable outflow corresponding to the reservoir's storage volume. It is incorporated into the optimization model as a constraint (Equation (4)). During the optimization process, reservoir storage is updated at each time, and the corresponding maximum outflow is applied to ensure feasible and physically realistic solutions.

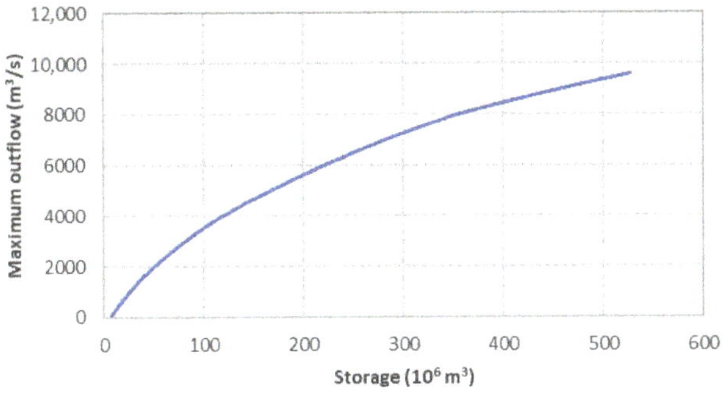

Figure 3. Reservoir volume–outflow relationship curve.

This study employs the Interior Point Optimizer (IPOPT) to solve the reservoir flood operation optimization problem. Reservoir flood control represents a large-scale, nonlinear, and constrained optimization challenge. IPOPT is an open-source solver [38] specifically developed to address such problems efficiently. It uses a primal-dual interior-point algorithm combined with a filter line-search strategy [39]. Inequality constraints are handled through the use of logarithmic barrier functions, introducing a barrier parameter that transforms the original problem into a sequence of equality-constrained sub-problems. As the barrier parameter is progressively reduced, these sub-problems are solved iteratively [40]. For each iteration, IPOPT applies a Newton-type algorithm with a line search that involves solving large, sparse, symmetric, and potentially indefinite linear systems. This approach ensures robust convergence and high computational performance, particularly in large-scale dynamic optimization tasks [41]. In this study, the optimization framework is implemented in Pyomo 6.7.3 [42], a Python-based algebraic modeling language that facilitates clear and intuitive representation of mathematical models. Pyomo serves as the interface for defining the optimization problem and transferring it to IPOPT for solution.

3.2. GRU-Based Reservoir Operation Model

The Gated Recurrent Unit (GRU), introduced by Cho et al. [43], is an advanced variant of the traditional Recurrent Neural Network (RNN). GRUs share structural similarities with Long Short-Term Memory (LSTM) networks, particularly in their use of gating mechanisms to regulate the flow of information. However, GRUs offer a more streamlined architecture by reducing the number of gates, thereby simplifying computations while maintaining comparable performance [44]. Several studies have shown that GRUs can deliver performance on par with LSTMs, but with improved training efficiency and faster execution [45]. For instance, Gao et al. [45] developed both LSTM- and GRU-based models for short-term runoff forecasting and found GRUs to be preferable due to their reduced training time and comparable predictive accuracy.

As illustrated in Figure 4, the GRU architecture enhances the memory capacity of standard RNNs while improving training stability and efficiency [44]. By employing gating mechanisms and hidden units, GRUs mitigate the vanishing gradient problem often encountered in traditional RNNs. Each GRU unit adaptively captures temporal dependencies across different time scales, making it well-suitable for sequential data modeling.

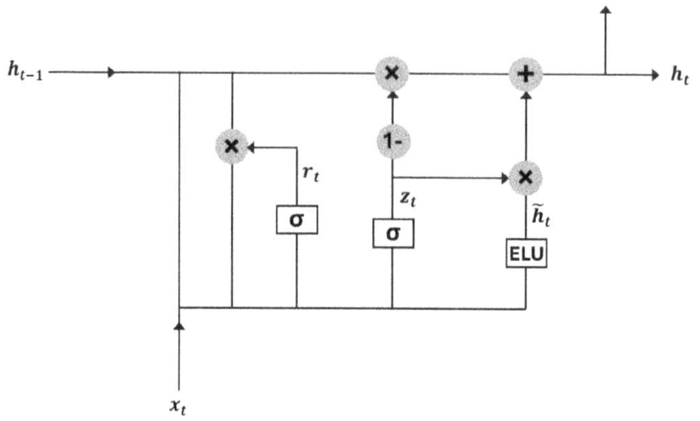

Figure 4. Structure schematic of a GRU cell.

At each time step, the GRU processes two inputs: the current input x_t and the previous hidden state h_{t-1}, which encapsulates information from prior time steps. The resulting output y_t is derived from the hidden node, while the updated hidden state h_t is forwarded to the next time step [46].

Unlike LSTM, which uses three gates (input, forget, and output), the GRU simplifies the architecture by using only two gates. One is the update gate (z_t) that determines the extent to which past information should be retained in the current state. A higher value indicates greater retention. The other is the reset gate (r_t) that controls the degree to which past information is forgotten. A lower value results in greater forgetting of previous state.

These gates enable GRUs to capture long-term dependencies while reducing model complexity and computational cost. The mathematical formation of the GRU is provided in Equations (5)–(9) [47,48]:

$$r_t = \sigma(w_r \cdot [h_{t-1}, x_t] + b_r) \quad (5)$$

$$z_t = \sigma(w_z \cdot [h_{t-1}, x_t] + b_z) \quad (6)$$

$$\tilde{h}_t = \mathrm{ELU}\left(w_{\tilde{h}} \cdot [r_t * h_{t-1}, x_t]\right) \quad (7)$$

$$h_t = (1 - z_t) * h_{t-1} + z_t * \tilde{h}_t \quad (8)$$

$$y_t = \sigma(w_o \cdot h_t) \quad (9)$$

where σ denotes the sigmoid activation function, which bounds the output between 0 and 1. Additionally, the Exponential Linear Unit (ELU) activation function is employed to introduce nonlinearity, allowing for negative values and enhancing convergence speed by increasing neuron diversity and smoothing activation dynamics.

In this study, the GRU model is employed to simulate reservoir operations under flood conditions. The model is trained to estimate reservoir outflows using input features such as inflow hydrographs, precipitation data, and optimized reservoir storage and release values derived from a reservoir operation optimization model. The modeling is conducted at an hourly resolution during selected flood events.

To train the model, the dataset is divided into predictor variables (inputs) and a target variable (output). The primary predictor is reservoir inflow, which captures the upstream hydrological response. Precipitation is also included, as it directly influences inflow and storage dynamics. Another key input variable is the reservoir's storage volume, a commonly used determinant of release decisions. Storage-based operational rules typically aim to satisfy multiple objectives, including water supply, flood control, irrigation, and hydropower generation [18]. The target variable is the hourly reservoir outflow at the current time step. With this structured input-output setup, the GRU-based reservoir operation model is developed to enhance real-time decision-making during flood events.

3.3. Scenarios

To evaluate the impact of different input combinations and application of DI, six scenarios are tested, as summarized in Table 2. The selection of input variables and their time lags is known to significantly influence the performance of machine/deep learning models [49,50]. Scenario 1 uses reservoir inflow and storage at time step $t-1$ as input variables. Scenario 2 adds precipitation at $t-1$ to examine its influence on GRU-based reservoir operation modeling. Scenarios 3 and 4 extend Scenarios 1 and 2, respectively, by integrating lagged observations over the past three hours into the GRU input, thereby incorporating DI. Scenarios 5 and 6 additionally include short-term forecasts (up to two days ahead) of inflow and precipitation, considering the importance of forecasting in operational decision-making [51,52]. Forecasts are assumed to be perfectly accurate and are generated from observed data.

Table 2. Input variable configurations for scenario analysis.

Scenarios	Input Variables		
	Inflow	Storage	Precipitation
1	I_{t-1}	S_{t-1}	-
2	I_{t-1}	S_{t-1}	P_{t-1}
3	$I_{t-3}, I_{t-2}, I_{t-1}$	$S_{t-3}, S_{t-2}, S_{t-1}$	-
4	$I_{t-3}, I_{t-2}, I_{t-1}$	$S_{t-3}, S_{t-2}, S_{t-1}$	$P_{t-3}, P_{t-2}, P_{t-1}$
5	$I_{t-3}, I_{t-2}, I_{t-1}, I_t, I_{t+1}, I_{t+2}$	$S_{t-3}, S_{t-2}, S_{t-1}$	-
6	$I_{t-3}, I_{t-2}, I_{t-1}, I_t, I_{t+1}, I_{t+2}$	$S_{t-3}, S_{t-2}, S_{t-1}$	$P_{t-3}, P_{t-2}, P_{t-1}, P_t, P_{t+1}, P_{t+2}$

4. Results and Discussion

Optimal reservoir storage and release trajectories for 42 historical flood events are generated using IPOPT, based on inflow hydrographs and initial storage conditions. These optimal values serve as training targets for GRU model development. The dataset is split into training (52%), validation (25%), and testing (23%) sets (Table 3).

Table 3. Data partitioning for GRU model training, validation and testing.

Training		Validation		Test	
Event	Max Inflow (m^3/s)	Event	Max Inflow (m^3/s)	Event	Max Inflow (m^3/s)
1	6524	4	3009	2	4149
3	3258	18	5846	5	3364
6	5792	19	4918	7	5442
8	14,818	22	7095	12	12,214
9	12,082	23	10,648	16	3543
10	6070	26	14,101	17	3005

Table 3. *Cont.*

Training		Validation		Test	
Event	Max Inflow (m^3/s)	Event	Max Inflow (m^3/s)	Event	Max Inflow (m^3/s)
11	3641	29	4546	20	3857
13	3534	33	3997	25	4895
14	8792	36	3492	34	5784
15	4211	39	5467	37	8729
21	5711	42	3658	38	6216
24	3767				
27	4050				
28	4898				
30	5257				
31	3368				
32	7433				
35	5443				
40	3932				
41	4798				

In this study, hyperparameter tuning focuses on three key parameters: window size, number of hidden nodes, and batch size. Fixed hyperparameters include dropout rate (0.2), learning rate (0.001), number of hidden layers (1), loss function (MSE), optimizer (RMSprop), and activation function (ELU). The number of hidden nodes is tuned within the range of 1–100. Early stopping with a maximum of 10,000 epochs is applied, and the number of epochs is not considered as a tuning parameter. Hyperparameter configurations are summarized in Table 4.

Table 4. Hyperparameter configurations considered for tuning and optimization.

Hyperparameter	Interval
Window size	[4, 6, 8]
Batch size	[8, 32]
Number of hidden nodes	[1, 100]

GRU model performance under different input scenarios and hyperparameter settings is presented in Table 5. To assess the impact of batch size and window size, Root Mean Squared Error (RMSE) and Nash-Sutcliffe Efficiency (NSE) values calculated using Equation (10) and Equation (11), respectively, are compared across multiple configurations. The number of hidden nodes is optimized in each setting using the RMSprop optimizer. Figure 5 shows the RMSE values using bar charts (e.g., S1_BS8_WS4 refers to Scenario 1 with batch size 8 and window size 4).

$$\mathrm{RMSE} = \sqrt{\frac{1}{n}\sum_{i=1}^{n}\left(Q_{opt,\ i} - Q_{sim,\ i}\right)^2} \tag{10}$$

$$\mathrm{NSE} = 1 - \frac{\sum_{i=1}^{n}\left(Q_{opt,i} - Q_{sim,\,i}\right)^2}{\sum_{i=1}^{n}\left(Q_{opt,i} - \overline{Q}_{opt,\,i}\right)^2} \quad (11)$$

where $Q_{opt,\,i}$ is the ith optimized reservoir outflow, $Q_{sim,i}$ is the ith GRU-simulated outflow, and n is the number of data points.

Table 5. RMSE and NSE values of GRU models for reservoir operation simulation.

Scenarios	GRU Hyperparameters			RMSE			NSE		
	Batch Size	Window Size	Number of Hidden Nodes	Training	Validation	Test	Training	Validation	Test
1	8	4	100	380.3	411.2	714.8	0.44	0.21	0.17
		6	84	364.0	378.4	712.7	0.49	0.34	0.18
		8	54	157.9	202.1	344.0	0.90	0.81	0.81
	32	4	90	439.0	417.8	717.5	0.26	0.19	0.16
		6	98	137.9	376.8	671.0	0.93	0.34	0.27
		8	84	154.4	251.6	414.4	0.91	0.71	0.72
2	8	4	60	345.3	397.1	694.9	0.54	0.27	0.21
		6	46	360.9	379.7	690.4	0.50	0.33	0.23
		8	22	125.0	294.5	521.9	0.94	0.60	0.56
	32	4	100	356.4	396.9	702.3	0.51	0.27	0.20
		6	10	356.6	398.0	700.5	0.51	0.26	0.20
		8	54	127.1	355.9	633.7	0.94	0.41	0.35
3	8	4	26	133.3	118.5	202.2	0.93	0.93	0.93
		6	26	155.8	187.6	310.9	0.91	0.84	0.84
		8	14	165.4	174.5	324.1	0.89	0.86	0.82
	32	4	62	145.4	204.6	366.2	0.92	0.81	0.78
		6	10	163.2	394.9	679.0	0.90	0.28	0.26
		8	16	167.7	216.8	379.1	0.89	0.78	0.76
4	8	4	20	125.9	160.7	262.5	0.94	0.88	0.89
		6	10	155.6	172.4	250.2	0.91	0.86	0.90
		8	10	173.8	226.0	375.6	0.88	0.76	0.76
	32	4	46	133.0	202.5	343.4	0.93	0.81	0.81
		6	20	153.4	210.4	366.8	0.91	0.79	0.78
		8	12	155.8	240.6	391.4	0.91	0.73	0.74
5	8	4	24	113.2	160.2	262.8	0.95	0.88	0.89
		6	16	144.7	148.5	261.7	0.92	0.90	0.89
		8	12	171.0	170.3	293.9	0.89	0.87	0.86
	32	4	20	124.3	160.9	262.2	0.94	0.88	0.89
		6	20	143.8	236.9	435.9	0.92	0.74	0.69
		8	24	174.1	187.0	318.5	0.88	0.84	0.83

Table 5. *Cont.*

Scenarios	GRU Hyperparameters			RMSE			NSE		
	Batch Size	Window Size	Number of Hidden Nodes	Training	Validation	Test	Training	Validation	Test
6	8	4	10	88.4	299.4	530.6	0.97	0.58	0.55
		6	50	123.2	200.2	323.9	0.94	0.81	0.83
		8	24	142.8	213.3	340.0	0.92	0.79	0.81
	32	4	14	109.5	249.2	434.1	0.95	0.71	0.70
		6	48	124.8	211.3	340.1	0.94	0.79	0.81
		8	40	144.5	225.9	365.7	0.92	0.76	0.78

Models with DI (Scenarios 3 and 4) trained with a batch size of 8 consistently outperform those with a batch size of 32. This indicates that smaller batch sizes may offer improved learning stability and responsiveness in DI-based model, likely due to their ability to better capture high-frequency variability in short hydrological sequences. In scenarios without DI (scenarios 1 and 2), a window size of 8 yields the best performance. In contrast, DI-based models (Scenarios 3 and 4) achieve optimal performance with window sizes of 4 and 6, respectively. For DI-based models incorporating forecasted variables (Scenarios 5 and 6), a window size of 6 provides the highest accuracy. These results indicate that GRU models with DI can achieve superior performance using shorter window sizes compared to those without DI. This finding is consistent with previous work by [30], which showed that DI enables models to capture short-term hydrologic dynamics more effectively by explicitly providing lagged observations as input features.

The best model performance is observed in S3_BS8_WS4, with a minimum RMSE of 202.2, an NSE of 0.93, and a coefficient of determination (R^2) of 0.89, followed by S4_BS8_WS6, which records a RMSE of 250.19, NSE of 0.90, and R^2 of 0.90. GRU models incorporating DI (scenarios 3 and 4) consistently outperform those without DI (scenarios 1 and 2), emphasizing the importance of explicitly incorporating recent observations in model training. This feature is especially critical for real-time flood operation, where quick decision-making is essential. Interestingly, S3_BS8_WS4, which relies solely on past inflow and storage observations, outperforms S4_BS8_SW6, which includes past precipitation as an input. This suggests that adding past precipitation does not enhance model accuracy, potentially due to data redundancy, as inflow and storage already encapsulate much of the system's hydrological behavior.

The inclusion of forecasted values in Scenarios 5 and 6 leads to improved accuracy in certain model configurations. However, the highest performance is achieved by the DI model utilizing only historical inflow and storage. These outcomes suggest that while forecasted data can provide additional information, its utility is limited and does not significantly enhance model accuracy. This may be due to the short lead times of forecasted inputs, which can compromise their reliability. Consequently, models based on recent historical data offer more robust and interpretable solutions, which are critical for real-time applications like reservoir operation.

(a)

(b)

Figure 5. Performance comparison of GRU-based reservoir operation models under various hyperparameter settings. (**a**) Batch size = 8; (**b**) Batch size = 32.

Despite the satisfactory performance of the models, further improvements are possible. In this study, several hyperparameters were manually tuned or selected, and no systematic hyperparameter optimization was performed. Future studies could benefit from more systematic hyperparameter optimization methods to further improve model performance. Nevertheless, the objective of this study is not solely to minimize RMSE or maximize NSE, but rather to identify key input variables for reservoir outflow prediction and evaluate the impact of DI and forecasted forcing on GRU model performance. Thus, the findings provide insights into effective model design for reservoir outflow prediction under real-time constraints.

Figure 6 presents the performance of the S3_BS8_WS4 model, comparing observed, optimized, and GRU-simulated outflows during the test periods. The results show that the optimized release hydrographs consistently demonstrate reduced peak discharges compared to observed historical operation. This supports the premise outlined in the Introduction that optimization-based approaches can effectively mitigate downstream flood risk. The GRU-simulated outflows closely align with the optimized trajectories, demonstrating the model's effectiveness in replicating optimal release patterns. Furthermore, the hourly reservoir storage, calculated based on the simulated outflows, remains within the operational storage bounds and closely tracks the optimized storage throughout the test periods, as shown in Figure 7. These findings validate the capability of the GRU model, particularly when trained with DI, to emulate optimization-derived control strategies under dynamic hydrologic conditions. This confirms the feasibility of integrating optimization outputs with deep learning models for real-time reservoir operation support, offering a promising approach to enhance the resilience and effectiveness of flood management systems.

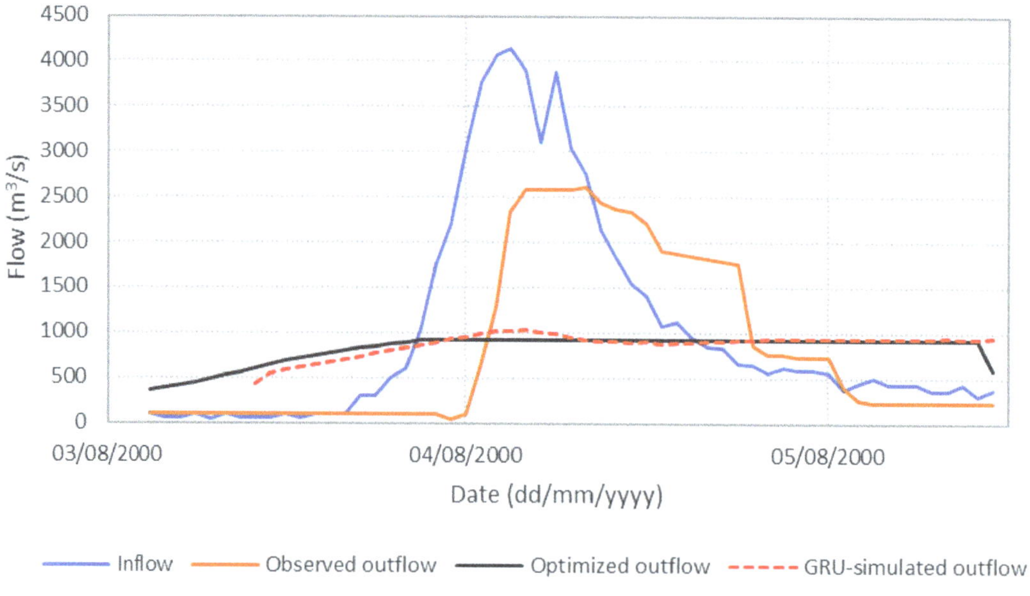

(a)

Figure 6. *Cont.*

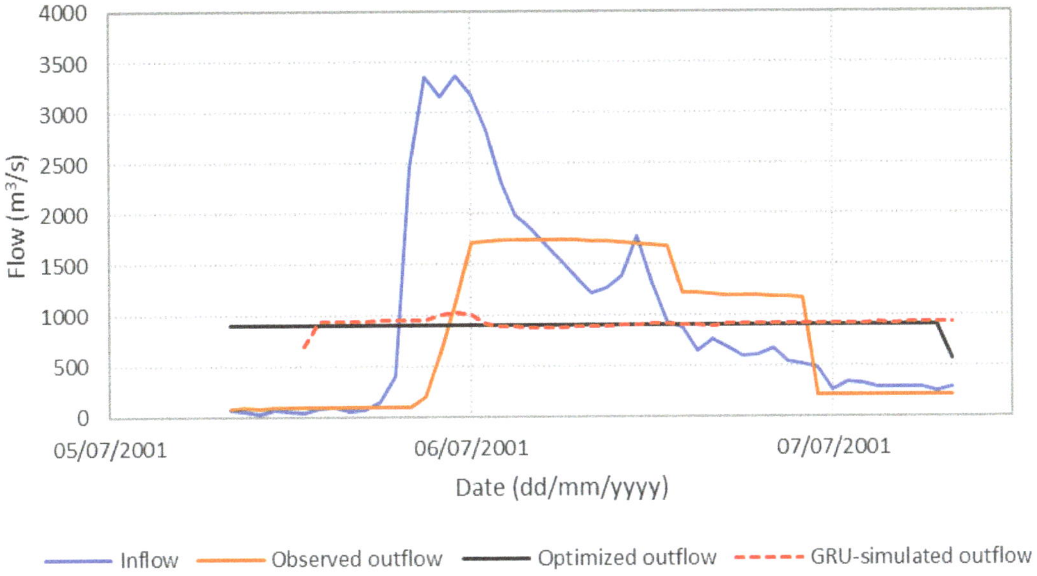

(**b**)

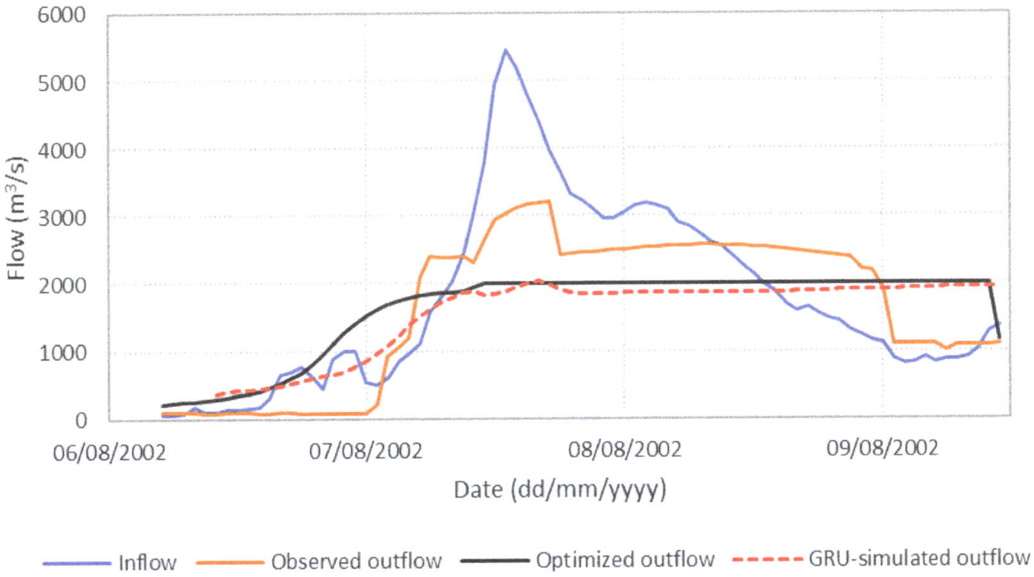

(**c**)

Figure 6. *Cont.*

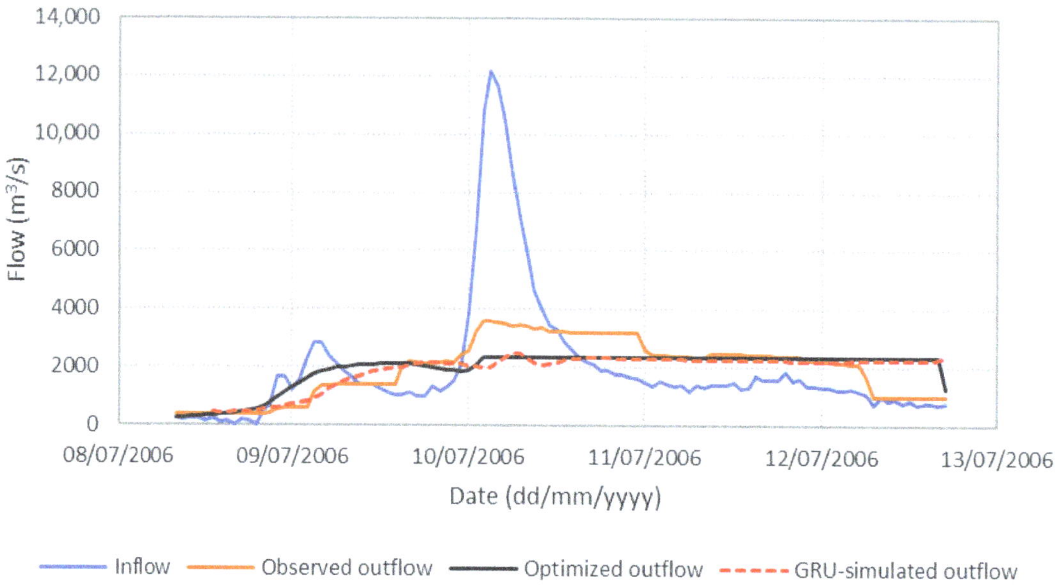

(**d**)

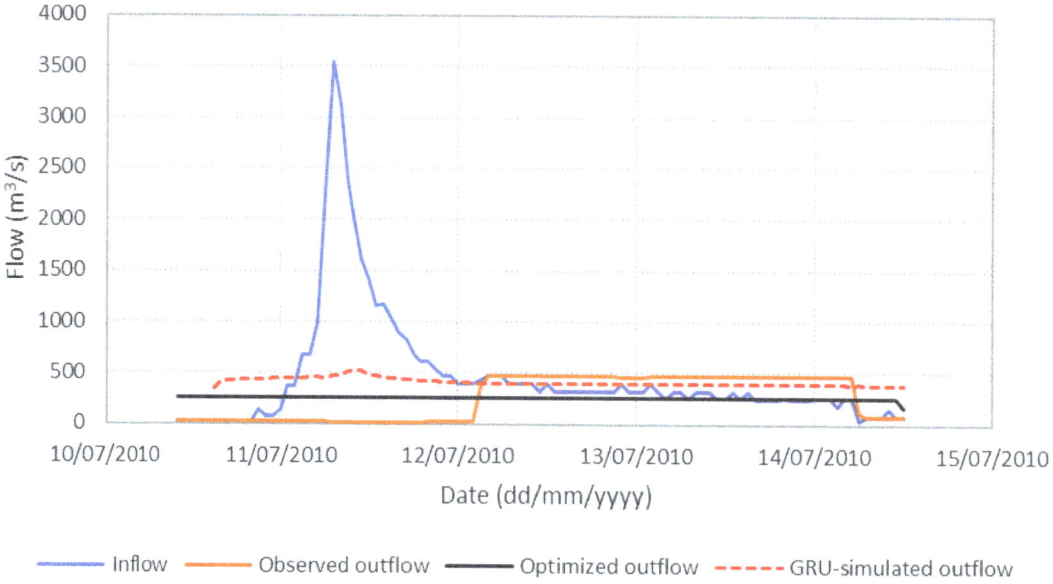

(**e**)

Figure 6. *Cont.*

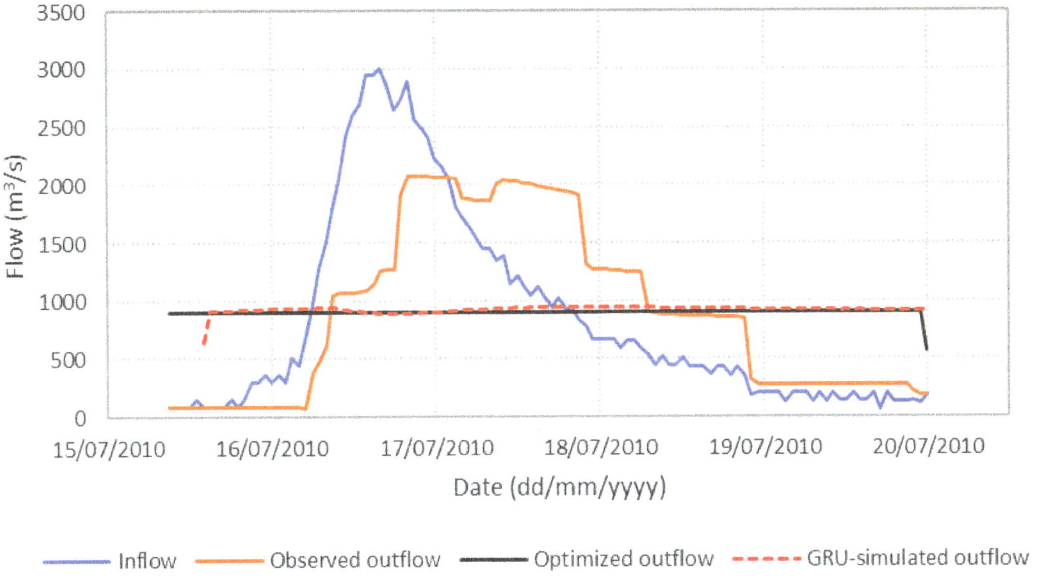

(f)

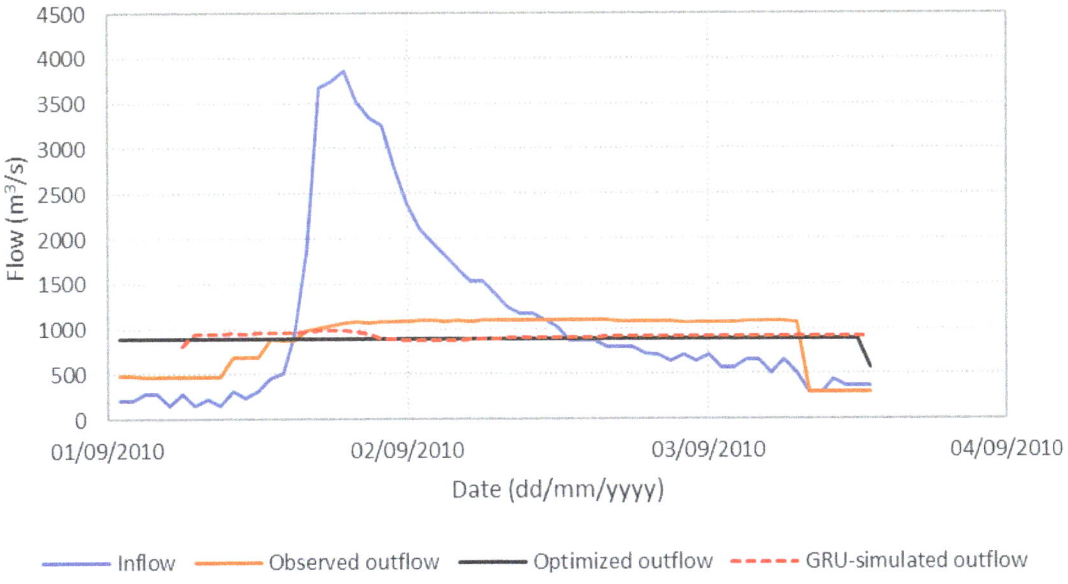

(g)

Figure 6. *Cont.*

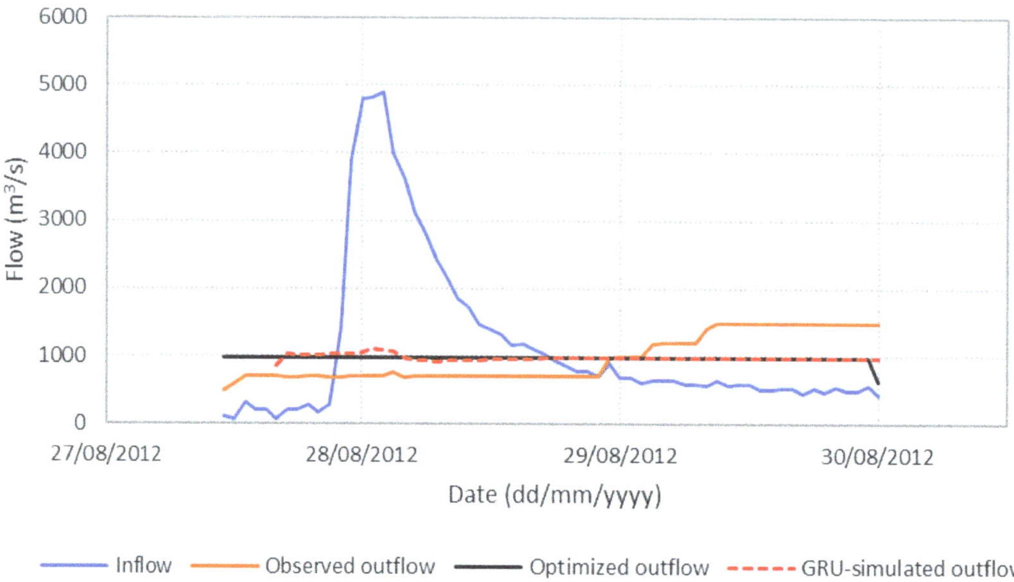

(**h**)

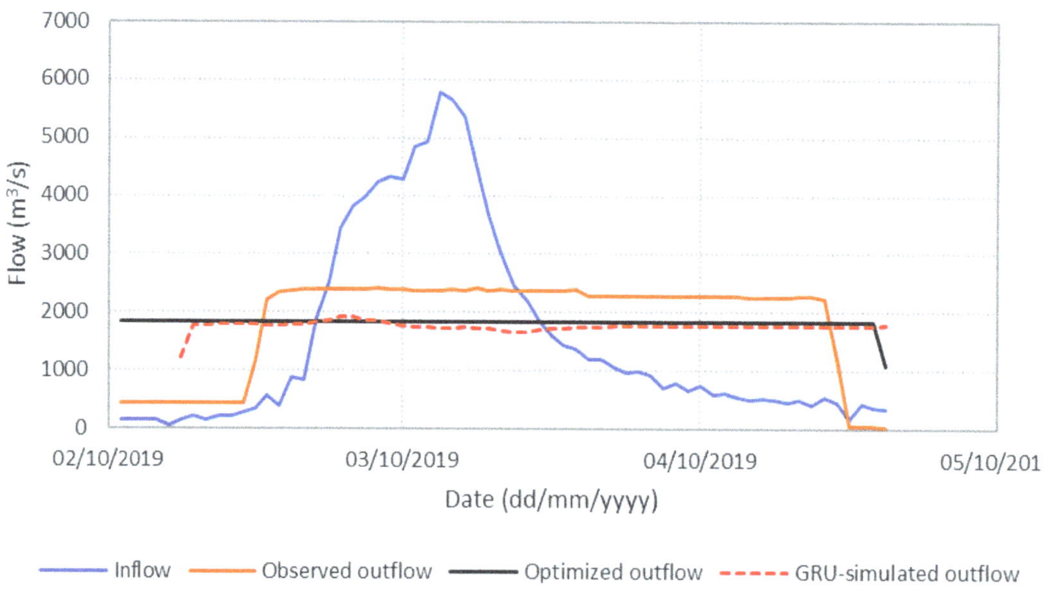

(**i**)

Figure 6. *Cont.*

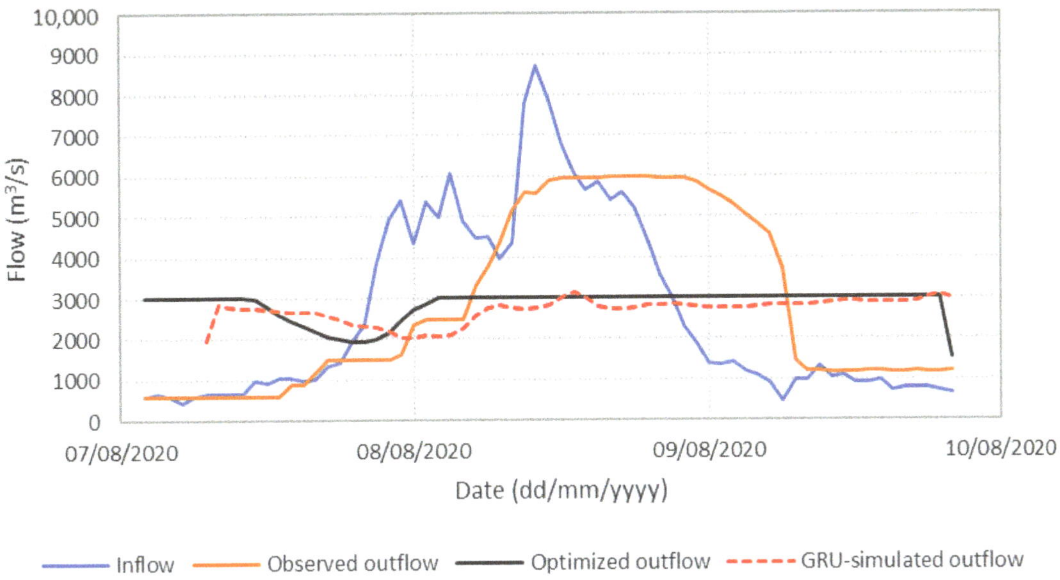

(**j**)

Figure 6. Comparison of observed, optimized, and GRU-simulated reservoir outflows. (**a**) Flood event 2; (**b**) Flood event 5; (**c**) Flood event 7; (**d**) Flood event 12; (**e**) Flood event 16; (**f**) Flood event 17; (**g**) Flood event 20; (**h**) Flood event 25; (**i**) Flood event 34; (**j**) Flood event 37.

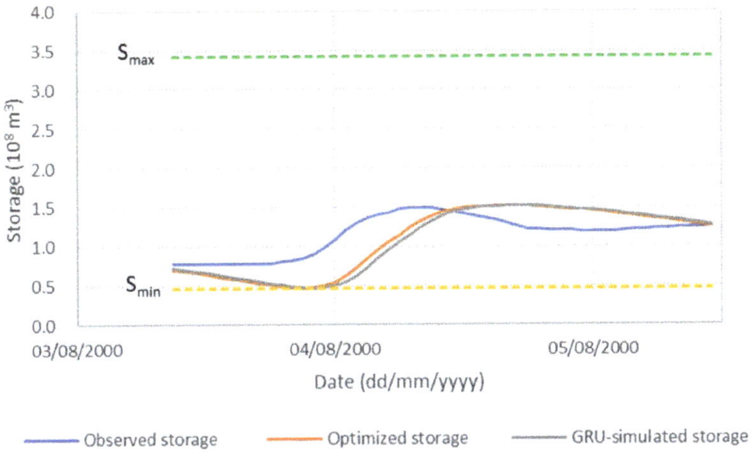

(**a**)

Figure 7. *Cont.*

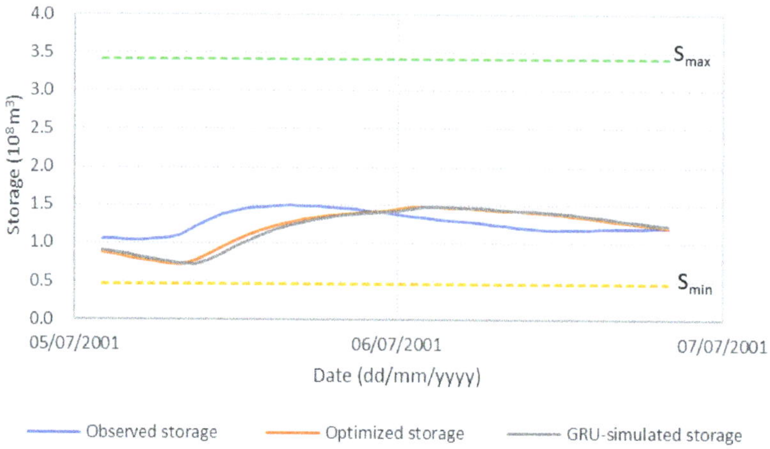

(**b**)

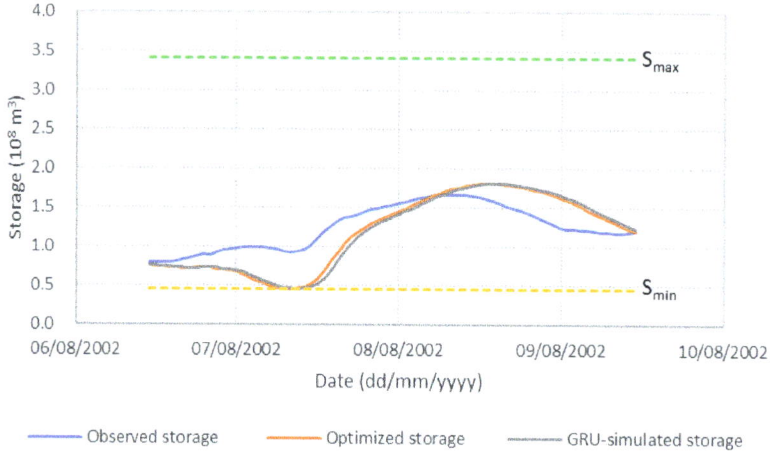

(**c**)

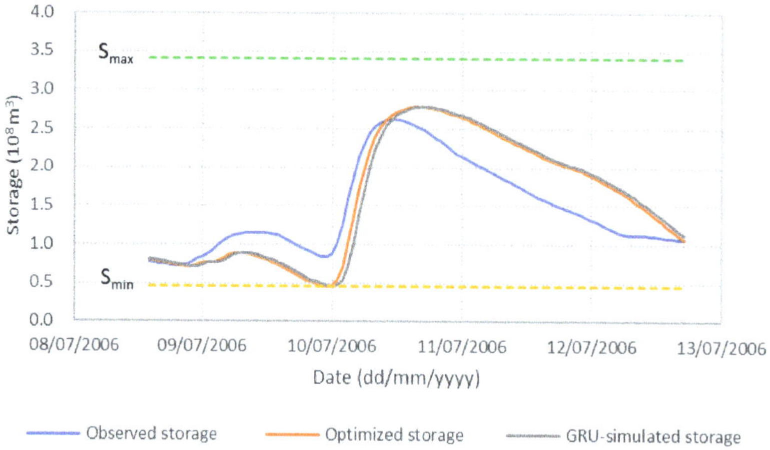

(**d**)

Figure 7. *Cont.*

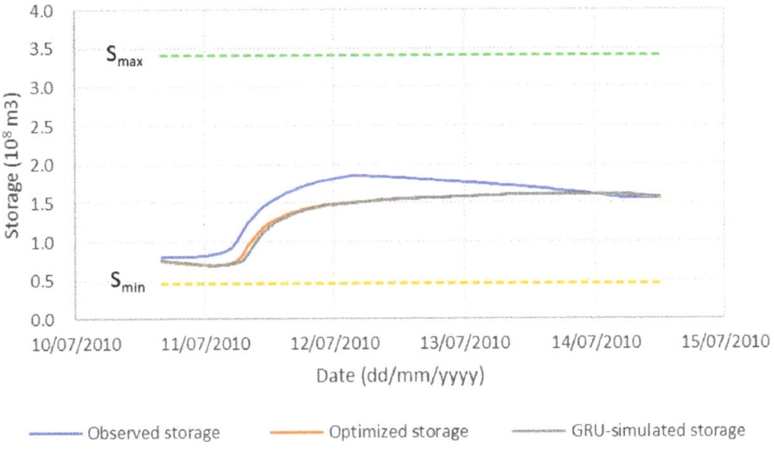

(e)

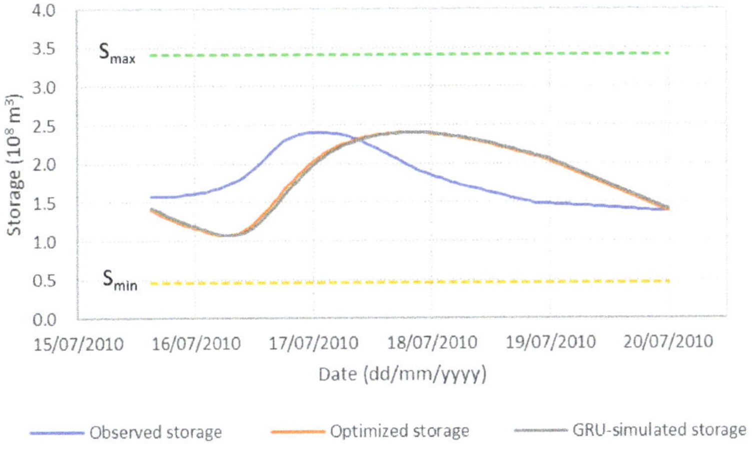

(f)

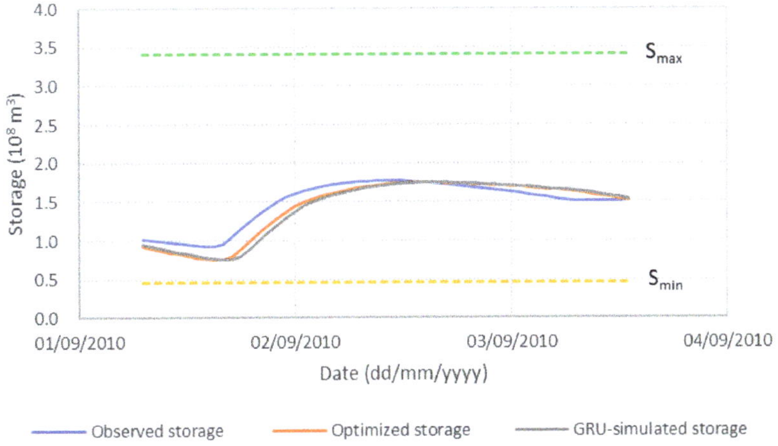

(g)

Figure 7. *Cont.*

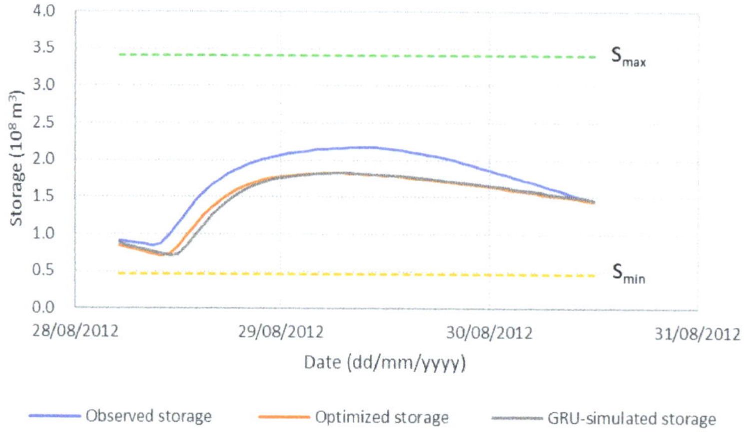

(**h**)

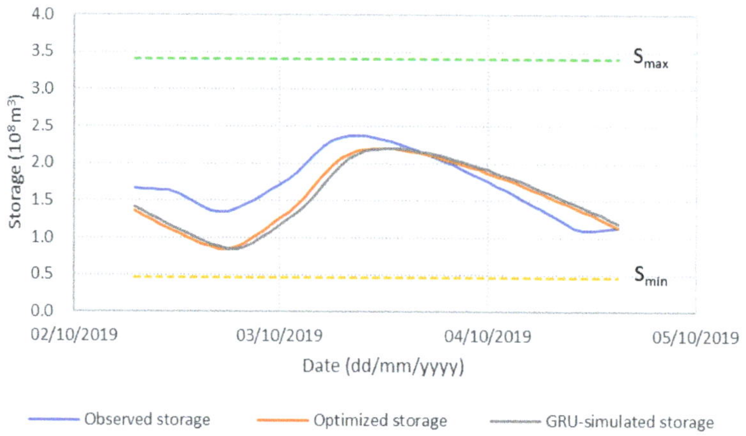

(**i**)

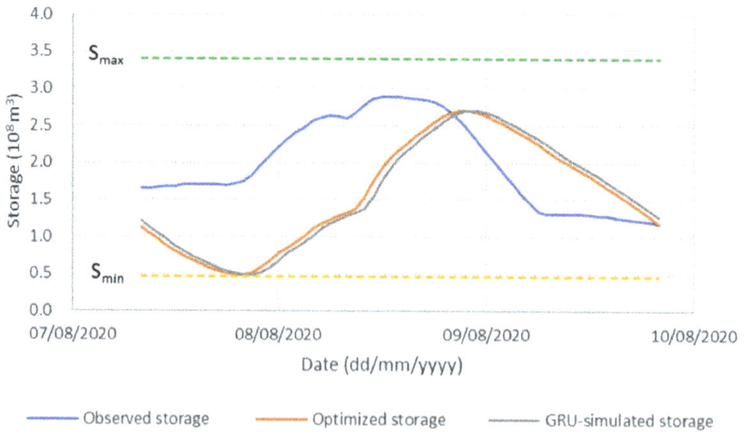

(**j**)

Figure 7. Comparison of observed, optimized, and GRU-simulated reservoir storage levels. (**a**) Flood event 2; (**b**) Flood event 5; (**c**) Flood event 7; (**d**) Flood event 12; (**e**) Flood event 16; (**f**) Flood event 17; (**g**) Flood event 20; (**h**) Flood event 25; (**i**) Flood event 34; (**j**) Flood event 37.

5. Conclusions

This study first applies IPOPT, a deterministic optimization method, to derive detailed operational rules of reservoir operation during flood periods. The peak releases in the optimized hydrographs consistently remain lower than those observed in historical operations. These optimized reservoir storage and release values, along with the corresponding inflow hydrographs, are then used to train a machine learning-based reservoir operation model.

This study employs the GRU model to integrate lagged observations for enhancing reservoir release forecasting at an hourly scale. Scenario analysis highlights the strength of GRU with DI in simulating reservoir releases and reveals the importance of input variable selection. By applying the DI procedure, the model incorporates recent observations to dynamically update hydrological conditions and predict reservoir releases with high accuracy. GRU models with DI consistently outperform those without, as DI directly provides short-term memory inputs that improve accuracy in memory-dependent systems, enhance interpretability, and suit real-time decision-making. Given its demonstrated effectiveness, the DI approach offers promising applications for real-time reservoir operation for flood control. The inclusion of precipitation or forecasted forcing does not improve model performance. Given that forecasted data often contains significant uncertainties that influence reservoir operations, the accuracy of the model incorporating forecasted forcings—as in Scenarios 5 and 6—may be lower than those relying solely on observed data. The GRU model learns to approximate the mathematical processes underlying both the hydrologic system and the DI procedure. The findings of this study provide a real-time operation tool for simulating optimal reservoir releases during flood events. With the same framework, it is possible to incorporate additional relevant variables, depending on the specific problem context.

The proposed GRU-DI model demonstrates the ability to simulate reservoir release dynamics and leverage observations to improve forecasting. However, the internal mechanisms by which GRU processes hydrological information remain difficult to interpret. Continued efforts in explainable machine learning are encouraged to reveal the GRU's internal representations, particularly regarding how it learns reservoir release behavior. This understanding could inform the development of more transparent and effective hydrologic models. Additionally, further studies may examine the effects of varying the time scale of lagged observations used in DI, which could further enhance model performance in reservoir operation.

Author Contributions: Conceptualization, L.L. and K.S.J.; methodology, L.L.; software, L.L.; validation, L.L. and K.S.J.; formal analysis, L.L.; investigation, L.L.; resources, L.L.; data curation, L.L.; writing—original draft preparation, L.L.; writing—review and editing, K.S.J.; visualization, L.L.; supervision, K.S.J.; project administration, K.S.J.; funding acquisition, K.S.J. All authors have read and agreed to the published version of the manuscript.

Funding: This work was supported by the Korea Environmental Industry and Technology Institute (KEITI) through R&D Program for Innovative Flood Protection Technologies against Climate Crisis, funded by Korea Ministry of Environment (MOE) (grant number: 2022003460001).

Data Availability Statement: Data are contained within the article.

Conflicts of Interest: The authors declare no conflicts of interest.

References

1. Yoo, C.; Jun, C.; Zhu, J.; Na, W. Evaluation of dam water-supply capacity in Korea using the water-shortage index. *Water* **2021**, *13*, 956. [CrossRef]
2. Zhang, D.; Peng, Q.; Lin, J.; Wang, D.; Liu, X.; Zhuang, J. Simulating reservoir operation using a recurrent neural network algorithm. *Water* **2019**, *11*, 865. [CrossRef]
3. MyWater. Available online: https://www.water.or.kr/ (accessed on 21 March 2025).
4. Udall, S.L. *Design of Small Dams United States Department of the Interior*; Bureau of Reclamation: Washington, DC, USA, 1961.
5. Labadie, J.W. Optimal operation of multireservoir systems: State-of-the-art review. *J. Water Resour. Plan. Manag.* **2004**, *130*, 93–111. [CrossRef]
6. Zhu, F.; Zhong, P.-A.; Sun, Y.; Yeh, W.W.G. Real-time optimal flood control decision making and risk propagation under multiple Uncertainties. *Water Resour. Res.* **2017**, *53*, 10635–10654. [CrossRef]
7. Ding, W.; Wei, G.; Zhou, H. Improving flood resilience through optimal reservoir operation. *J. Hydrol.* **2023**, *620*, 129494. [CrossRef]
8. Yakowitz, S. Dynamic programming applications in water resources. *Water Resour. Res.* **1982**, *18*, 673–696. [CrossRef]
9. Malekmohammadi, B.; Zahraie, B.; Kerachian, R. A real-time operation optimization model for flood management in river-reservoir systems. *Nat. Hazards* **2010**, *53*, 459–482. [CrossRef]
10. Haddeland, I.; Skaugen, T.; Lettenmaier, D.P. Anthropogenic impacts on continental surface water fluxes. *Geophys. Res. Lett.* **2006**, *33*, L08406. [CrossRef]
11. Mohanty, A.; Sahoo, B.; Kale, R.V. A coupled optimized hedging rule-based reservoir operation and hydrodynamic model framework for riverine flood risk management. *Water Res.* **2025**, *279*, 123443. [CrossRef]
12. Chen, R.; Wang, D.; Mei, Y.; Lin, Y.; Lin, Z.; Zhang, Z.; Zhuang, S.; Zhu, J.; Kam, J.; Wu, Y.; et al. A knowledge-guided LSTM reservoir outflow model and its application to streamflow simulation in reservoir-regulated basins. *J. Hydrol.* **2025**, *658*, 133164. [CrossRef]
13. Sigvaldson, O.T. A simulation model for operating a multipurpose multireservoir system. *Water Resour. Res.* **1976**, *12*, 263–278. [CrossRef]
14. Uyasal, G.; Sensoy, A.; Sorman, A.A.; Akgun, T.; Gezgin, T. Basin/reservoir system integration for real time reservoir operation. *Water Resour. Manag.* **2016**, *30*, 1653–1668. [CrossRef]
15. Yates, D.; Sieber, J.; Purkey, D.; Huber-Lee, A. WEAP21-A demand-, priority-, and preference-driven water planning model: Part 1: Model characteristics. *Water Int.* **2005**, *30*, 487–500. [CrossRef]
16. Klipsch, J.D.; Hurst, M.B. *HEC-ResSim Reservoir System Simulation User's Manual, Version 3.0*; USACE: Davis, CA, USA, 2007; p. 512.
17. Zhang, D.; Lin, J.; Peng, Q.; Wang, D.; Yang, T.; Sorooshian, S.; Liu, X.; Zhuang, J. Modeling and simulating of reservoir operation using the artificial neural network, support vector regression, deep learning algorithm. *J. Hydrol.* **2018**, *565*, 720–736. [CrossRef]
18. Yang, S.; Yang, D.; Chen, J.; Zhao, B. Real-time reservoir operation using recurrent neural networks and inflow forecast from a distributed hydrological model. *J. Hydrol.* **2019**, *579*, 124229. [CrossRef]
19. Saavedra Valeriano, O.C.; Koike, T.; Yang, K.; Graf, T.; Li, X.; Wang, L.; Han, X. Decision support for dam release during floods using a distributed biosphere hydrological model driven by quantitative precipitation forecasts. *Water Resour. Res.* **2010**, *46*, W10544. [CrossRef]
20. Allwi, M.F.; Jaafar, O.; Hamzah, F.M.; Abdullah, S.M.S.; El-shafie, A. Review on application of artificial intelligence methods for dam and reservoir-hydro-environment models. *Environ. Sci. Pollut. Res.* **2018**, *25*, 13446–13469. [CrossRef]
21. Soria-Lopez, A.; Sobrido-Pouso, C.; Mejuto, J.C.; Astray, G. Assessment of different machine learning methods for reservoir outflow forecasting. *Water* **2023**, *15*, 3380. [CrossRef]
22. Yassin, F.; Razavi, S.; Elshamy, M.; Davison, B.; Sapriza-Azuri, G.; Wheater, H. Representation and improved parameterization of reservoir operation in hydrological and land-surface models. *Hydrol. Earth Syst. Sci.* **2019**, *23*, 3735–3764. [CrossRef]
23. Wannasin, C.; Brauer, C.C.; Uijlenhoet, R.; Torfs, P.J.J.F.; Weerts, A.H. Machine learning for real-time reservoir operation simulation: Comparing input variables and algorithms for the Sirikit Reservoir, Thailand. *J. Hydroinform.* **2024**, *26*, 3151. [CrossRef]
24. Niu, W.-J.; Feng, Z.-K.; Feng, B.-F.; Min, Y.-W.; Cheng, C.-T.; Zhou, J.-Z. Comparison of multiple linear regression, artificial neural network, extreme learning machine, and support vector machine in deriving operation rule of hydropower reservoir. *Water* **2019**, *11*, 88. [CrossRef]
25. Yang, T.; Zhang, L.; Kim, T.; Hong, Y.; Zhang, D.; Peng, Q. A large-scale comparison of artificial intelligence and data mining (AI&DM) techniques in simulating reservoir releases over the Upper Colorado Region. *J. Hydrol.* **2021**, *602*, 126723.

26. Chen, Y.; Li, D.; Zhao, Q.; Cai, X. Developing a generic data-driven reservoir operation model. *Adv. Water Resour.* **2022**, *167*, 104274. [CrossRef]
27. Qie, G.; Zhang, Z.; Getahun, E.; Mamer, E.A. Comparison of machine learning models performance on simulating reservoir outflow: A case study of two reservoir in Illinois, USA. *J. Am. Water Resour. Assoc.* **2022**, *12*, JAWR-21-0019-P. [CrossRef]
28. Zarei, M.; Bozorg-Haddad, O.; Baghban, S.; Delpasand, M.; Goharian, E.; Loaiciga, H.A. Machine-learning algorithms for forecast-informed reservoir operation (FIRO) to reduce flood damages. *Sci. Rep.* **2021**, *11*, 24295. [CrossRef]
29. Wannasin, C.; Brauer, C.; Uijlenhoet, R.; Weerts, A. Modelling and reforecasting real-time reservoir operation and outflow with neural networks: Case study of the multi-purpose Sirikit reservoir in Thailand. In Proceedings of the EGU General Assembly 2023, Vienna, Austria, 24–28 April 2023.
30. Feng, D.; Fang, K.; Shen, C. Enhancing streamflow forecast and extracting insights using Long-Short Term Memory networks with data integration at continental scales. *Water Resour. Res.* **2020**, *56*, e2019WR026793. [CrossRef]
31. Nearing, G.S.; Klotz, D.; Frame, J.M.; Gauch, M.; Gilon, O.; Kratzert, F.; Sampson, A.K.; Shalev, G.; Sella, N. Technical note: Data assimilation and autoregression for using near-real-time streamflow observations in long short-term memory networks. *Hydrol. Earth Syst. Sci.* **2022**, *26*, 5493–5513. [CrossRef]
32. Song, Y.; Tsai, W.-P.; Gluck, J.; Rhoades, A.; Zarzycki, C.; McCrary, R.; Lawson, K.; Shen, C. LSTM-based data integration to improve snow water equivalent prediction and diagnose error sources. *J. Hydrometeorol.* **2024**, *25*, 223–237. [CrossRef]
33. Jung, W.; Park, N.; Lee, S. Current status of conflicts in the downstream area of Namgang Dam and solutions for resolution. *Water Future* **2021**, *54*, 46–54.
34. Chong, S.; Lee, H.; An, K.-G. Predicting taste and odor compounds in a shallow reservoir using a three-dimensional hydrodynamic ecological model. *Water* **2018**, *10*, 1396. [CrossRef]
35. Rye, Y.; Moon, H.; Kim, J.; Kim, T.-J.; Boo, K.-O.; Guan, B.; Kamae, Y.; Mei, W.; Park, C.; Son, S.-W.; et al. A multi-inventory ensemble analysis of the effects of atmospheric rivers on precipitation and streamflow in the Namgang-Dam basin in Korea. *Water Resour. Res.* **2021**, *57*, e2021WR030058. [CrossRef]
36. Guo, Y. Hydrologic design of urban flood control detention ponds. *J. Hydrol. Eng.* **2001**, *6*, 472–479. [CrossRef]
37. Guo, Y.; Baetz, B.W. Sizing of rainwater storage units for green building applications. *J. Hydrol. Eng.* **2007**, *12*, 197–205. [CrossRef]
38. Gustafsson, O.; Vigerske, S.; COIN-OR Foundation I. Interior Point Optimizer (IPOPT). 2024. Available online: https://github.com/coin-or/Ipopt (accessed on 2 April 2025).
39. Wächter, A.; Biegler, L.T. On the implementation of an interior-point filter lin-search algorithm for large-scale nonlinear programming. *Math. Program.* **2006**, *106*, 25–27. [CrossRef]
40. Kyriakidis, L.; Mendez, M.A.; Bähr, M. A hybrid algorithm based on Bayesian optimization and Interior Point OPTimizer for optimal operation of energy conversion systems. *Energy* **2024**, *312*, 133416. [CrossRef]
41. Biegler, L.T.; Zavala, V.M. Large-scale nonlinear programming using IPOPT: An integrating framework for enterprise-wide dynamic optimization. *Comput. Chem. Eng.* **2009**, *33*, 575–582. [CrossRef]
42. Hart, W.E.; Watson, J.-P.; Woodruff, D.L. Pyomo: Modeling and solving mathematical programs in Python. *Math. Program. Comput.* **2011**, *3*, 219–260. [CrossRef]
43. Cho, K.; van Merrienboer, B.; Bahdanau, D.; Bengio, Y. On the Properties of Neural Machine Translation: Encoder-Decoder Approaches. 2014. Available online: https://arxiv.org/abs/1409.1259 (accessed on 25 April 2025).
44. Ranjbarzadeh, R.; Caputo, A.; Tirkolaee, E.B.; Ghoushchi, S.J.; Bendechache, M. Brain tumor segmentation of MRI images: A comprehensive review on the application of artificial intelligence tools. *Comput. Biol. Med.* **2023**, *152*, 106405. [CrossRef]
45. Gao, S.; Huang, Y.; Zhang, S.; Han, J.; Wang, G.; Zhang, M.; Lin, Q. Short-term runoff prediction with GRU and LSTM networks without requiring time step optimization during sample generation. *J. Hydrol.* **2020**, *589*, 125188. [CrossRef]
46. Guo, Z.; Yang, C.; Wang, D.; Liu, H. A novel deep learning model integrating CNN and GRU to predict particulate matter concentrations. *Process Saf. Environ. Prot.* **2023**, *173*, 604–613. [CrossRef]
47. Ji, L.; Fu, C.; Ju, Z.; Shi, Y.; Wu, S.; Tao, L. Short-term canyon wind speed prediction based on CNN-GRU transfer learning. *Atmosphere* **2022**, *13*, 813. [CrossRef]
48. Liu, C.; Yang, X.; Peng, S.; Zhang, Y.; Peng, L.; Zhong, R.Y. Spark analysis based on the CNN-GRU model for WEDM process. *Micromachines* **2021**, *12*, 702. [CrossRef]
49. Jain, A.; Patel, H.; Nagalapatti, L.; Gupta, N.; Mehta, S.; Guttula, S.; Mujumdar, S.; Afzal, S.; Sharma Mittal, R.; Munigala, V. Overview and importance of data quality for machine learning tasks. In Proceedings of the 26th ACM SIGKDD International Conference on Knowledge Discovery & Data Mining, Virtual Event, NY, USA, 6–10 July 2020; pp. 3561–3562.
50. Khatun, A.; Nisha, M.N.; Chatterjee, S.; Sridhar, V. A novel insight on input variable and time lag selection in daily streamflow forecasting using deep learning models. *Environ. Model. Softw.* **2024**, *179*, 106126. [CrossRef]

51. Anghileri, D.; Voisin, N.; Castelletti, A.; Pianosi, F.; Nijssen, B.; Lettenmaier, D.P. Value of long-term streamflow forecasts to reservoir operations for water supply in snow-dominated river catchments. *Water Resour. Res.* **2016**, *52*, 4209–4225. [CrossRef]
52. Yao, H.; Georgakakos, A. Assessment of Folsom Lake response to historical and potential future climate scenarios: 2. Reservoir management. *J. Hydrol.* **2001**, *249*, 176–196. [CrossRef]

water

MDPI

Review

A Bibliometric-Systematic Literature Review (B-SLR) of Machine Learning-Based Water Quality Prediction: Trends, Gaps, and Future Directions

Jeimmy Adriana Muñoz-Alegría [1,*], Jorge Núñez [2], Ricardo Oyarzún [2,3,4], Cristian Alfredo Chávez [5], José Luis Arumí [3,6] and Lien Rodríguez-López [7]

1 Doctorate Program in Energy, Water and Environment, University of La Serena, La Serena 1700000, Chile
2 Department of Mining Engineering, University of La Serena, La Serena 1700000, Chile; jhnunez@userena.cl (J.N.); royarzun@userena.cl (R.O.)
3 Water Research Center for Agriculture and Mining (CRHIAM), ANID FONDAP Center, Universidad de Concepción, Concepción 4070411, Chile; jarumi@udec.cl
4 Center for Advanced Studies in Arid Zones (CEAZA), La Serena 1700000, Chile
5 Department of Mechanical Engineering, University of La Serena, La Serena 1700000, Chile; cchavez@userena.cl
6 Department of Water Resources, Universidad de Concepción, Chillán 3812120, Chile
7 Faculty of Engineering, Universidad San Sebastián, Concepción 4030000, Chile; lien.rodriguez@uss.cl
* Correspondence: jeimmy.munoz@userena.cl

Abstract: Predicting the quality of freshwater, both surface and groundwater, is essential for the sustainable management of water resources. This study collected 1822 articles from the Scopus database (2000–2024) and filtered them using Topic Modeling to create the study corpus. The B-SLR analysis identified exponential growth in scientific publications since 2020, indicating that this field has reached a stage of maturity. The results showed that the predominant techniques for predicting water quality, both for surface and groundwater, fall into three main categories: (i) ensemble models, with Bagging and Boosting representing 43.07% and 25.91%, respectively, particularly random forest (RF), light gradient boosting machine (LightGBM), and extreme gradient boosting (XGB), along with their optimized variants; (ii) deep neural networks such as long short-term memory (LSTM) and convolutional neural network (CNN), which excel at modeling complex temporal dynamics; and (iii) traditional algorithms like artificial neural network (ANN), support vector machines (SVMs), and decision tree (DT), which remain widely used. Current trends point towards the use of hybrid and explainable architectures, with increased application of interpretability techniques. Emerging approaches such as Generative Adversarial Network (GAN) and Group Method of Data Handling (GMDH) for data-scarce contexts, Transfer Learning for knowledge reuse, and Transformer architectures that outperform LSTM in time series prediction tasks were also identified. Furthermore, the most studied water bodies (e.g., rivers, aquifers) and the most commonly used water quality indicators (e.g., WQI, EWQI, dissolved oxygen, nitrates) were identified. The B-SLR and Topic Modeling methodology provided a more robust, reproducible, and comprehensive overview of AI/ML/DL models for freshwater quality prediction, facilitating the identification of thematic patterns and research opportunities.

Keywords: water quality; machine learning; explainable artificial intelligence (XAI); Bibliometric-Systematic Literature Review (B-SLR); Topic Modeling

1. Introduction

Surface and groundwater pollution is a problem of global concern, as it can negatively affect freshwater availability, a key aspect for the conservation of ecosystems and the health of the population. This is aggravated by the strong pressure exerted by a growing demand for fresh water to sustain population growth and associated economic development [1–3]. Consequently, access to reliable sources of clean water directly influences the capacity to ensure water security, particularly given that nearly 2.2 billion people lack safe drinking water [4].

Aquatic conditions are shaped by multiple variables, including weather patterns [5], seasonal hydrological fluctuations [6], geological features [7], and anthropogenic influences such as land use. Traditionally, assessments have relied on in situ sensors or laboratory analysis of biological and physicochemical parameters. More recently, predictive approaches have incorporated physically based simulation tools like QUAL2K [8–10] and WASP [11,12], alongside statistical techniques such as multivariate analysis [13,14], and linear regression [15,16].

Within this landscape, the application of artificial intelligence (AI) over recent decades has significantly advanced predictive techniques for aquatic system assessment. AI transcends basic sensor-based data collection [17], by processing, interpreting, and modeling the complex, multidimensional relationships inherent in water systems [18,19]. This approach allows it to dynamically predict the state of water quality [20], supporting proactive and optimized resource management. Notably, the integration of machine learning (ML) [20–23] and deep learning (DL) [24–30] models have strengthened the path toward meeting the Sustainable Development Goals (SDGs) related to water [31]. These prediction models can handle large amounts of data and adapt to complex relationships. In fact, the use of ML for water quality studies is highlighted as an example of the United Nations Global Acceleration Framework, for the scope of the SDG-6 [32,33].

A key advantage of applying AI/ML/DL to water quality prediction lies in their ability to anticipate variations in physicochemical and biological parameters using historical data, remote sensing, or hydrological models—thus enhancing environmental management and evidence-based decision-making. Two main approaches are commonly used: (a) prediction of individual water quality parameters (WQP) [10,34–37] and (b) integration of these parameters into a composite indicator, traditionally known as the water quality index (WQI) [38–42]. For the first approach, one example is the prediction of total phosphorus concentrations in Taihu Lake, China, using the Complete Ensemble Empirical Mode Decomposition with Adaptive Noise (CEEMDAN) and long short-term memory (LSTM) algorithms, combined with Shapley Additive Explanations (SHAP) for interpretability [43]. Another study [44], applied the random forest (RF) method to predict 14 physicochemical parameters from historical data in the Loa River, Chile. Regarding the second approach, ML models were used to estimate WQI in the Bug River, Ukraine [45]. High predictive performance was also reported in Kerala, India [46], where various ML models—including extreme gradient boosting (XGB), support vector regression (SVR), artificial neural network (ANN), and random forest regression (RFR)—were evaluated to predict the entropy-weighted water quality index (EWQI) and assess groundwater quality.

The aforementioned studies are examples of how AI/ML/DL methods have been consolidated as water quality prediction techniques, given their robustness and efficiency to extract patterns of complex systems, as water systems, making these techniques an interesting alternative to conventional water quality modeling techniques [47–49]. Indeed, systematic reviews of the literature at a global level show the high interest of researchers in the use of AI/ML/DL for water quality prediction. For example, the authors in [50] reviewed 876 articles within the period 2015–2022, showing that the United States, England, Iran, India, and China have emerged as major contributors to the field of water quality prediction with ML and DL. Likewise, the authors in [51] reviewed 249 articles on water quality using Internet of Things (IoT) models and machine learning. In another study [52], 253 articles were reviewed, finding that DL optimizes the processing of large volumes of data through parallel computing, facilitating the effective prediction of water quality, although its success depends on the quality of the data used. Finally, in [53], the literature was reviewed in terms of groundwater quality prediction studies using AI/ML/DL, finding that the ANN, Adaptive Neuro-Fuzzy Inference System (ANFIS) and support vector machine (SVM) techniques have proven to be efficient and accurate tools for such purposes.

Despite the intrinsic value of the aforementioned literature review studies, it is important to highlight a key conceptual distinction between the two main review approaches. One is known as bibliometric analysis, that is, a quantitative method that allows the identification of emerging trends, thematic networks, and research gaps [54–56]. Another corresponds to a systematic review of the literature, a rigorous and structured method used to collect and evaluate all available evidence on a research question [57–59]. Although the two approaches complement each other, and together they form a promising methodology with which to develop robust reviews of the scientific literature, to date, no work, to the best of our knowledge, has addressed the study of the state of the art on the application of AI/ML/DL for water quality prediction through the integration of both approaches. Thus, such a dual-approach review corresponds to the main objective of this work, as this contribution proposes to analyze the state of the art on the application of AI/ML/DL for water quality prediction in the 2000–2024 time frame, using the Bibliometric-Systematic approach Literature Review (B-SLR) [60]. We hope that the results of this work can guide researchers interested in the study, assessment and prediction of water quality using AI/ML/DL techniques.

2. Materials and Methods

We employed a structured framework for conducting systematic literature reviews, known as the Bibliometric-Systematic Literature Review (B-SLR), following the guidelines outlined in [60]. This approach was used to assess the current status, emerging trends, and research gaps in the application of AI/ML/DL techniques to freshwater (surface and groundwater) quality studies. The methodology B-SLR combines the bibliometric analysis (BA) [56,61,62] with the systematic literature review (SLR) [63–65]. Thus, the B-SLR approach facilitates the broadening of topic scope, expanding the domain of knowledge available to researchers working in the field [59]. According to the adopted methodology, this work was developed in three sequential stages: (a) data collection, (b) bibliometric analysis, and (c) a systematic review of the literature, as illustrated in Figure 1.

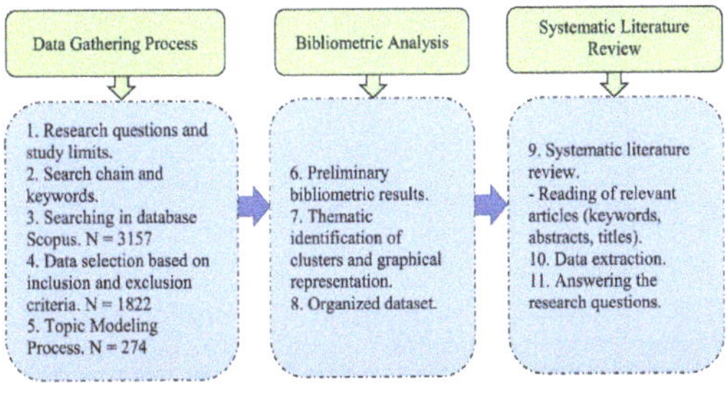

Figure 1. Stages of the B-SLR methodology. Stage 1: data collection: defining keywords, search terms, applying inclusion/exclusion criteria, and filtering irrelevant results using topic modeling. Stage 2: bibliometric analysis: identifying trends, thematic clusters, and performance metrics. Stage 3: systematic review: extracting and synthesizing the evidence to answer the research questions.

2.1. Data Gathering Process

This stage aims to identify the research topic and the structural framework to establish the research questions, keywords, search strings, database, time scope of the research, techniques to carry out the search and analysis tools, where each stage generates a result that serves as an input for the next stage.

The research questions raised in this study seek to identify a set of patterns behind the application of AI/ML/DL techniques in water quality prediction. Table 1 presents the guiding research questions of this study with their corresponding justification, following the guidelines of the systematic review of the literature [64].

Table 1. Research questions defined in this study.

Research Questions	Justification
RQ1. What are the most commonly used AI/ML/DL algorithms for predicting water quality	To establish a general overview of the research topic.
RQ2. Which AI/ML/DL algorithm provides the most accurate estimation of water quality?	To identify knowledge gaps in AI/ML/DL prediction models.
RQ3. What limitations have been identified in water quality prediction using AI/ML/DL techniques?	To uncover potential research opportunities and future work
RQ4. What emerging variants currently exist in AI/ML/DL models for estimating water quality?	To identify current trends in AI/ML/DL techniques for water quality prediction.
RQ5. What are the key water quality indicators used to assess natural water sources?	To review and understand the factors that determine water quality.

The bibliographic references were obtained from the Scopus database (https://www.scopus.com/home.uri), which offers a comprehensive and global view of scientific production. Scopus is one of the most used sources in bibliometric analysis and is recognized for its reliability and high data quality, especially in academic research related to water resources and water quality [17,55,56,66–68].The temporal scope was defined from January 2000 to December 2024. The structured search string used to obtain the first consolidated set of documents (N = 3157) and the characteristic code line generated after applying the inclusion/exclusion criteria to create a second consolidated set of documents (N = 1822) are presented in Table 2. The inclusion and exclusion criteria presented in Table 3 allow determining the eligibility of the primary studies for inclusion in the bibliographic review and are useful for extracting relevant information and answering the research questions posed.

Table 2. Search strategy applied to Scopus database.

Search Strategy	Total Documents
Search chain: "water" AND "quality" AND "prediction" AND "machine" AND "learning" OR "water" AND "quality" AND "prediction" AND "artificial" AND "intelligence" OR "water" AND "quality" AND "prediction" AND "deep" AND "learning"	3157
Research code line: (TITLE-ABS-KEY (water AND quality AND prediction AND machine AND learning) OR TITLE-ABS-KEY (water AND quality AND prediction AND artificial AND intelligence) OR TITLE-ABS-KEY (water AND quality AND prediction AND deep AND learning)) AND PUBYEAR > 1999 AND PUBYEAR < 2025 AND (LIMIT-TO (DOCTYPE, "re") OR LIMIT-TO (DOCTYPE, "ar")) AND (LIMIT-TO (LANGUAGE, "English")) AND (LIMIT-TO (SUBJAREA, "ENVI") OR LIMIT-TO (SUBJAREA, "ENGI") OR LIMIT-TO (SUBJAREA, "EART") OR LIMIT-TO (SUBJAREA, "MULT"))	1822

Table 3. Inclusion and exclusion criteria used in the B-SLR methodology.

Inclusion Criteria
Publications classified as "Research Article" or "Review"
The study must be published in English to ensure accessibility and comprehension.
Publications within the subject areas of Environmental Science, Engineering, Earth and Planetary Sciences, and Multidisciplinary
Articles containing the keywords specified in the search string
Exclusion Criteria
Conference proceedings, books, book chapters, theses and reports were excluded.
Publications that have not undergone a formal peer review process, such as preprints, unpublished reports, or unreviewed gray literature.
Research whose primary focus is not on the application of AI/ML/DL for the prediction of freshwater quality (surface and groundwater)

The bibliographic references were exported from Scopus in CSV format and subsequently imported into the R programming environment [62]. To filter the 1822 original articles selected for the B-SLR analysis, a comprehensive list of 29,144 Scimago journals (https://www.scimagojr.com/) for the year 2023 was incorporated, with the objective of retaining only those studies published in journals classified within quartiles Q1 and Q2, in accordance with the recommendations described in reference [60]. This procedure involved excluding publications prior to 2014, indicating that such studies were published in journals indexed outside Q1 and Q2, thereby establishing by default an analysis period between 2015 and 2024. In addition, a systematic cleaning process was carried out, which included the removal of duplicate rows, incomplete records in the fields of Index Keywords, Author Keywords, DOI, and Abstract, as well as the elimination of entries with duplicate DOIs and false positives [69]. These publications were not aligned with the objectives of the present study.

The automation of this process was achieved through the implementation of an iterative text-mining approach based on Topic Modeling, whose procedure is illustrated in Figure 2. For this procedure, we used the TextMiner package in R as a specialized tool for the construction and analysis of topic models [70]. The process operates under the assumption that each document is a combination of several topics and that each topic is a set of words that frequently co-occur [71]. The algorithm identifies these groups of coexisting words to infer the latent topics within the text [72,73].

2.2. Bibliometric Analysis (BA)

The bibliometric analysis was conducted using the Bibliometric package in the R programming language version 4.4.1 [62]. This package has been widely applied in bibliometric studies related to hydrology [55,74] and water quality [75,76]. The analysis included the generation and visualization of maps and graphs corresponding to performance analysis [17,55,61] and science mapping [56,77,78].

2.3. Systematic Literature Review (SLR)

A comprehensive reading stage was conducted for the 276 selected articles, conducted using the Systematic Literature Review (SLR) approach described in [59], in order to extract key data and information to address the research questions. This methodology enabled the identification, collection, analysis, and synthesis of relevant information to answer the research questions posed [79]. Its robustness stems primarily from the transparency of its implementation process, which ensures the reproducibility of the review [80,81].

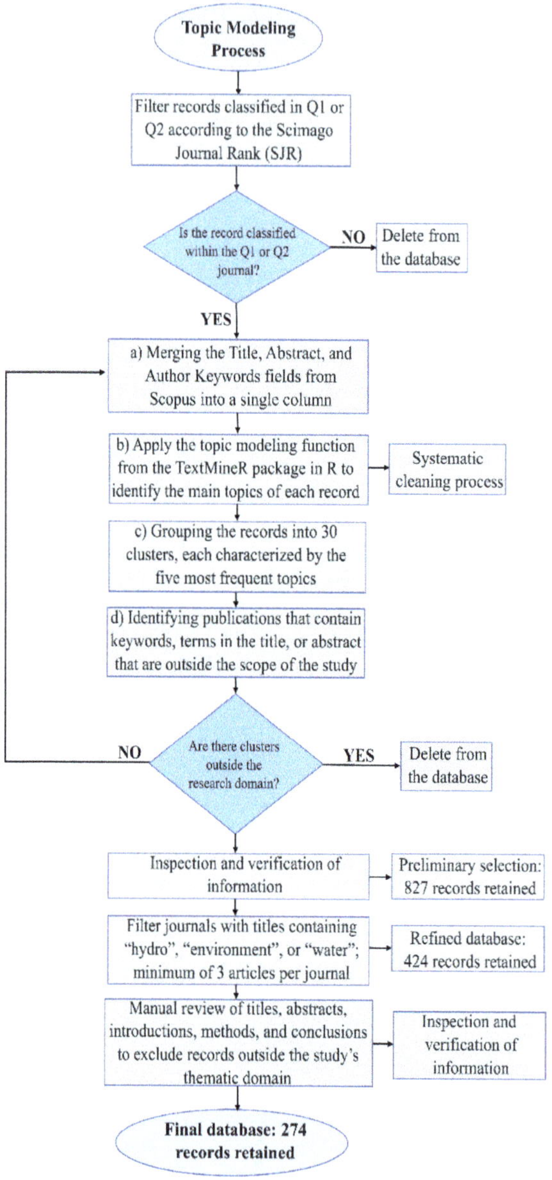

Figure 2. Flowchart of the Topic Modeling Process.

Each study was classified as either a Review or a Research article. Within the Research category, articles were further classified according to the type of water body under study: surface water or groundwater. A particular case involved two Review articles [82,83] that addressed both systems (groundwater and surface water); this condition did not affect the classification by water body type, as these studies were not part of the Research category.

3. Results

3.1. Bibliometric Performance Analysis

The annual scientific output on water quality prediction using AI/ML/DL has increased in the study period, as shown in Figure 3, exhibiting an exponential increase, especially marked since 2020. It should be noted that the annual production of the domain field shows an exponential growth pattern between 2015 and 2024. Consequently, it was decided to apply an exponential model, in order to more accurately represent both the growth acceleration and the current development phase of this research area.

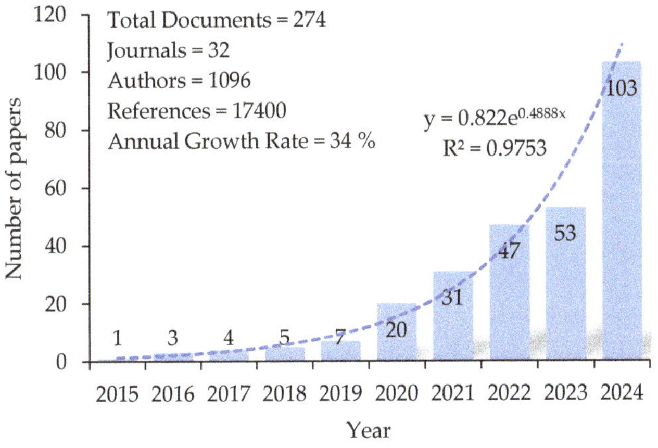

Figure 3. Annual publication output on freshwater quality prediction using AI/ML/DL.

The growing interest in research on the prediction of water quality using AI/ML/DL is within a global context driven by climate change, and increased pollution and demand for fresh water [5,84]. This evolution indicates that the analyzed domain has passed an initial stage of exploration, positioning itself in a phase of maturity. Likewise, it is possible to expect an even more significant development in the coming years, anticipating a continuous strengthening of both interest and research in this field [85,86].

3.1.1. Journals

In this context, the H-index is a metric that allows for the evaluation of scientific research productivity and journal impact. On the other hand, the Total Citations (TC) refers to the cumulative number of times a publication has been cited by other researchers in academic publications, revealing the scope and influence of the research from complementary perspectives. Global Citations (GC) refers to the total number of times an article has been cited in the Scopus database.

When ranking the top 10 of the 32 sources based on the H-index and Total Citations (TC) [87], it is observed that the journals Water (Switzerland), Journal of Hydrology, Environmental Science and Pollution Research, Science of the Total Environment and Water Research are the publications with the highest number of papers (first 5) within the reference collection (Table 4).

3.1.2. Most Cited Documents

The most cited documents within the analyzed collection of bibliographic references were identified. Both global and local citation counts were considered to assess the reach and influence of the research from complementary perspectives. In this context, local citations reflect the impact of a document within the specific dataset under analysis, while global citations represent the total number of times each article has been cited in the Scopus database.

Table 4. Journals Local impact.

ID	Journals	H Index	TC
1	Water (Switzerland)	19	1568
2	Journal of Hydrology	18	2231
3	Environmental Science and Pollution Research	14	803
4	Water Research	10	823
5	Science of the Total Environment	9	737
6	Journal of Environmental Management	7	193
7	International Journal of Environmental Research and Public Health	6	103
8	Environmental Monitoring and Assessment	5	115
9	Hydrological Processes	5	82
10	Process Safety and Environmental Protection	5	147

Table 5 provides the top 10 most cited publications and corresponding (leading) authors in terms of local and global citations. The referred papers reveal important contributions to research on water quality prediction using AI/ML/DL techniques.

Table 5. Top ten Local citations and Global citations.

Ranking	First Author	Year	LC [1]	GC [2]	Reference
1	Rahim Barzegar	2020	32	330	[88]
2	Rahim Barzegar	2016	15	149	[89]
3	Jun Yung Ho	2019	13	101	[90]
4	Amir Hamzeh Haghiabi	2018	13	290	[37]
5	Xiaoliang Ji	2017	11	10	[91]
6	Elham Fijani	2019	10	146	[92]
7	Sani Isah Abba	2020	9	91	[93]
8	Muhammed Sit	2020	9	273	[82]
9	Roohollah Noori	2015	9	66	[94]
10	Bachir Sakaa	2022	7	66	[95]

[1] LC = Local citation. [2] GC = Global citation.

According to Table 3, the most cited reference corresponds to the work of Barzegar [88] entitled "Short-term water quality variable prediction using a hybrid CNN–LSTM deep learning model", which investigates the prediction of water quality in Lake Prespa, Greece, and proposes a hybrid CNN-LSTM model to predict dissolved oxygen (DO; mg/L) and chlorophyll-a (Chl-a; μg/L).

Second, the study [89] compares different machine learning methods (such as artificial neural networks and support vector machines) for the prediction of water quality parameters of northwest Iran's Aji-Chay River, demonstrating the effectiveness of these techniques in improving environmental monitoring. Its high number of global citations reflects the interest in using water quality prediction to support the transition toward more sustainable water resources management.

Third, the work of [90] examined the application of a decision tree model for predicting the Water quality index (WQI) in the Klang River. The research successfully showed that the number of water quality parameters needed for monitoring can be reduced while maintaining prediction accuracy above a 75% benchmark. Finally, the most recent study of the list [95], proposes a hybrid approach that combines the Random Forest algorithm with an improved version of the SMO algorithm for support vector machines. Applied to the Saf-Saf River Basin, this model improved WQI prediction, highlighting the growing role of optimized hybrid approaches in watershed management.

3.2. *Bibliometric Science Mapping*

3.2.1. Network Analysis of the Co-Occurrence of Authors' Keywords

The network analysis of co-occurrence of author keywords presented in Figure 4 reveals a conceptual structure derived from studies on machine learning and water quality management. Each node represents a keyword, with its size indicating the degree of connectivity (number of links to other terms). Connecting lines reflect co-occurrence relationships, and their thickness represents the strength of those associations. Nodes are grouped into three thematic communities, distinguished by color: group 1 (blue) and group 2 (orange), enabling the identification of conceptual subdomains within the field.

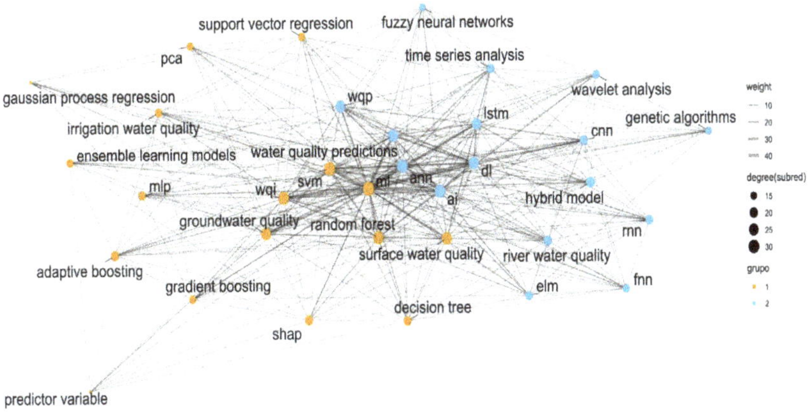

Figure 4. Network analysis co-occurrence of the author's keywords.

Central terms such as support vector machine, deep learning, water quality predictions, and water quality management, along with technical acronyms like ML (machine learning), AI (artificial intelligence), SHAP (SHapley Additive exPlanations), ANN, CNN (convolutional neural networks), LSTM (long short-term memory), and random forest, exhibit high connectivity. This suggests their integrative role in the analyzed literature and highlights their methodological relevance in the development of predictive and explanatory models applied to water systems. The visualization supports the identification of thematic clusters, methodological relationships, and emerging areas at the intersection of artificial intelligence and environmental management.

3.2.2. Word Cloud

Another method for identifying the frequency of the most common terms used by the authors in the collection corresponds to the word cloud generated from the keywords. Figure 5 shows a predominant interest in "water quality," which serves as the central axis of the study corpus. Its association with indicators such as "WQI," "WQP," and water bodies like "groundwater quality" and "surface water" reaffirms the environmental orientation of the studies. From a methodological standpoint, there is a predominant interest in techniques such as "random forest," "deep learning," "ML," and "ANN," along with specific architectures like "LSTM," "CNN," "RNN," and "SVM," used in modeling and prediction processes of hydrological parameters.

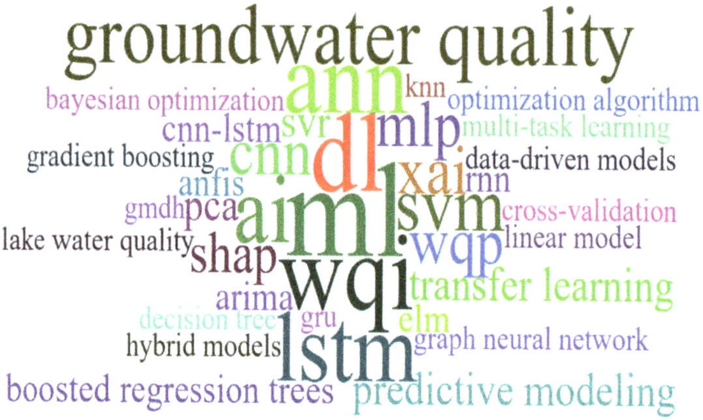

Figure 5. Word cloud of water quality prediction by AI/ML/DL. The size of the words reflects their frequency and the color is used only to differentiate them within the group of words.

Complementing this trend is the appearance of terms like "hybrid model," "ensemble learning," and "transfer learning," reflecting the adoption of integrated strategies and techniques for transferring pretrained representations aimed at improving model generalization and efficiency. Finally, "XAI" and "SHAP" reveal an interest in model explainability, pointing toward advanced analytical systems for the management and evaluation of water sustainability.

3.2.3. Thematic Map with Authors' Keywords

Figure 6 presents a thematic map derived from bibliometric analysis, grouping clusters associated with distinct research subtopics in the field of water quality prediction using AI, ML, and DL. This map enables the evaluation of each topic in terms of centrality and maturity, identifying established, emerging, and declining areas. In the quadrant of Basic themes, terms like "water quality prediction," "deep learning," and "LSTM" reflect the consolidation of deep neural networks as a methodological foundation. Notably, LSTM networks have been highlighted in multiple studies for their strength in modeling complex temporal dynamics in water quality parameter prediction [30,88,96,97]. Furthermore, hybrid models and the ANFIS technique have provided enhanced precision and flexibility in complex scenarios, particularly in surface water contexts [89,98–100].

In the quadrant of Motor themes, terms such as "WQI," "ANN," "AI," and "ML" stand out, indicating that artificial neural networks and machine learning models have been extensively studied and applied, yielding significant results. Conversely, "transfer learning" emerges as a nascent exploratory line within water quality prediction, although several studies [6,96,101–104] have demonstrated its potential to improve predictions in data-scarce environments.

In the quadrant of Niche themes, the terms "data-driven models," "lake water quality," and "prediction intervals" suggest that lake water analysis and data-driven modeling represent growing niches within water-related research, though they have yet to become dominant in AI/ML/DL-based water quality prediction.

In the quadrant of Emerging or declining themes, terms such as "support vector machine," "feature selection," and "gaussian process regression" indicate that these methods

(SVR, GPR), although historically robust and widely used in this field, may be losing relevance compared to more modern approaches such as deep learning. Additionally, procedures like feature selection may have already transitioned into standard practice within AI/ML/DL applications. Meanwhile, "extreme gradient boosting" and "SHAP" are associated with advanced ML techniques but have not yet reached the prominence of core topics. The emerging discipline known as Explainable Artificial Intelligence (XAI), or Interpretable Machine Learning (iML), seeks to address the challenges posed by the opacity of black-box techniques in AI/ML/DL. Its application has expanded in recent years across various engineering domains [85,105–107], and has recently been implemented in the field of water quality, particularly in groundwater studies [108,109].

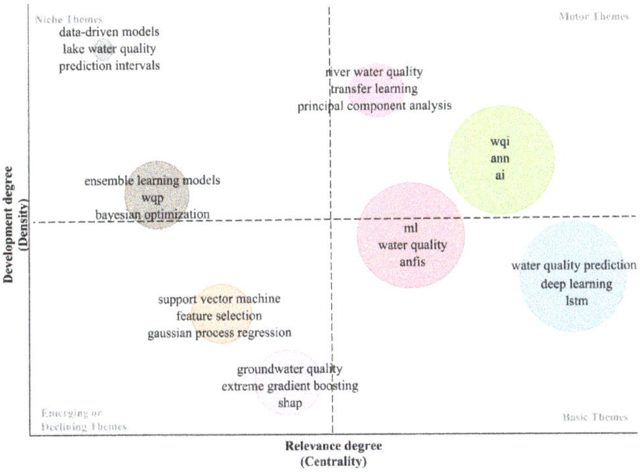

Figure 6. Thematic map based on the author's keyword. The colors represent the different clusters of related terms that form a thematic group, and the size of each circle indicates the frequency of occurrence or density of publications related to that group.

3.2.4. Thematic Evolution and Trend Topics with Keywords Plus

The thematic evolution in the BA is useful to visualize more clearly the change and evolution of research topics over time, allowing the identification of windows of opportunity for future research work. To have a more global view of the thematic evolution and current trends in water quality prediction using AI/ML/DL, Figure 7 presents the thematic evolution based on the Keywords Plus. These words are automatically generated by the BA algorithms from sources such as the titles of the references cited in the documents, allowing for to assessment, from a complementary perspective, of the areas that are related to the central axis of the research search. Thus, the application of models based on ANN and decision support techniques is maintained over time, while models based on SVM have been less constant in recent years.

In recent years, cutting-edge ML and DL techniques have increasingly been applied to environmental challenges such as pollution, climate change, water quality, and water availability. These approaches are often integrated with Geographic Information Systems (GIS) technologies [1,2,110–112], enhancing spatial analysis capabilities.

While GIS is now widely used in the field of water resources, it has yet to become a dominant tool for predicting water quality through AI/ML/DL methods. Notably, interest in developing predictive models for surface water quality has remained consistent throughout the study period. Since 2015, ML has emerged as the leading approach, demonstrating strong applicability in environmental monitoring, wastewater treatment, and water quality assessment, indicating a sustained research focus on water resources management.

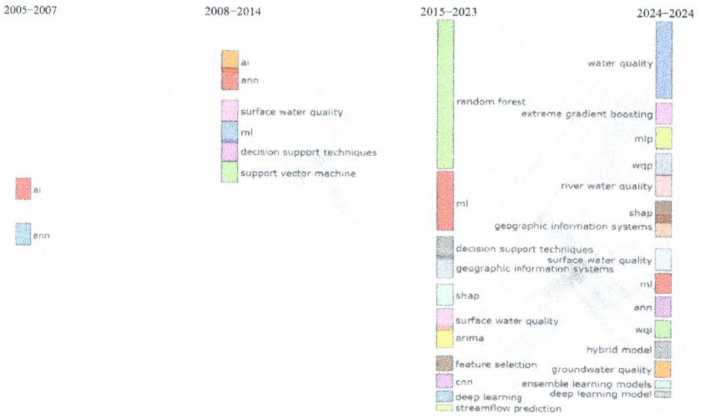

Figure 7. Thematic evolution of AI/ML/DL applications in water quality prediction.

3.2.5. Social Structure

Figure 8 presents the map of main collaborations between countries, which reveals a highly interconnected structure of scientific cooperation. Countries such as China and the United States stand out, acting as central nodes, evidencing their leadership in the production of knowledge and in the articulation of international efforts. This dynamic reflects not only the shared interest in addressing the challenges of water quality prediction but also the growing need to integrate data, methodologies, and interdisciplinary approaches on a global scale. Countries such as India, Australia, the United Kingdom, Italy, and Germany stand out for their collaborative connections as well.

Figure 8. Collaboration with country authors. The thickness of each line represents the intensity of collaboration between two countries and the colors represent the dominant clusters of regional or thematic collaboration.

However, the map also reveals a striking absence of Latin American countries, suggesting a potential asymmetry in scientific visibility. This underrepresentation may be partially attributed to the methodological filter applied in this study, which included only publications indexed in high-impact Q1 and Q2 journals. While this criterion ensures scientific rigor, it may inadvertently exclude valuable regional research that, due to structural or editorial limitations, does not reach these publication venues. As a result, the map not only visualizes collaboration intensity but also reflects broader epistemic inequalities in global scientific discourse.

3.3. Systematic Literature Reviews Results

The systematic review allowed us to classify each work into different categories according to: Approach (Review or Research articles), body of water of the study, that is, Surface water (rivers, reservoirs and lakes) or Groundwater, as presented in Figure 9. Figure 9a shows the classification of the 274 articles collected according to their type: Research Articles and Review Articles. The disparity between the two approaches suggests that, although research in the area is growing rapidly, a significant gap persists in terms of consolidating knowledge through systematic reviews of the state of the art. This represents a valuable opportunity to strengthen the field through studies that integrate and synthesize existing findings (such as ours).

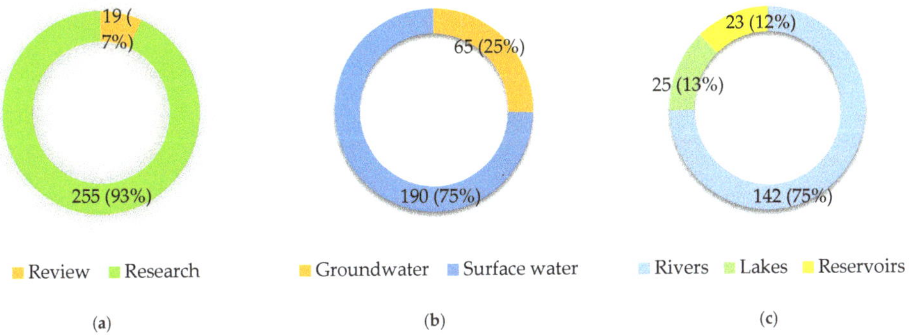

Figure 9. Classification of publications in water quality prediction according to (**a**) Approach: Article research versus Review. (**b**) Body of water: Underground versus Surface, (**c**) Surface water: River, lake, reservoir.

Within Research articles, 75% (n = 190) correspond to studies focused on surface water quality, while the remaining 25% (n = 65) address groundwater quality issues (Figure 9b). This bias towards the study of surface waters can be attributed to the relevance of watershed monitoring for the sustainable management of water resources. This monitoring allows us to understand the complex interactions between biological, chemical, physical, and environmental factors that determine water quality, as well as to anticipate the appearance of alterations to water quality. This information is crucial for informed decision-making and long-term planning, which are essential for preserving the health of water bodies and the ecosystems that depend on them [113].

In Figure 9c, we present the distribution of works considering only surface water bodies, i.e., rivers, lakes, and reservoirs. It is shown a major focus in rivers, which could be because they are particularly exposed to pollution caused by anthropogenic activities [39,114]. In addition, during periods of drought, rivers are more easily and rapidly affected by the reduction in the flow, affecting the quality of the water in terms of its physical, chemical and biological properties [115]. The study [57], made a significant contribution with the exhaustive review of the literature in the period 2000–2020 on the modeling of river water quality in the world, reviewing 209 research articles from Scopus journals. The results showed that most of the study areas are Asian countries, such as China, Iran, India, Malaysia, Taiwan, Korea, Iraq, Bangladesh and Thailand, accounting for more than 50% of research papers. These results are consistent with the findings of the current work for the period 2015–2024. Based on the classification shown in Figure 8, this study delves into the two water bodies that are currently of greatest interest in the application of AI/ML/DL techniques for water quality prediction: rivers and groundwater. For this reason, a detailed description of the literature for them is presented below.

3.3.1. Prediction of River Water Quality Using AI/ML/DL

Focusing on the publications on water quality prediction in rivers, Table 6 presents the characterization of a random sample of the database, equivalent to 40% (n = 57).

Table 6. Characterization of predictive models in a representative sample of river water quality prediction studies (n = 57).

ID	River	Algorithm	Approach	Reference
1	Yangtze River, China	CNN-LSTM	WQP	[116]
2	Delaware River Basin, USA	XGB, RF, KNN	WQP	[117]
3	Sheshui River in Wuhan, China	RF, SSA-CNN-LSTM	WQP	[114]
4	Vaigai, Madurai, and Tamil Nadu Rivers, India	Optimization algorithm and LSTM	WQP	[118]
5	Upper Red River Basin (URRB), USA	TL, FFNNs	WQP	[102]
6	The South Platte River, Colorado, USA	EBM, SWEBM	WQP	[119]
7	Cauvery River, India	AO-SVM	WQI	[120]
8	Indian Rivers	DT, RF, GBT, ANN, SVM	WQP	[42]
9	Cauvery River, India	CNN	WQP	[121]
10	Han River, South Korea	RF, SVR, XGB, LGB, and a hybrid model. SHAP, LIME	WQI	[122]
11	Tanjiang River, China	SVR	WQP	[123]
12	Des Moines, Iowa, and Cedar Rivers, Iowa, USA	LSTM, GRU	WQP	[124]
13	Mahanadi River, India	LSTM, GRU, XGB	WQI	[125]
14	Oyster River, New Hampshire, USA	CNN-LSTM	WQP	[126]
15	Li River and Liu River, China	SSA, GRU, SHAP	WQP	[127]
16	Fujian River Network, China	WA-LSTM-TL	WQP	[103]
17	Euphrates River, Iraq	RC, DR, REPT, AR	WQP	[128]
18	Ohio River, USA	LSTM	WQP	[111]
19	USA Rivers	RF	WQP	[129]
20	Xiaofu River, China	LSTM	WQP	[130]
21	Lijiang River, China	BPNN, SVR, GRU	WQP	[131]
22	Drinking water quality, South Korea	LSTM, GRU	WQP	[132]
23	Indian, Rivers *	DT, LR, Ridge, Lasso, SVR, RF, ETR, ANN	WQI	[133]
24	Júcar River, Spain	RF, XGB, SHAP	WQP	[134]
25	Bullfrog River, Tampa, Florida USA	SVM, RF, XGB, ANN, SHAP	WQP	[135]
26	Talar River, Iran	EN, AMT, REPT	WQP	[136]
27	Wadi Saf-Saf River, Algeria	SMO-SVM, RF	WQI	[95]
28	Pearl River, China	CEEMDAN -LSTM	WQP	[137]
29	Fuyang River, China	RF, PLS	WQP	[138]
30	Synthetic dataset, Wabash River, USA	SVMR	WQP	[139]
31	Yamuna River, India	LSTM, SVR, CNN-LSTM	WQP	[30]
32	Kelantan River, Malaysia	KNN, ANN, DT, RF, GB	WQP	[140]
33	Langat River, Malaysia	ANN	WQP	[141]
34	Mid-Atlantic and Pacific Northwest USA, River Basin	SVR, XGB	WQP	[142]
35	Santiago-Guadalajara River, Mexico	SLR, MLR	WQI	[143]
36	Danube, Tisa, and Sava Rivers, Vojvodina Province, Serbia	Naïve Bayes algorithm	WQI	[144]
37	Yamuna River, India	ANFIS–GP, ANFIS–SC	WQP	[145]
38	Fanno Creek in Oregon, USA	DRNN, SVM, ANN	WQP	[146]
39	Dongjiang River, China	WT-MLR, WT-SVM, WT-ANN, WT-RF	WQP	[147]
40	Klang and Penang Rivers, Malaysia	MLP, SVM, RF, BDT	WQI	[148]
41	Nakdong River, South Korea	CEEMDAN, CSA, MARS	WQP	[149]
42	Luan River, Tangshan China	1-DRCNN *, BiGRU	WQP	[150]
43	Tyhume, Bloukrans, Buffalo Rivers Province of South Africa	ANN, MLP, RBF	WQP	[151]
44	Kinta River, Malaysia	EANN-GA, EANN, FFNN, NNE	WQI	[152]
45	Xin'anjiang River, China	CNN-LSTM, CEEMDAN	WQP	[153]

Table 6. Cont.

ID	River	Algorithm	Approach	Reference
46	The Juhe River, Sanhe China	PSO-DBN-LSSVR	WQP	[24]
47	Burnett River, Australia	kPCA, RNN, FFNN, SVR, GRNN	WQP	[154]
48	Nakdong River, South Korea	CNN-LSTM	WQP	[155]
49	Sefid Rud River, Iran	W-MGGP, GEP, DWT	WQP	[156]
50	Talar River, Iran	RF, RFC	WQI	[157]
51	Yangtze River, Jiangsu, China	IABC-BP	WQP	[158]
52	Klang River, Malaysia	DT	WQI	[90]
53	Langat River, Malaysia	MLP-FFA	WQP	[159]
54	Tireh River, Iran	ANN, GMDH, SVM	WQP	[37]
55	Danube Delta River, Romania	ANN, KNN, BPNN	WQI	[160]
56	Sefidrood River, Iran	SVM	WQP	[94]
57	Aji-Chay River, Iran	ANN, ANFIS, WT	WQP	[89]

* Indian water quality data from Kaggle, 1-DRCNN: One-dimensional residual convolutional neural networks.

Recent advances in AI/ML/DL techniques have significantly enhanced the prediction of river water quality, as demonstrated by the 57 studies summarized in Table 4. These investigations employ a range of algorithmic approaches, including CNN-LSTM, XGB, RF, LSTM, SVM, and hybrid models to estimate key indicators such as the water quality parameter (WQP) and the water quality index (WQI) across rivers in Asia, America, Europe, and Africa. Artificial neural networks (ANNs) have played a prominent role in this domain, owing to their capacity to model complex nonlinear relationships and anticipate fluctuations in water quality, as confirmed by studies [135,140]. In [42], ANN-based models were evaluated alongside gradient boosted trees (GBT), decision trees (DTs), support vector machines (SVM), and random forests (RFs) for predicting the WQI of the Indian river system, with GBT achieving the highest performance. Similarly, ref. [157] implemented four standalone decision tree models and twelve hybrid configurations to estimate the WQI of the Talar River in Iran. The Bagging-Random Tree (BA-RT) approach yielded the most accurate results, reinforcing the superiority of hybrid models over simpler alternatives, as also noted in [161].

The integration of optimization algorithms has proven to significantly enhance the accuracy of machine learning (ML) models in predicting river water quality. For instance, studies [24,95] employed Particle Swarm Optimization (PSO) and Sequential Minimal Optimization (SMO) to strengthen models such as PSO-DBN-LSSVR for the Juhe River in China and SMO-SVM for the Wadi Saf-Saf River in Algeria, respectively. Additionally, the application of Wavelet Transform (WT) has been key in identifying relevant variables and reducing noise, as demonstrated in studies on the Aji-Chay River (Iran) [89], Fujian (China) [103], Dongjiang (China) [147], and Sefid Rud (Iran) [156]. These enhancements have led to improved predictive performance in models such as WT-SVM, WT-RF, and WT-ANN. Another widely applicable technique is XGB, which has been successfully implemented in basins such as the Delaware River (USA) [117], the Han River (South Korea), and regions of the northwestern United States [142], due to its ability to capture complex interactions among predictors and deliver high predictive accuracy.

Hybrid architectures that integrate Convolutional Neural Networks (CNNs) with LSTM units have also demonstrated strong performance in river water quality prediction, as shown in studies on the Yangtze River [116,158]. In [114], the SSA-CNN-LSTM model was applied to the Sheshui River, achieving effective integration of temporal and spatial patterns for estimating the water quality parameter (wQp). Other studies have explored variants such as Gated Recurrent Units (GRUs) [124,127], Bidirectional GRU (BiGRU) networks [150], and hybrid approaches like ANFIS–GP and ANFIS–SC [145] to address nonlinear modeling challenges. In this context, emerging techniques such as Transfer Learning (TL) have shown considerable promise by enabling the reuse of previously acquired knowledge in new environments, thereby enhancing predictive performance, as demonstrated in studies on the Fujian river system in China [102,103].

Finally, the geographical diversity of the studies listed in Table 4 spans a wide range of river systems, including the Yangtze [116,158], Sheshui [114], Tanjiang [123], Li and Liu [127], Pearl [137], Fuyang [138], Xiaofu [130], Lijiang [131], Juhe [24], Euphrates [128,145], Júcar [134], Yamuna [30,145], Langat [141], Klang [148], Sefid Rud [156], Talar [136], Sefidrood [94], Aji-Chay [89], Nakdong [149,155], Oyster River [126], Upper Red River Basin [102], Danube [144], Burnett [154], Kelantan [140], and Bullfrog [135]. These rivers span diverse climatic zones from tropical and temperate to arid and mountainous and reflect a broad spectrum of hydrological and geological contexts. This territorial breadth highlights the versatility of AI/ML/DL models in adapting to varied environmental conditions, reinforcing their value as effective tools for the sustainable management of water resources.

3.3.2. Prediction of Groundwater Quality Using AI/ML/DL

Regarding groundwater, Table 7 presents a random sample of 40% (n = 26) of the publications on groundwater quality prediction found in our database, spanning various regions of the world. Diverse environmental, hydrogeological, socioeconomic, and technological factors have shaped the development of advanced computational approaches for predicting groundwater quality.

These approaches range from traditional statistical models and standalone machine learning algorithms (e.g., decision trees, support vector machines, etc.) to hybrid, ensemble methods and SHAP-enhanced models. They also include the integration of geospatial tools, remote sensing data, and cloud-based platforms for data processing and visualization. The most frequently studied contaminants are nitrates, arsenic, fluoride, heavy metals, and total dissolved solids. In addition to studies focused on individual water quality parameters (WQPs), there is research incorporating integrative water quality indices such as the water quality index (WQI), irrigation water quality index (IWQI), entropy-weighted water quality index (EWQI), and groundwater quality index (GWQI).

Table 7. Characterization of predictive models in a representative sample of groundwater quality prediction studies (n = 26).

ID	Region	Parameters	Algorithm	Reference
1	Madrid, Spain	Nitrate concentrations	DT, RF, AdaBoost, ExT	[162]
2	Songyuan City, China	Strontium (Sr^{2+})	GAN, KNN, GPR	[163]
3	Mekong Delta regi6n, Vietnam	Salinity levels	Bagging, CatBoost, ExT, HGB, XGB, DT, RF, LightGBM, KNN, SHAP	[164]
4	Eden Valley, Cumbria, North West England	Nitrate concentrations	DT, XGB, RF, KNN, SHAP	[165]
5	Kerala, India	EWQI	XGB, SVR, ANN, RF	[46]
6	The Mitidja plain, northern Algeria	IWQI	LSTM	[166]
7	Groundwater dataset	Salinity levels	GMDH algorithm	[167]

Table 7. *Cont.*

ID	Region	Parameters	Algorithm	Reference
8	Tamil Nadu, India	IWQI	SVM, ANN, LRM, RT, GPR, BRT	[168]
9	North China Plain, Beijing	Arsenic (As) and fluoride (F−) concentrations	XGB, RF, SVM,	[169]
10	Eastern India	WQI	MLP-ANN	[170]
11	Raipur district, Chhattisgarh, India	WQI	ANN-LR	
12	Midwestern United States	Redox Conditions	GBM, XGB, RF	[171]
13	Hawasinah catchment Wilayat Al-Khaburah, Oman	TDS	CatBoost regression, ETR, Bagging regression	[172]
14	Vehari, Punjab Province of Pakistan	WQI	ANN, RF, LR	[173]
15	Northeast of Tamil Nadu, India	WQI	GB, RF, DT, KNN, MLP, XGB, SVR	[174]
16	Qom City, Iran	Nitrate concentration	KNN, SVR, RF	[175]
17	Savar, Dhaka district, Bangladesh	GWQI *	LR, SVM, ANN	[176]
18	Al Qunfudhah, Saudi Arabia	WQI	CNN, XGB, SHAP	[177]
19	Fars Province, Iran	WQI	RF, BRT, MnLR	[178]
20	Wendeng District, China	WQI	LSTM	[179]
21	Taiwan Groundwater Pollution Monitoring Standard	Heavy Metal Concentrations	SVR, KNN, MLP, GBR, LIME, SHAP	[180]
22	Middle Black Sea Region of Turkey	WQP	CNN, RF, XGB, DNN	[181]
23	Noida, Uttar Pradesh, India	WQP	MLR, SVR, DT	[182]
24	The Akot basin, Akola district of Maharashtra, India	IWQI	ANN, LSTM, MLR	[183]
25	North Carolina, USA	Nitrate concentrations	RF	[184]
26	Dezful Aquifer, Iran	TDS	RF	[185]

The wide variety of AI/ML/DL models highlights a transition from traditional approaches such as ANN, LSTM, Multilayer Perceptron (MLP) and CNN, to more sophisticated models such as XGB and LightGB. These models are generally applied in studies that require modeling nonlinear relationships or time series, and have been widely used to estimate water quality indices as well. From the reading of the studies listed in Table 5, it is possible to identify a current trend in the use of ensemble-type algorithms, such as Random Forest, Gradient Boosting and Bagging, which are considered robust methods capable of handling the high dimensionality and multicollinearity of predictive models [46,164].

Other ensemble algorithms such as CatBoost, Bagging, and Extra Trees have also been frequently used to predict specific WQP such as nitrates [162,165,176,184], salinity levels [164,173], metals [163,180], as well as water quality indices such as WQI, IWQI, EWQI, and GWQ. These algorithms allow multiple hydrochemical variables to be handled, and reliable predictions to be constructed in complex environments. Regression models have also been widely used in this area, such as the use of Multinomial Logistic Regression (MnLR), which allows water quality to be classified into multiple categories, being useful in environmental risk and zoning studies [178].

Among the emerging models, Generative Adversarial Network (GAN) [163] and Group Method of Data Handling (GMDH) [167] stand out. They are generally used in contexts with complex dynamics, particularly with scarce data, and be particularly useful in predictions of Sr^{2+} and salinity levels [163]. In addition, explanatory models such as SHAP and LIME represent a significant advance in the interpretability and explainability of predictive models applied to groundwater quality [180]. These techniques, typical of the field of XAI, are important because they allow the outputs of complex algorithms to be broken down into quantifiable contributions of each input variable, allowing one to understand how much each factor influences the final prediction. For example, in [164] SHAP was applied to assess the impact of physicochemical factors on salinity levels in multiple aquifers, revealing key spatial patterns using interpretable maps. Similarly, [165] used SHAP in a nitrate prediction model to identify the main geoenvironmental pollution-related drivers in a UK aquifer, integrating interpretation and prediction into a unified framework.

In this sense, the integration of GIS technologies, together with XAI models, strengthens the capacity to spatially represent the results, detect risk areas and generate useful tools for water management. Together, the AI/ML/DL models allow the prediction of groundwater quality to be approached from an interdisciplinary, explanatory and applicable perspective, configuring themselves as key tools for water sustainability in complex environmental scenarios [108,170].

On the other hand, in the hydrogeological context, studies are conducted in regions characterized by a wide range of factors, including aquifers affected by saline intrusion (Vietnam, Iran), arid zones (Algeria, Saudi Arabia), and areas with carbonate lithology that induce water hardness (India, USA). Natural processes such as weathering, evapotranspiration, and the influence of volcanic emissions also contribute to the spatial and temporal variability of hydrochemical parameters [163,179,180]. These hydrogeological factors, combined with local climatic conditions, shape physically complex and dynamic environments that demand predictive models with high adaptive capacity. In this context, the suitability of groundwater quality for human consumption is conditioned by these factors, among them, variations in well depth, which can significantly alter water mineralization and contaminant mobility due to interactions with lithological formations, hydraulic gradients, and specific redox conditions [174,186]. Therefore, the applied algorithms must be capable of capturing aquifer heterogeneity by integrating geological and hydrodynamic variables that directly influence groundwater quality.

Finally, at the regional level, Asia leads scientific production, with strong representation from India, China, Iran, and Bangladesh. These countries face severe water stress, which justifies the scientific community's growing interest in groundwater quality and the development of advanced predictive models.

4. Answering the Research Questions

Consistent with the B-SLR framework adopted in this study, the driven research questions are answered below.

RQ1. What are the most widely used AI/ML/DL algorithms in water quality prediction?

The literature review showed that assembly models, such as Bagging and Boosting, have established themselves as predominant techniques in the prediction of both surface water and groundwater quality [187]. Their effectiveness is supported by studies that demonstrate their ability to reduce mean absolute errors and quadratic errors [188]. In addition, recent research has explored variants of these models, such as Grid Search Random Forest (GS-RF) and XGB, to optimize prediction accuracy on parameters such as turbidity and different nutrients [189], showing that algorithms such as CatBoost Regression (CBR) offer advantages in terms of stability and adaptability, particularly in handling heterogeneous datasets, minimizing overfitting, and maintaining consistent performance across varying environmental conditions and input configurations [173].

These assembly models have been successfully applied in diverse hydrological contexts, including urban rivers and Rural [57], and their use extends to the prediction of water quality indices in multivariable and complex scenarios [190]. Moreover, their ability to handle large volumes of data and correlated variables characterizes them as robust tools in environmental studies.

On the other hand, models based on ANNs have evolved considerably in recent years, incorporating more flexible architectures and optimization techniques such as genetic algorithms, wavelet transforms and hybrid strategies inspired by nature, improving performance against nonlinear and highly noisy datasets [109,161]. Its applicability has extended from the prediction of individual parameters to multivariable estimates of water quality, including dissolved oxygen and organic pollutants. Also, models based on fuzzy logic have gained relevance due to their ability to handle uncertainty (inherent in environmental data). Indeed, research in Saudi Arabia, India, and China has shown that fuzzy logic, when integrated with neural networks or evolutionary algorithms, exhibits suitable performances in regions with fragmented or scattered data [100,191,192]. Techniques such as Extreme Learning Machine (ELM), in combination with RF, have also been used to extend their applicability in environments with high hydrological variability [127].

Additionally, algorithms such as SVM, SVR, and RF remain highly popular due to their effectiveness in estimating water quality parameters. These methods offer non-parametric solutions that learn directly from observed data, facilitating uncertainty management and contextualized interpretation of pollution patterns [26].

The bibliometric analysis also reflects a sustained increase in integrating XAI techniques in research oriented to water quality prediction, to interpret interpreting the internal behavior of complex models and facilitating transparent decision-making. Techniques such as SHAP and LIME have been implemented to overcome the "black box" of ML/DL algorithms, allowing researchers to assess and visualize the influence of each predictor variable on the results [85,107,193]. XAI has been shown to improve model reliability and provide useful explanations for water quality management [108,194,195]. A prominent example is [196], in which an interpretable learning framework based on SHAP and RF is developed, applied to hydrodynamic scenarios. This approach allowed us to understand the impact of environmental variables on water quality, reinforcing the usefulness of XAI in complex studies of aquatic systems.

Figure 10a summarizes the predictive models identified in this study, classifying and organizing the AI/ML/DL algorithms most commonly used in freshwater quality prediction studies, both for surface water and groundwater systems. The percentage of the models with the highest applicability in predicting water quality in both surface and groundwater is presented in Figure 10b.

Finally, building reliable predictive models involves following a rigorous workflow that includes database consolidation, raw data preprocessing, proper predictive algorithm selection, model training, and validation. Recent literature underscores that each stage is critical to ensure the accuracy and robustness of the predictive model [47,197–199]. In particular, the selection of the algorithm must be aligned with the nature of the data and the specific objectives of a given study, ensuring interpretable, adaptable and relevant results for environmental decision-making.

RQ2. Which AI/ML/DL algorithm allows a better estimate of water quality?

The literature review identified a wide range of AI/ML/DL models developed both as standalone approaches and hybrid frameworks to enhance water quality prediction. Defining a single robust predictive model remains a challenge for researchers. However, certain algorithms stand out for their efficiency. For instance, the light gradient boosting machine (LightGBM) has emerged as a highly effective option due to its ability to process large datasets, fast training speed, and optimized architecture designed to minimize computational cost while maximizing predictive accuracy. Recent studies have reported accuracies exceeding 90% in the estimation of physicochemical parameters [200–202]. This technique is particularly notable for its consistent performance in water bodies with high variability.

The XGB algorithm has demonstrated remarkable performance in classifying both surface and groundwater quality, achieving accuracy levels close to 89% [203,204]. Its integration with XAI techniques such as SHAP enables the interpretation of key indicators such as zinc, nitrates, and chlorides, thereby improving the transparency of predictive models [203]. On the other hand, study [205] evaluated the XGB model alongside the ETR and GBR regression models to predict water quality in the Weihe River Basin, China. The study addressed the challenges of monitoring NH_3-N, TP, COD, and DO using multispectral images by integrating meteorological data (temperature, precipitation, evapotranspiration, and wind speed) and land use types as covariates in the models, which were assessed under three scenarios: remote sensing; remote sensing + meteorology; and remote sensing + meteorology + land use. Methods such as Correlation Analysis (CA), Feature Importance Analysis (FIA), and SHAP were employed to select sensitive spectral bands and evaluate the contribution of multi-source data to improve model accuracy. The results showed that the GBR and ETR models were more robust and transferable, enabling the generation of spatial inversion maps with accurate concentrations of WQPs.

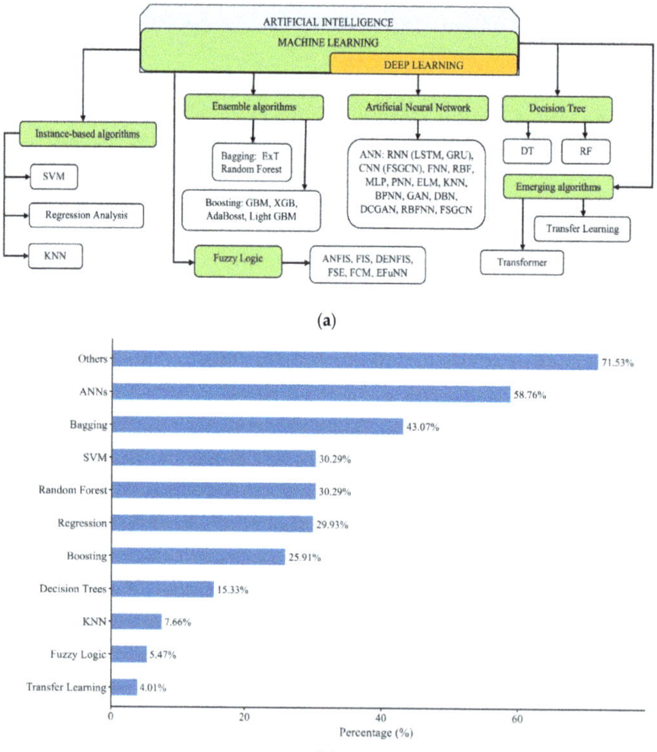

Figure 10. ML/DL models most used in water quality prediction: (**a**) Classification; (**b**) Percentage of applicability.

Time-series-based models, particularly LSTM architectures and their hybrid variants such as LSTM-CNN and CEEMDAN-LSTM, have achieved accuracies ranging from 90% to 93% by capturing complex dynamic patterns [6,43,88,116,155,206]. The incorporation of Transfer Learning further enhances performance by enabling efficient adaptation under data limitations [101,103,104]. Although MLP models have a simpler structure, they yield accuracies between 85% and 89% in sequential scenarios. Optimizing their parameters through genetic algorithms has improved precision in hydrologically realistic contexts [197,207–209].

Finally, hybrid approaches that combine multiple algorithms have reached accuracies above 92%, standing out for their adaptability to outlier values and their generalization capacity [157,210]. In summary, although no single model consistently outperforms across all contexts, algorithms that integrate interpretability, deep architecture, and hybrid strategies have delivered more accurate and reliable results. The selection of the most suitable model depends on the type of data, the analytical objectives, and the specific conditions of the water system under study.

RQ3. What limitations have been identified in the use of AI/ML/DL for water quality prediction?

A relevant aspect identified in this work is the study of spatial and temporal variations in the evaluation of water quality in the context of missing data, primarily due to missing data caused by measurement system failures, operational errors, environmental phenomena, and the non-continuous sampling frequency of water quality data. These limitations constrain the availability of reliable datasets for classification and evaluation purposes. In this context, Time Series Analysis (TSA) models combined with machine learning approaches, particularly long short-term memory (LSTM) networks have proven highly effective in addressing these challenges and estimating future values of water quality parameters (WQP) or water quality indices (WQI) based on historical data [211–213].

It was observed that DL models face various challenges compared to traditional physical models for water quality prediction. These include the complexity of internal structure and parameter adjustment, reliance on large datasets for effective training, and a lack of physical constraints, which can make it difficult to explain prediction results. Similarly, obtaining high-reliability data for certain water quality parameters can be difficult, limiting the applicability of deep learning approaches [214]. However, these challenges have begun to be overcome with the development of hybrid models and the use of interpretable approaches [25,88,114,215,216]. In general, the accuracy of the predictions of AI/ML/DL models is influenced by the availability and quality of historical data, and the models developed can be sensitive to variations in environmental conditions over time.

Recent literature has shown that some of these limitations can be mitigated through the use of LSTM models combined with Transfer Learning (TL) techniques [101,103], particularly instance-based approaches such as TrAdaBoost [96]. This model inherits the strengths of both LSTM and TL, offering powerful capabilities to capture long-term dependencies in time series and the flexibility to leverage related knowledge from complete datasets to fill large-scale consecutive data gaps [96]. Notably, prediction performance can be further enhanced by applying wavelet transforms to suppress noise in time-series signals, serving as an optimization mechanism for predictive modeling [89]. In this regard, model performance depends heavily on how variables are cleaned, transformed, and selected. For this reason, optimization algorithms such as Particle Swarm Optimization (PSO) [217,218], among others [114,118,219], are widely used.

Finally, variations in model performance caused by seasonal changes, extreme events, or point-source contamination which can significantly affect prediction accuracy have been addressed using Generative Adversarial Networks (GANs). These networks enable the simulation of abnormal or extreme conditions that are poorly represented in real datasets [163]. While data scarcity and climate variability driven by extreme events have posed significant challenges to water quality prediction using AI/ML/DL techniques, the recent literature reveals a growing interest in bridging this gap—an essential step toward ethical and transparent water resource management. In this context, study [220] introduced the Multi-Scale Weighted-Slope Regression (MS-WSR) model, which demonstrated strong robustness in predicting water levels across six lakes, achieving high accuracy and adaptability under climate variability.

From an ethical and regulatory standpoint, AI-based models have emerged as powerful yet ambivalent tools in water treatment, conservation, and management. While they offer transformative potential to address pressing water challenges, but its use raises concern about its considerable water consumption and potential environmental impact. These issues can be mitigated through the integration of renewable energy sources reducing both water footprints and greenhouse gas emissions, and the implementation of water reuse systems that lessen reliance on freshwater and minimize wastewater-related effects [221]. Furthermore, the lack of explainability in certain models may undermine trust in the predictions that inform transparent decision-making in water governance. Emerging technologies such as XAI offer promising solutions to issues of opacity and bias, paving the way for more sustainable and accountable water resource management.

RQ4. What emerging variants currently exist in AI/ML/DL models for estimating water quality?

A current trend in water resource management using AI/ML/DL is the development of hybrid algorithms, which have gained relevance as a strategy to improve the accuracy of water quality parameter estimation. For instance, models based on Variational mode decomposition optimized by the sparrow search algorithm (SSA-VMD), combined with Bidirectional Gated Recurrent Units (BiGRU), have achieved over 96% efficiency in the case of Qiandao Lake, China [222]. These hybrid approaches enable the capture of both nonlinear patterns and temporal dynamics in hydro-environmental data.

In the context of time series prediction for water quality assessment, Transformer models have shown notable advantages over LSTM networks. This architecture overcomes the limitations of previous models by effectively capturing long-term dependencies and correlations between distant points in time series—an essential capability for predicting variables such as pH, turbidity, and dissolved oxygen. Unlike LSTMs, which process data sequentially, the Transformer simultaneously analyzes the entire historical dataset, significantly reducing training time, enhancing the extraction of relevant features, and improving overall modeling efficiency [26]. Finally, there is growing interest in enhancing the interpretability of machine learning models, given their "black box" nature. In this regard, Explainable AI (XAI) techniques have been implemented to identify the relative impact of water quality parameters on model predictions. The recent literature highlights successful applications of SHAP in models estimating salinity and dissolved oxygen, and in the prediction of heavy metal concentrations, contributing to more informed decision-making in water management [164,165,223].

RQ5. What are the key water quality indicators used to assess natural water sources?

Water quality assessment in natural sources relies on a set of physicochemical and biological parameters that characterize both the environmental status and the suitability of water for various uses. According to the scientific literature reviewed, the most frequently used parameters in predictive studies employing AI/ML/DL techniques include dissolved oxygen (DO), biochemical oxygen demand (BOD), chemical oxygen demand (COD), total suspended solids (TSS), temperature, pH, electrical conductivity (EC), chlorophyll-a (Chl-a), nitrates, phosphates, and coliform bacteria. These indicators are considered conventional and are regulated by international environmental standards due to their relevance in characterizing both surface and groundwater bodies.

Beyond these essential parameters, there is a growing trend in the literature toward incorporating complementary variables such as heavy metals (e.g., arsenic, copper, lead), nutrients (NO_3^-, NO_2^-, NH_4^+, PO_4^{3-}), and trace metals (Fe, Mn, Zn, Cu, Cr), which enhance the discrimination between different water quality categories. Additionally, the integration of hydrogeological, meteorological, land use, and socioeconomic variables has proven valuable in enriching predictive models by capturing the influence of external factors on water body dynamics [129,224,225]. In this regard, meteorological, land use, socioeconomic, and hydrogeological variables [226], for example, help illustrate how human activities can alter the export of chemical elements through changes in vegetation cover, ultimately affecting water quality [227].

5. Contribution and Future Work

The scientific literature review conducted in this article, covering the period 2015–2024, using the integrated bibliometric-systematic literature review (B-SLR) methodology and Topic Modeling, delved into AI/ML/DL techniques for predicting freshwater quality (surface and groundwater). The review confirmed a state of maturity in the field, characterized by exponential growth in scientific publications since 2020. It also allowed us to identify key authors in the field, research gaps, emerging trends, and models with high predictive performance. The main contribution of this study lies in its methodological approach, which allowed us to identify the most relevant works that apply AI/ML/DL models in studies related to surface and groundwater quality.

The results of this research confirmed the findings of the study [57], which received more than 452 citations and reviewed predictive models for river water quality during the period 2000–2020. However, by applying the B-SLR methodology, the present study significantly expanded the scope of the analysis, updating the literature to 2024, identifying emerging trends, and incorporating studies on freshwater groundwater as a potential source of drinking water. This broader and more integrative approach allowed for a deeper thematic mapping of the field, revealing underexplored areas and offering a more complete understanding of current scientific advances. Regarding groundwater quality, the study [109] from the period (1994–2022) highlighted that ANNs are the most widely used in water quality modeling, with nitrate being the most studied parameter.

Complementing this knowledge, the present work identifies a transition from more traditional approaches, such as ANNs, to more sophisticated models, such as XGB, LightGB, and hybrid explainable architectures, such as XGB, LSTM-CNN, and SHAP. In addition, promising emerging approaches are identified, such as Generative Adversarial Networks (GANs) and Group Data Handling (GMDH), used in data-scarce contexts. Transfer Learning (TL), applied for knowledge reuse and performance improvement in data-constrained environments; and Transformer architectures, which have been shown to outperform LSTM networks in time series prediction.

As future research, we propose expanding the spectrum of water quality parameters considered in predictive models, incorporating emerging contaminants such as microplastics, pharmaceuticals, and persistent organic compounds (POCs), which are of current concern [195] in water management. Furthermore, a promising research gap could explore the scalability of Transformer models in real-time water quality monitoring systems, especially in resource-limited settings.

Finally, we recognize that using citations as an indicator of quality has limitations. The lack of distinction between open access and subscription articles can bias the results, given that visibility and accessibility influence citation frequency. Furthermore, citation practices influenced by national or institutional affinities can distort the relationship between citation counts and intrinsic quality.

The total number of citations reflects the overall scholarly impact of a body of work, yet it may be misleading when disproportionately driven by a few highly cited publications or diluted through extensive co-authorship, obscuring individual contributions [87]. Future research should consider incorporating complementary metrics that evaluate the quality of cited works, alongside approaches that integrate contextual impact and explainability indicators. Such enhancements would enable a more equitable and nuanced assessment of scientific value, particularly in disciplines where visibility does not necessarily correlate with methodological rigor.

6. Conclusions

This research demonstrates that the Bibliometric-Systematic Literature Review (B-SLR) approach constitutes a robust methodology for analyzing the state of the art in highly dynamic scientific fields, such as water quality prediction using AI/ML/DL techniques. By integrating the structured rigor of a systematic review with the analytical depth of bibliometric analysis, the B-SLR enabled the identification of domain-specific trends, thematic mapping of knowledge, assessment of scientific impact, and detection of gaps in the literature—offering a more comprehensive, precise, and context-aware understanding of the field.

The findings reveal that ensemble models (e.g., Bagging, Boosting), deep neural networks (LSTM, CNN, MLP), and hybrid approaches have overcome the limitations of conventional methods, delivering greater accuracy, adaptability, and the ability to handle incomplete or nonlinear data. The integration of explainable artificial intelligence (XAI) techniques, such as SHAP and LIME, has facilitated the development of more transparent and reliable models, enhanced result interpretation and supported informed decision-making. Accordingly, the selection of the optimal predictive model depends on multiple factors, including the type of water body, geographic context, data availability, and the specific objectives of the study. In this regard, hybrid and interpretable models emerge as the most promising alternatives for addressing current challenges in water quality prediction.

Regarding the methodological approach employed in this study, the application of B-SLR enabled the refinement of an initial database of 1822 articles into a final corpus of 274 highly relevant publications, through automated procedures and manual validation. This process ensured the quality and relevance of the analyzed studies, reinforcing the reliability of the results obtained. Furthermore, a detailed classification was achieved, covering the most frequently used algorithms, the types of water bodies studied, key quality indicators, and the methodological limitations faced by predictive models. Ultimately, this work establishes a replicable and scalable methodological foundation for future research.

Author Contributions: Conceptualization, J.A.M.-A., J.N., R.O., C.A.C., J.L.A. and L.R.-L.; methodology, J.A.M.-A.; software, J.A.M.-A. and J.N.; validation, J.A.M.-A. and J.N.; formal analysis, J.A.M.-A., J.N., R.O., C.A.C., J.L.A. and L.R.-L.; investigation, J.A.M.-A., J.N. and R.O.; resources, J.N. and R.O.; data curation, J.A.M.-A. and J.N.; writing—original draft preparation, J.A.M.-A., J.N. and R.O.; writing—review and editing, J.A.M.-A., J.N., R.O., C.A.C., J.L.A. and L.R.-L.; visualization, J.A.M.-A. and J.N.; supervision, J.N. and R.O.; project administration, R.O.; funding acquisition R.O., J.N. and J.L.A. All authors have read and agreed to the published version of the manuscript.

Funding: This research was supported by DIDULS Regular PR2553851 Project of the University of La Serena and ANID/FONDAP/1523A0001. The APC was sponsored by the CRHIAM Water Center, Universidad de Concepción, Chillán, Chile.

Data Availability Statement: No new data were created or analyzed in this study. Data sharing is not applicable to this article.

Acknowledgments: Ricardo Oyarzún and Jorge Núñez acknowledges the financial support of DIDULS/ULS, through the project PR2553851 (University of La Serena, Chile). José Luis Arumí and Ricardo Oyarzún acknowledges the financial support of the Water Research Center CRHIAM: ANID/FONDAP/1523A0001.

Conflicts of Interest: The authors declare no conflicts of interest.

Abbreviations

The following abbreviations are used in this manuscript:

AMT	Alternating Model Tree
ANFIS	Adaptive Neuro-Fuzzy Inference System
ANFIS–GP	Adaptive Neuro Fuzzy Inference System–Grid Partitioning
ANFIS–SC	ANFIS–Subtractive Clustering
ANN	Artificial Neural Network
AO-SVM	Aquila Optimization Support Vector Machine
AR	Additive Regression
AdaBoost	Adaptive Boosting
BDT	Boosted Decision Tree
B-SLR	Bibliometric-Systematic Literature Review
BiGRU	Bi-directional Gated Recurrent Units
BMEF	Bayesian Maximum Entropy-based Fusion
BNN	Bayesian Neural Network
BOD	Biochemical Oxygen Demand
BPNN	Backpropagation Neural Network
BRT	Boosted Regression Trees
CA	Correlation Analysis
Chl-a	Chlorophyll-a
CART	Classification and Regression Tree
CatBoost	Categorical Boosting
CBR	CatBoost Regression
CEEMD	Complete Ensemble Empirical Mode Decomposition with Adaptive Noise
CNN	Convolutional Neural Network
COD	Chemical Oxygen Demand
CSA	Crow Search Algorithm
DL	Deep Learning
DBN	Deep Belief Network
DCGAN	Deep Convolutional Generative Adversarial Network
DENFIS	Dynamic Evolving Neural-Fuzzy Inference System
DNN	Deep Neural Network
DO	Dissolved Oxygen
DR	Discretization Regression
DRNN	Deep Recurrent Neural Network
DT	Decision Tree
DWT	Discrete Wavelet Transform
EANN	Emotional Artificial Neural Network
EANN-GA	Emotional Artificial Neural Network–Genetic Algorithm
EBM	Ensemble Bagged Machine
EC	Electrical Conductivity
EFuNN	Evolving Fuzzy Neural Network
ELM	Extreme Learning Machine
EN	Elastic Network
EWQI	Entropy-weighted Water Quality Index

ExT	Extra Trees
FIA	Feature Importance Analysis
ETR	Extra Tree Regression
FFNNs	Feedforward Neural Networks
FNN	Feed-forward Neural Network
FSGCN	Functional-Structural Sub-Region Graph Convolutional Network
FFA	Firefly Algorithm
GAN	Generative Adversarial Network
GB	Gradient Boosting
GBM	Gradient Boosting Machine
GBR	Gradient Boosting Regression
GBT	Gradient Boosted Trees
GC	Global Citations
GEP	Gene Expression Programming
GMDH	Group Method of Data Handling
GNB	Gaussian Naïve Bayes
GPR	Gaussian Process Regression
GRNN	Generalized Regression Neural Network
GRU	Gated Recurrent Unit
GS-RF	Grid Search-Random Forest
GS-SVR	Grid Search-Support Vector Regression
GWQI	Groundwater Quality Index
HGB	Histogram Gradient Boosting
IABC-BPNN	Improved Artificial Bee Colony–Backpropagation
iML	Interpretable Machine Learning
IoT	Internet of Things
IWQI	Irrigation Water Quality Index
KNN	K-Nearest Neighbors
kPCA	Kernal Principal Component Analysis
LC	Local Citation
LIME	Local Interpretable Model-agnostic Explanations
LR	Logistic Regression
LRM	Logistic Regression Model
LSSVR	Least Squares Support Vector Regression
LSTM	Long Short-Term Memory
LightGBM	Light Gradient Boosting Machine
LGB	Light Gradient Boosting
MARS	Multivariate Adaptive Regression Spline
ML	Machine Learning
MLR	Multiple Linear Regression
MLRF	Multi-label Classification Through Random Forest
MLP	Multi-Layer Perceptron
MnLR	Multinomial Logistic Regression
MS-WSR	Multi-Scale Weighted-Slope Regression
NNE	Neural Network Ensemble
PCA	Principal Component Analysis
PCOs	Persistent Organic Compounds
PLS	Partial Least Squares
PNN	Probabilistic Neural Network
PSO	Particle Swarm Optimization
RBF	Radial Basis Function
RBFNN	Radial Basis Function Neural Network
RC	Random Committee

REPT	Reduced Error Pruning Tree
RF	Random Forest
RFR	Random Forest Regression
RFC	Randomizable Filtered Classification
RNN	Recurrent Neural Network
RR	Ridge Regression
RQ	Research Questions
RT	Regression Tree
SDGs	Sustainable Development Goals
SHAP	SHapley Additive exPlanations
SLR	Simple Linear Regression
SMO	Sequential Minimal Optimization
SMO-SVM	Sequential Minimal Optimization-Support Vector Machine
SSA-CNN-LSTM	Sparrow Search Algorithm-Convolutional Neural Network-Long Short-Term Memory
SSA-VMD	Sparrow search algorithm-Variational mode decomposition
SVM	Support Vector Machines
SVR	Support Vector Regression
SVMR	Support Vector Machine Regression
SWEBM	Stochastic Weighted Ensemble Bagged Machine
TC	Total Citations
TDS	Total Dissolved Solids
TL	Transfer Learning
TSS	Total Suspended Solids
WA	Wavelet Analysis
W-MGGP	Wavelet-Multigene Genetic Programming
WQI	Water Quality Index
WQP	Water Quality Parameters
WT	Wavelet Transform
WT-ANN	Wavelet Transform-Artificial Neural Network
XAI	eXplainable Artificial Intelligence
XGB	eXtreme Gradient Boosting

References

1. Ahmed, W.; Mohammed, S.; El-Shazly, A.; Morsy, S. Tigris River water surface quality monitoring using remote sensing data and GIS techniques. *Egypt. J. Remote Sens. Sci.* **2023**, *26*, 816–825. [CrossRef]
2. Gaagai, A.; Aouissi, H.A.; Bencedira, S.; Hinge, G.; Athamena, A.; Heddam, S.; Gad, M.; Elsherbiny, O.; Elsayed, S.; Eid, M.H.; et al. Application of Water Quality Indices, Machine Learning Approaches, and GIS to Identify Groundwater Quality for Irrigation Purposes: A Case Study of Sahara Aquifer, Doucen Plain, Algeria. *Water* **2023**, *15*, 2. [CrossRef]
3. Rahaman, M.H.; Sajjad, H.; Hussain, S.; Roshani; Masroor, M.; Sharma, A. Surface water quality prediction in the lower Thoubal river watershed, India: A hyper-tuned machine learning approach and DNN-based sensitivity analysis. *J. Environ. Chem. Eng.* **2024**, *12*, 112915. [CrossRef]
4. UNESCO; UN-Water; World Water Assessment Programme. *The United Nations World Water Development Report 2023: Partnerships and Cooperation for Water*; UNESCO: Paris, France, 2023. Available online: http://digitallibrary.un.org/record/4007797 (accessed on 26 September 2025).
5. Gao, J.; Zhu, S.; Li, D.; Jiang, H.; Deng, G.; Wen, Y.; He, C.; Cao, Y. Bibliometric analysis of climate change and water quality. *Hydrobiologia* **2023**, *850*, 3441–3459. [CrossRef]
6. Pyo, J.; Pachepsky, Y.; Kim, S.; Abbas, A.; Kim, M.; Kwon, Y.S.; Ligaray, M.; Cho, K.H. Long short-term memory models of water quality in inland water environments. *Water Res. X* **2023**, *21*, 100207. [CrossRef]
7. Hussein, E.E.; Baloch, M.Y.J.; Nigar, A.; Abualkhair, H.F.; Aldawood, F.K.; Tageldin, E. Machine Learning Algorithms for Predicting the Water Quality Index. *Water* **2023**, *15*, 3540. [CrossRef]
8. Kamal, N.A.; Muhammad, N.S.; Abdullah, J. Scenario-based pollution discharge simulations and mapping using integrated QUAL2K-GIS. *Environ. Pollut.* **2020**, *259*, 113909. [CrossRef]

9. Mummidivarapu, S.K.; Rehana, S.; Rao, Y.R.S. Mapping and assessment of river water quality under varying hydro-climatic and pollution scenarios by integrating QUAL2K, GEFC, and GIS. *Environ. Res.* **2023**, *239*, 117250. [CrossRef]
10. Sarafaraz, J.; Kaleybar, F.A.; Karamjavan, J.M.; Habibzadeh, N. Predicting river water quality: An imposing engagement between machine learning and the QUAL2Kw models (case study: Aji-Chai, river, Iran). *Results Eng.* **2024**, *21*, 101921. [CrossRef]
11. Chueh, Y.-Y.; Fan, C.; Huang, Y.-Z. Copper concentration simulation in a river by SWAT-WASP integration and its application to assessing the impacts of climate change and various remediation strategies. *J. Environ. Manag.* **2021**, *279*, 111613. [CrossRef]
12. Prajapati, S.; Sabokruhie, P.; Brinkmann, M.; Lindenschmidt, K.-E. Modelling Transport and Fate of Copper and Nickel across the South Saskatchewan River Using WASP—TOXI. *Water* **2023**, *15*, 265. [CrossRef]
13. Alam, R.; Ahmed, Z.; Seefat, S.M.; Nahin, K.T.K. Assessment of surface water quality around a landfill using multivariate statistical method, Sylhet, Bangladesh. *Environ. Nanotechnol. Monit. Manag.* **2021**, *15*, 100422. [CrossRef]
14. Isaac, R.; Siddiqui, S.; Higgins, P.; Paul, A.S.; Lawrence, N.A.; Lall, A.S.; Khatoon, A.; Singh, A.; Majeed, P.A.; Massey, S.; et al. Assessment of seasonal impacts on Water Quality in Yamuna river using Water Quality Index and Multivariate Statistical approaches. *Waste Manag. Bull.* **2024**, *2*, 145–153. [CrossRef]
15. Fernandes, C.P.; Fonseca, A.R.; Pacheco, F.A.L.; Fernandes, L.F.S. Water quality predictions through linear regression—A brute force algorithm approach. *MethodsX* **2023**, *10*, 102153. [CrossRef] [PubMed]
16. Galoie, M.; Motamedi, A.; Fan, J.; Moudi, M. Prediction of water quality under the impacts of fine dust and sand storm events using an experimental model and multivariate regression analysis. *Environ. Pollut.* **2023**, *336*, 122462. [CrossRef]
17. Pandey, D.K.; Hunjra, A.I.; Bhaskar, R.; Al-Faryan, M.A.S. Artificial intelligence, machine learning and big data in natural resources management: A comprehensive bibliometric review of literature spanning 1975–2022. *Resour. Policy* **2023**, *86*, 104250. [CrossRef]
18. Li, X.; Su, J.; Wang, H.; Boczkaj, G.; Mahlknecht, J.; Singh, S.V.; Wang, C. Bibliometric analysis of artificial intelligence in wastewater treatment: Current status, research progress, and future prospects. *J. Environ. Chem. Eng.* **2024**, *12*, 113152. [CrossRef]
19. Gonzales-Inca, C.; Calle, M.; Croghan, D.; Haghighi, A.T.; Marttila, H.; Silander, J.; Alho, P. Geospatial Artificial Intelligence (GeoAI) in the Integrated Hydrological and Fluvial Systems Modeling: Review of Current Applications and Trends. *Water* **2022**, *14*, 2211. [CrossRef]
20. Aliaga-Alvarado, M.; Gómez-Escalonilla, V.; Martínez-Santos, P. Identification of non-conventional groundwater resources by means of machine learning in the Aconcagua basin, Chile. *J. Hydrol. Reg. Stud.* **2023**, *49*, 101502. [CrossRef]
21. Khoi, D.N.; Quan, N.T.; Linh, D.Q.; Nhi, T.T.; Thuy, N.T.D. Using Machine Learning Models for Predicting the Water Quality Index in the La Buong River, Vietnam. *Water* **2022**, *14*, 1552. [CrossRef]
22. Mallya, G.; Hantush, M.M.; Govindaraju, R.S. A Machine Learning Approach to Predict Watershed Health Indices for Sediments and Nutrients at Ungauged Basins. *Water* **2023**, *15*, 586. [CrossRef]
23. Ghobadi, F.; Kang, D. Application of Machine Learning in Water Resources Management: A Systematic Literature Review. *Water* **2023**, *15*, 620. [CrossRef]
24. Yan, J.; Gao, Y.; Yu, Y.; Xu, H.; Xu, Z. A Prediction Model Based on Deep Belief Network and Least Squares SVR Applied to Cross-Section Water Quality. *Water* **2020**, *12*, 1929. [CrossRef]
25. Zhou, Y. Real-time probabilistic forecasting of river water quality under data missing situation: Deep learning plus post-processing techniques. *J. Hydrol.* **2020**, *589*, 125164. [CrossRef]
26. Tripathy, K.; Mishra, A.K. Deep learning in hydrology and water resources disciplines: Concepts, methods, applications, and research directions. *J. Hydrol.* **2024**, *628*, 130458. [CrossRef]
27. Zheng, Y.; Wei, J.; Zhang, W.; Zhang, Y.; Zhang, T.; Zhou, Y. An ensemble model for accurate prediction of key water quality parameters in river based on deep learning methods. *J. Environ. Manag.* **2024**, *366*, 121932. [CrossRef]
28. Chellaiah, C.; Anbalagan, S.; Swaminathan, D.; Chowdhury, S.; Kadhila, T.; Shopati, A.K.; Shangdiar, S.; Sharma, B.; Amesho, K.T.T. Integrating deep learning techniques for effective river water quality monitoring and management. *J. Environ. Manag.* **2024**, *370*, 122477. [CrossRef]
29. Prasad, D.V.V.; Venkataramana, L.Y.; Kumar, P.S.; Prasannamedha, G.; Harshana, S.; Srividya, S.J.; Harrinei, K.; Indraganti, S. Analysis and prediction of water quality using deep learning and auto deep learning techniques. *Sci. Total Environ.* **2022**, *821*, 153311. [CrossRef]
30. Khullar, S.; Singh, N. Water quality assessment of a river using deep learning Bi-LSTM methodology: Forecasting and validation. *Environ. Sci. Pollut. Res.* **2022**, *29*, 12875–12889. [CrossRef]
31. United Nations Department of Economic and Social Affairs. *The Sustainable Development Goals Report 2023*, Special Edition; United Nations Department of Economic and Social Affairs: New York, NY, USA, 2023. Available online: https://unstats.un.org/sdgs/report/2023/The-Sustainable-Development-Goals-Report-2023.pdf (accessed on 4 June 2025).

32. UN-Water. *The Sustainable Development Goal 6 Global Acceleration Framework*; UN-Water: Geneva, Switzerland, 2020. Available online: https://unsceb.org/sdg-6-global-acceleration-framework (accessed on 4 June 2025).
33. de Carvalho Marques, M.; Mohamed, A.A.; Feitosa, P. Sustainable development goal 6 monitoring through statistical machine learning—Random Forest method. *Clean. Prod. Lett.* **2025**, *8*, 100088. [CrossRef]
34. Rodríguez-López, L.; Bustos Usta, D.; Bravo Alvarez, L.; Duran-Llacer, I.; Lami, A.; Martínez-Retureta, R.; Urrutia, R. Machine Learning Algorithms for the Estimation of Water Quality Parameters in Lake Llanquihue in Southern Chile. *Water* **2023**, *15*, 1994. [CrossRef]
35. Ansari, T.; Nigar, N.; Faisal, H.M.; Shahzad, M.K. AI for clean water: Efficient water quality prediction leveraging machine learning. *Water Pract. Technol.* **2024**, *19*, 1986–1996. [CrossRef]
36. Zamani, M.G.; Nikoo, M.R.; Niknazar, F.; Al-Rawas, G.; Al-Wardy, M.; Gandomi, A.H. A multi-model data fusion methodology for reservoir water quality based on machine learning algorithms and bayesian maximum entropy. *J. Clean. Prod.* **2023**, *416*, 137885. [CrossRef]
37. Haghiabi, H.; Nasrolahi, A.H.; Parsaie, A. Water quality prediction using machine learning methods. *Water Qual. Res. J.* **2018**, *53*, 3–13. [CrossRef]
38. Satish, N.; Anmala, J.; Varma, M.R.R.; Rajitha, K. Performance of Machine Learning, Artificial Neural Network (ANN), and stacked ensemble models in predicting Water Quality Index (WQI) from surface water quality parameters, climatic and land use data. *Process Saf. Environ. Prot.* **2024**, *192*, 177–195. [CrossRef]
39. Farzana, S.Z.; Paudyal, D.R.; Chadalavada, S.; Alam, M.J. Temporal Dynamics and Predictive Modelling of Streamflow and Water Quality Using Advanced Statistical and Ensemble Machine Learning Techniques. *Water* **2024**, *16*, 2107. [CrossRef]
40. Uddin, M.G.; Nash, S.; Diganta, M.T.M.; Rahman, A.; Olbert, A.I. Robust machine learning algorithms for predicting coastal water quality index. *J. Environ. Manag.* **2022**, *321*, 115923. [CrossRef]
41. Shamsuddin, I.S.; Othman, Z.; Sani, N.S. Water Quality Index Classification Based on Machine Learning: A Case from the Langat River Basin Model. *Water* **2022**, *14*, 2939. [CrossRef]
42. Singh, S.; Das, A.; Sharma, P. Predictive modeling of water quality index (WQI) classes in Indian rivers: Insights from the application of multiple Machine Learning (ML) models on a decennial dataset. *Stoch. Environ. Res. Risk Assess.* **2024**, *38*, 3221–3238. [CrossRef]
43. Yao, J.; Chen, S.; Ruan, X. Interpretable CEEMDAN-FE-LSTM-transformer hybrid model for predicting total phosphorus concentrations in surface water. *J. Hydrol.* **2024**, *629*, 130609. [CrossRef]
44. Flores, V.; Bravo, I.; Saavedra, M. Water Quality Classification and Machine Learning Model for Predicting Water Quality Status—A Study on Loa River Located in an Extremely Arid Environment: Atacama Desert. *Water* **2023**, *15*, 2868. [CrossRef]
45. Masood, A.; Niazkar, M.; Zakwan, M.; Piraei, R. A Machine Learning-Based Framework for Water Quality Index Estimation in the Southern Bug River. *Water* **2023**, *15*, 3543. [CrossRef]
46. Aju, D.; Achu, A.L.; Mohammed, M.; Raicy, M.C.; Gopinath, G.; Reghunath, R. Groundwater quality prediction and risk assessment in Kerala, India: A machine-learning approach. *J. Environ. Manag.* **2024**, *370*, 122616. [CrossRef] [PubMed]
47. Zhu, M.; Wang, J.; Yang, X.; Zhang, Y.; Zhang, L.; Ren, H.; Wu, B.; Ye, L. A review of the application of machine learning in water quality evaluation. *Eco-Environ. Health* **2022**, *1*, 107–116. [CrossRef]
48. del Castillo, F.; Garibay, M.V.; Díaz-Vázquez, D.; Yebra-Montes, C.; Brown, L.E.; Johnson, A.; Garcia-Gonzalez, A.; Gradilla-Hernández, M.S. Improving river water quality prediction with hybrid machine learning and temporal analysis. *Ecol. Inform.* **2024**, *82*, 102655. [CrossRef]
49. Yan, T.; Zhou, A.; Shen, S.-L. Prediction of long-term water quality using machine learning enhanced by Bayesian optimisation. *Environ. Pollut.* **2023**, *318*, 120870. [CrossRef]
50. Liu, C.; Xu, J.; Li, X.; Yu, Z.; Wu, J. Water resource forecasting with machine learning and deep learning: A scientometric analysis. *Artif. Intell. Geosci.* **2024**, *5*, 100084. [CrossRef]
51. Jayaraman, P.; Nagarajan, K.K.; Partheeban, P.; Krishnamurthy, V. Critical review on water quality analysis using IoT and machine learning models. *Int. J. Inf. Manag. Data Insights* **2024**, *4*, 100210. [CrossRef]
52. Li, W.; Zhao, Y.; Zhu, Y.; Dong, Z.; Wang, F.; Huang, F. Research progress in water quality prediction based on deep learning technology: A review. *Environ. Sci. Pollut. Res.* **2024**, *31*, 26415–26431. [CrossRef]
53. Nordin, N.F.C.; Mohd, N.S.; Koting, S.; Ismail, Z.; Sherif, M.; El-Shafie, A. Groundwater quality forecasting modelling using artificial intelligence: A review. *Groundw. Sustain. Dev.* **2021**, *14*, 100643. [CrossRef]
54. Li, X.; Li, Y.; Li, G. A scientometric review of the research on the impacts of climate change on water quality during 1998–2018. *Environ. Sci. Pollut. Res.* **2020**, *27*, 14322–14341. [CrossRef]
55. Bose, S.; Mazumdar, A.; Basu, S. Evolution of groundwater quality assessment on urban area- a bibliometric analysis. *Groundw. Sustain. Dev.* **2023**, *20*, 100894. [CrossRef]

56. Donthu, N.; Kumar, S.; Mukherjee, D.; Pandey, N.; Lim, W.M. How to conduct a bibliometric analysis: An overview and guidelines. *J. Bus. Res.* **2021**, *133*, 285–296. [CrossRef]
57. Tiyasha; Tung, T.M.; Yaseen, Z.M. A survey on river water quality modelling using artificial intelligence models: 2000–2020. *J. Hydrol.* **2020**, *585*, 124670. [CrossRef]
58. Cojbasic, S.; Dmitrasinovic, S.; Kostic, M.; Sekulic, M.T.; Radonic, J.; Dodig, A.; Stojkovic, M. Application of machine learning in river water quality management: A review. *Water Sci. Technol.* **2023**, *88*, 2297–2308. [CrossRef] [PubMed]
59. Kitchenham, B.; Brereton, O.P.; Budgen, D.; Turner, M.; Bailey, J.; Linkman, S. Systematic literature reviews in software engineering—A systematic literature review. *Inf. Softw. Technol.* **2009**, *51*, 7–15. [CrossRef]
60. Marzi, G.; Balzano, M.; Caputo, A.; Pellegrini, M.M. Guidelines for Bibliometric-Systematic Literature Reviews: 10 steps to combine analysis, synthesis and theory development. *Int. J. Manag. Rev.* **2025**, *27*, 81–103. [CrossRef]
61. Lim, W.M.; Kumar, S.; Donthu, N. How to combine and clean bibliometric data and use bibliometric tools synergistically: Guidelines using metaverse research. *J. Bus. Res.* **2024**, *182*, 114760. [CrossRef]
62. Aria, M.; Cuccurullo, C. bibliometrix: An R-tool for comprehensive science mapping analysis. *J. Informetr.* **2017**, *11*, 959–975. [CrossRef]
63. Mohammed, M.A.; De-Pablos-Heredero, C.; Botella, J.L.M. A Systematic Literature Review on the Revolutionary Impact of Blockchain in Modern Business. *Appl. Sci.* **2024**, *14*, 11077. [CrossRef]
64. Lahami, M.; Maalej, A.J.; Krichen, M. A systematic literature review on dynamic testing of blockchain oriented software. *Sci. Comput. Program.* **2025**, *240*, 103211. [CrossRef]
65. Rousso, B.Z.; Bertone, E.; Stewart, R.; Hamilton, D. A systematic literature review of forecasting and predictive models for cyanobacteria blooms in freshwater lakes. *Water Res.* **2020**, *182*, 115959. [CrossRef]
66. Baas, J.; Schotten, M.; Plume, A.; Côté, G.; Karimi, R. Scopus as a curated, high-quality bibliometric data source for academic research in quantitative science studies. *Quant. Sci. Stud.* **2020**, *1*, 377–386. [CrossRef]
67. Baarimah, O.A.; Bazel, A.M.; Alaloul, S.W.; Alazaiza, D.Y.M.; Al-Zghoul, M.T.; Almuhaya, B.; Khan, A.; Mushtaha, W.A. Artificial intelligence in wastewater treatment: Research trends and future perspectives through bibliometric analysis. *Case Stud. Chem. Environ. Eng.* **2024**, *10*, 100926. [CrossRef]
68. Gusenbauer, M. Searchsmart.org: Guiding researchers to the best databases and search systems for systematic reviews and beyond. *Res. Synth. Methods* **2024**, *15*, 1200–1213. [CrossRef] [PubMed]
69. van Dinter, R.; Tekinerdogan, B.; Catal, C. Automation of systematic literature reviews: A systematic literature review. *Inf. Softw. Technol.* **2021**, *136*, 106589. [CrossRef]
70. *textmineR: Functions for Text Mining and Topic Modeling*, version 3.0.5; CRAN: Venna, Austria, 2021. Available online: https://cran.r-project.org/package=textmineR (accessed on 29 August 2025).
71. Chen, X.; Xie, H.; Tao, X.; Xu, L.; Wang, J.; Dai, H.; Wang, L.F. A topic modeling-based bibliometric exploration of automatic summarization research. *WIREs Data Min. Knowl. Discov.* **2024**, *14*, e1540. [CrossRef]
72. Cobelli, N.; Blasi, S. Combining topic modeling and bibliometric analysis to understand the evolution of technological innovation adoption in the healthcare industry. *Eur. J. Innov. Manag.* **2024**, *27*, 127–149. [CrossRef]
73. Xu, S.; Hao, L.; An, X.; Yang, G.; Wang, F. Emerging research topics detection with multiple machine learning models. *J. Informetr.* **2019**, *13*, 100983. [CrossRef]
74. Kassem, A.; Sefelnasr, A.; Ebraheem, A.A.; Sherif, M. Seawater intrusion physical models: A bibliometric analysis and review of mitigation strategies. *J. Hydrol.* **2024**, *634*, 131135. [CrossRef]
75. Phiri, Z.; Moja, N.T.; Nkambule, T.T.I.; de Kock, L.-A. Utilization of biochar for remediation of heavy metals in aqueous environments: A review and bibliometric analysis. *Heliyon* **2024**, *10*, e25785. [CrossRef]
76. Liu, H.; Kong, F.; Yin, H.; Middel, A.; Zheng, X.; Huang, J.; Xu, H.; Wang, D.; Wen, Z. Impacts of green roofs on water, temperature, and air quality: A bibliometric review. *Build. Environ.* **2021**, *196*, 107794. [CrossRef]
77. Biazatti, J.; Justi, A.C.A.; Souza, R.F.; de Carvalho Miranda, J.C. Soybean biorefinery and technological forecasts based on a bibliometric analysis and network mapping. *Environ. Dev.* **2024**, *52*, 101074. [CrossRef]
78. Batagelj, V.; Cerinšek, M. On bibliographic networks. *Scientometrics* **2013**, *96*, 845–864. [CrossRef]
79. Pandey, H.; Maraseni, T.N.; Apan, A.A. Enhancing systematic literature review adapting 'double diamond approach. *Heliyon* **2024**, *10*, e40581. [CrossRef] [PubMed]
80. Gusenbauer, M.; Gauster, S. How to search for literature in systematic reviews and meta-analyses: A comprehensive step-by-step guide. *Technol. Forecast. Soc. Chang.* **2025**, *212*, 123833. [CrossRef]
81. Petersen, K.; Vakkalanka, S.; Kuzniarz, L. Guidelines for conducting systematic mapping studies in software engineering: An update. *Inf. Softw. Technol.* **2015**, *64*, 1–18. [CrossRef]

82. Sit, M.; Demiray, B.Z.; Xiang, Z.; Ewing, G.J.; Sermet, Y.; Demir, I. A comprehensive review of deep learning applications in hydrology and water resources. *Water Sci. Technol.* **2020**, *82*, 2635–2670. [CrossRef]
83. Huang, R.; Ma, C.; Ma, J.; Huangfu, X.; He, Q. Machine learning in natural and engineered water systems. *Water Res.* **2021**, *205*, 117666. [CrossRef]
84. Tefera, G.W.; Ray, R.L.; Singh, V. Surface water quality under climate change scenarios in the Bosque watershed, Central Texas of United States. *Ecohydrol. Hydrobiol.* **2024**, *25*, 477–492. [CrossRef]
85. Ramya, S.; Srinath, S.; Tuppad, P. Comprehensive analysis of multiple classifiers for enhanced river water quality monitoring with explainable AI. *Case Stud. Chem. Environ. Eng.* **2024**, *10*, 100822. [CrossRef]
86. Sidek, L.M.; Mohiyaden, H.A.; Marufuzzaman, M.; Noh, N.S.M.; Heddam, S.; Ehteram, M.; Kisi, O.; Sammen, S.S. Developing an ensembled machine learning model for predicting water quality index in Johor River Basin. *Environ. Sci. Eur.* **2024**, *36*, 67. [CrossRef]
87. Hirsch, J.E. An index to quantify an individual's scientific research output. *Proc. Natl. Acad. Sci. USA* **2005**, *102*, 16569–16572. [CrossRef]
88. Barzegar, R.; Aalami, M.T.; Adamowski, J. Short-term water quality variable prediction using a hybrid CNN–LSTM deep learning model. *Stoch. Environ. Res. Risk Assess.* **2020**, *34*, 415–433. [CrossRef]
89. Barzegar, R.; Adamowski, J.; Moghaddam, A.A. Application of wavelet-artificial intelligence hybrid models for water quality prediction: A case study in Aji-Chay River, Iran. *Stoch. Environ. Res. Risk Assess.* **2016**, *30*, 1797–1819. [CrossRef]
90. Ho, J.Y.; Afan, H.A.; El-Shafie, A.H.; Koting, S.B.; Mohd, N.S.; Jaafar, W.Z.B.; Hin, L.S.; Malek, M.A.; Ahmed, A.N.; Mohtar, W.H.M.W.; et al. Towards a time and cost effective approach to water quality index class prediction. *J. Hydrol.* **2019**, *575*, 148–165. [CrossRef]
91. Ji, X.; Shang, X.; Dahlgren, R.A.; Zhang, M. Prediction of dissolved oxygen concentration in hypoxic river systems using support vector machine: A case study of Wen-Rui Tang River, China. *Environ. Sci. Pollut. Res.* **2017**, *24*, 16062–16076. [CrossRef]
92. Fijani, E.; Barzegar, R.; Deo, R.; Tziritis, E.; Skordas, K. Design and implementation of a hybrid model based on two-layer decomposition method coupled with extreme learning machines to support real-time environmental monitoring of water quality parameters. *Sci. Total Environ.* **2019**, *648*, 839–853. [CrossRef]
93. Abba, S.I.; Pham, Q.B.; Saini, G.; Linh, N.T.T.; Ahmed, A.N.; Mohajane, M.; Khaledian, M.; Abdulkadir, R.A.; Bach, Q. Implementation of data intelligence models coupled with ensemble machine learning for prediction of water quality index. *Environ. Sci. Pollut. Res.* **2020**, *27*, 41524–41539. [CrossRef]
94. Noori, R.; Yeh, H.-D.; Abbasi, M.; Kachoosangi, F.T.; Moazami, S. Uncertainty analysis of support vector machine for online prediction of five-day biochemical oxygen demand. *J. Hydrol.* **2015**, *527*, 833–843. [CrossRef]
95. Sakaa, B.; Elbeltagi, A.; Boudibi, S.; Chaffaï, H.; Islam, A.R.T.; Kulimushi, L.C.; Choudhari, P.; Hani, A.; Brouziyne, Y.; Wong, Y.J.; et al. Water quality index modeling using random forest and improved SMO algorithm for support vector machine in Saf-Saf river basin. *Environ. Sci. Pollut. Res.* **2022**, *29*, 48491–48508. [CrossRef] [PubMed]
96. Chen, Z.; Xu, H.; Jiang, P.; Yu, S.; Lin, G.; Bychkov, I.; Hmelnov, A.; Ruzhnikov, G.; Zhu, N.; Liu, Z.; et al. A transfer Learning-Based LSTM strategy for imputing Large-Scale consecutive missing data and its application in a water quality prediction system. *J. Hydrol.* **2021**, *602*, 126573. [CrossRef]
97. Jamshidzadeh, Z.; Ehteram, M.; Shabanian, H. Bidirectional Long Short-Term Memory (BILSTM)—Support Vector Machine: A new machine learning model for predicting water quality parameters. *Ain Shams Eng. J.* **2024**, *15*, 102510. [CrossRef]
98. Chen, T.-C. Application of wavelet theory to enhance the performance of machine learning techniques in estimating water quality parameters (case study: Gao-Ping River). *Water Sci. Technol.* **2023**, *87*, 1294–1315. [CrossRef]
99. Wang, K.; Liu, L.; Ben, X.; Jin, D.; Zhu, Y.; Wang, F. Hybrid deep learning based prediction for water quality of plain watershed. *Environ. Res.* **2024**, *262*, 119911. [CrossRef]
100. Manzar, M.S.; Benaafi, M.; Costache, R.; Alagha, O.; Mu'azu, N.D.; Zubair, M.; Abdullahi, J.; Abba, S.I. New generation neurocomputing learning coupled with a hybrid neuro-fuzzy model for quantifying water quality index variable: A case study from Saudi Arabia. *Ecol. Inform.* **2022**, *70*, 101696. [CrossRef]
101. Chen, X.; Sun, W.; Jiang, T.; Ju, H. Enhanced prediction of river dissolved oxygen through feature- and model-based transfer learning. *J. Environ. Manag.* **2024**, *372*, 123310. [CrossRef]
102. Khodkar, K.; Mirchi, A.; Nourani, V.; Kaghazchi, A.; Sadler, J.M.; Mansaray, A.; Wagner, K.; Alderman, P.D.; Taghvaeian, S.; Bailey, R.T.; et al. Stream salinity prediction in data-scarce regions: Application of transfer learning and uncertainty quantification. *J. Contam. Hydrol.* **2024**, *266*, 104418. [CrossRef]
103. Chen, S.; Huang, J.; Wang, P.; Tang, X.; Zhang, Z. A coupled model to improve river water quality prediction towards addressing non-stationarity and data limitation. *Water Res.* **2024**, *248*, 120895. [CrossRef]

104. Peng, L.; Wu, H.; Gao, M.; Yi, H.; Xiong, Q.; Yang, L.; Cheng, S. TLT: Recurrent fine-tuning transfer learning for water quality long-term prediction. *Water Res.* **2022**, *225*, 119171. [CrossRef]
105. Longo, L.; Brcic, M.; Cabitza, F.; Choi, J.; Confalonieri, R.; Del Ser, J.; Guidotti, R.; Hayashi, Y.; Herrera, F.; Holzinger, A.; et al. Explainable Artificial Intelligence (XAI) 2.0: A manifesto of open challenges and interdisciplinary research directions. *Inf. Fusion* **2024**, *106*, 102301. [CrossRef]
106. Núñez, J.; Cortés, C.B.; Yáñez, M.A. Explainable Artificial Intelligence in Hydrology: Interpreting Black-Box Snowmelt-Driven Streamflow Predictions in an Arid Andean Basin of North-Central Chile. *Water* **2023**, *15*, 3369. [CrossRef]
107. Madni, H.A.; Umer, M.; Ishaq, A.; Abuzinadah, N.; Saidani, O.; Alsubai, S.; Hamdi, M.; Ashraf, I. Water-Quality Prediction Based on H2O AutoML and Explainable AI Techniques. *Water* **2023**, *15*, 475. [CrossRef]
108. Alshehri, F.; Rahman, A. Coupling Machine and Deep Learning with Explainable Artificial Intelligence for Improving Prediction of Groundwater Quality and Decision-Making in Arid Region, Saudi Arabia. *Water* **2023**, *15*, 2298. [CrossRef]
109. Haggerty, R.; Sun, J.; Yu, H.; Li, Y. Application of machine learning in groundwater quality modeling—A comprehensive review. *Water Res.* **2023**, *233*, 119745. [CrossRef]
110. Ibrahim, H.; Yaseen, Z.M.; Scholz, M.; Ali, M.; Gad, M.; Elsayed, S.; Khadr, M.; Hussein, H.; Ibrahim, H.H.; Eid, M.H.; et al. Evaluation and Prediction of Groundwater Quality for Irrigation Using an Integrated Water Quality Indices, Machine Learning Models and GIS Approaches: A Representative Case Study. *Water* **2023**, *15*, 694. [CrossRef]
111. Rahat, S.H.; Steissberg, T.; Chang, W.; Chen, X.; Mandavya, G.; Tracy, J.; Wasti, A.; Atreya, G.; Saki, S.; Bhuiyan, M.A.E.; et al. Remote sensing-enabled machine learning for river water quality modeling under multidimensional uncertainty. *Sci. Total Environ.* **2023**, *898*, 165504. [CrossRef]
112. Huangfu, K.; Li, J.; Zhang, X.; Zhang, J.; Cui, H.; Sun, Q. Remote Estimation of Water Quality Parameters of Medium- and Small-Sized Inland Rivers Using Sentinel-2 Imagery. *Water* **2020**, *12*, 3124. [CrossRef]
113. O'Grady, J.; Zhang, D.; O'Connor, N.; Regan, F. A comprehensive review of catchment water quality monitoring using a tiered framework of integrated sensing technologies. *Sci. Total Environ.* **2021**, *765*, 142766. [CrossRef]
114. Bai, Y.; Peng, M.; Wang, M. A River Water Quality Prediction Method Based on Dual Signal Decomposition and Deep Learning. *Water* **2024**, *16*, 3099. [CrossRef]
115. Ahmed, N.; Othman, F.B.; Afan, H.A.; Ibrahim, R.K.; Fai, C.M.; Hossain, M.S.; Ehteram, M.; Elshafie, A. Machine learning methods for better water quality prediction. *J. Hydrol.* **2019**, *578*, 124084. [CrossRef]
116. Zhang, M.; Zhang, Z.; Wang, X.; Liao, Z.; Wang, L. The Use of Attention-Enhanced CNN-LSTM Models for Multi-Indicator and Time-Series Predictions of Surface Water Quality. *Water Resour. Manag.* **2024**, *38*, 6103–6119. [CrossRef]
117. Sadler, J.M.; Koenig, L.E.; Gorski, G.; Carter, A.M.; Hall, R.O., Jr. Evaluating a process-guided deep learning approach for predicting dissolved oxygen in streams. *Hydrol. Process.* **2024**, *38*, e15270. [CrossRef]
118. Rajagopal, S.; Ganesh, S.S.; Karthick, A.; Sampradeepraj, T. Environmental water quality prediction based on COOT-CSO-LSTM deep learning. *Environ. Sci. Pollut. Res.* **2024**, *31*, 54525–54533. [CrossRef] [PubMed]
119. Poursaeid, M.; Poursaeed, A.H.; Shabanlou, S. Water quality fluctuations prediction and Debi estimation based on stochastic optimized weighted ensemble learning machine. *Process Saf. Environ. Prot.* **2024**, *188*, 1160–1174. [CrossRef]
120. Vellingiri, J.; Kalaivanan, K.; Shanmugaiah, K.; Bai, F.J.J.S. AO-SVM: A machine learning model for predicting water quality in the cauvery river. *Environ. Res. Commun.* **2024**, *6*, 075025. [CrossRef]
121. Poluru, R.K.; Sundararajan, S.; Balakrishnan, S.; Rajagopal, M. Predicting nitrous oxide contaminants in Cauvery basin using region-based convolutional neural network. *Groundw. Sustain. Dev.* **2024**, *26*, 101194. [CrossRef]
122. Lin, Z.; Lim, J.Y.; Oh, J.-M. Innovative interpretable AI-guided water quality evaluation with risk adversarial analysis in river streams considering spatial-temporal effects. *Environ. Pollut.* **2024**, *350*, 124015. [CrossRef]
123. Liu, W.; Lin, S.; Li, X.; Li, W.; Deng, H.; Fang, H.; Li, W. Analysis of dissolved oxygen influencing factors and concentration prediction using input variable selection technique: A hybrid machine learning approach. *J. Environ. Manag.* **2024**, *357*, 120777. [CrossRef]
124. Saha, G.; Shen, C.; Duncan, J.; Cibin, R. Performance evaluation of deep learning based stream nitrate concentration prediction model to fill stream nitrate data gaps at low-frequency nitrate monitoring basins. *J. Environ. Manag.* **2024**, *357*, 120721. [CrossRef]
125. Singh, R.B.; Patra, K.C.; Pradhan, B.; Samantra, A. HDTO-DeepAR: A novel hybrid approach to forecast surface water quality indicators. *J. Environ. Manag.* **2024**, *352*, 120091. [CrossRef] [PubMed]
126. Hu, Y.; Liu, C.; Wollheim, W.M. Prediction of riverine daily minimum dissolved oxygen concentrations using hybrid deep learning and routine hydrometeorological data. *Sci. Total Environ.* **2024**, *918*, 170383. [CrossRef] [PubMed]
127. Wang, Z.; Wang, Q.; Liu, Z.; Wu, T. A deep learning interpretable model for river dissolved oxygen multi-step and interval prediction based on multi-source data fusion. *J. Hydrol.* **2024**, *629*, 130637. [CrossRef]

128. Al-Mukhtar, M.; Srivastava, A.; Khadke, L.; Al-Musawi, T.; Elbeltagi, A. Prediction of Irrigation Water Quality Indices Using Random Committee, Discretization Regression, REPTree, and Additive Regression. *Water Resour. Manag.* **2024**, *38*, 343–368. [CrossRef]
129. E, B.; Zhang, S.; Driscoll, C.T.; Wen, T. Human and natural impacts on the U.S. freshwater salinization and alkalinization: A machine learning approach. *Sci. Total Environ.* **2023**, *889*, 164138. [CrossRef]
130. Luo, L.; Zhang, Y.; Dong, W.; Zhang, J.; Zhang, L. Ensemble Empirical Mode Decomposition and a Long Short-Term Memory Neural Network for Surface Water Quality Prediction of the Xiaofu River, China. *Water* **2023**, *15*, 1625. [CrossRef]
131. Xu, R.; Wu, W.; Cai, Y.; Wan, H.; Li, J.; Zhu, Q.; Shen, S. Feature Extraction and Prediction of Water Quality Based on Candlestick Theory and Deep Learning Methods. *Water* **2023**, *15*, 845. [CrossRef]
132. Im, Y.; Song, G.; Lee, J.; Cho, M. Deep Learning Methods for Predicting Tap-Water Quality Time Series in South Korea. *Water* **2022**, *14*, 3766. [CrossRef]
133. Hoque, J.M.Z.; Aziz, N.A.A.; Alelyani, S.; Mohana, M.; Hosain, M. Improving Water Quality Index Prediction Using Regression Learning Models. *Int. J. Environ. Res. Public Health* **2022**, *19*, 13702. [CrossRef]
134. Dorado-Guerra, D.Y.; Corzo-Pérez, G.; Paredes-Arquiola, J.; Pérez-Martín, M.Á. Machine learning models to predict nitrate concentration in a river basin. *Environ. Res. Commun.* **2023**, *4*, 125012. [CrossRef]
135. Adedeji, C.; Ahmadisharaf, E.; Sun, Y. Predicting in-stream water quality constituents at the watershed scale using machine learning. *J. Contam. Hydrol.* **2022**, *251*, 104078. [CrossRef]
136. Khosravi, K.; Golkarian, A.; Melesse, A.M.; Deo, R.C. Suspended sediment load modeling using advanced hybrid rotation forest based elastic network approach. *J. Hydrol.* **2022**, *610*, 127963. [CrossRef]
137. Song, C.; Yao, L. A hybrid model for water quality parameter prediction based on CEEMDAN-IALO-LSTM ensemble learning. *Environ. Earth Sci.* **2022**, *81*, 262. [CrossRef]
138. Hou, Y.; Zhang, A.; Lv, R.; Zhao, S.; Ma, J.; Zhang, H.; Li, Z. A study on water quality parameters estimation for urban rivers based on ground hyperspectral remote sensing technology. *Environ. Sci. Pollut. Res.* **2022**, *29*, 63640–63654. [CrossRef] [PubMed]
139. Balson, T.; Ward, A.S. A machine learning approach to water quality forecasts and sensor network expansion: Case study in the Wabash River Basin, United States. *Hydrol. Process.* **2022**, *36*, e14619. [CrossRef]
140. Malek, H.A.; Yaacob, W.F.W.; Nasir, S.A.M.; Shaadan, N. Prediction of Water Quality Classification of the Kelantan River Basin, Malaysia, Using Machine Learning Techniques. *Water* **2022**, *14*, 1067. [CrossRef]
141. Rizal, N.N.M.; Hayder, G.; Yusof, K.A. Water Quality Predictive Analytics Using an Artificial Neural Network with a Graphical User Interface. *Water* **2022**, *14*, 1221. [CrossRef]
142. Weierbach, H.; Lima, A.R.; Willard, J.D.; Hendrix, V.C.; Christianson, D.S.; Lubich, M.; Varadharajan, C. Stream Temperature Predictions for River Basin Management in the Pacific Northwest and Mid-Atlantic Regions Using Machine Learning. *Water* **2022**, *14*, 1032. [CrossRef]
143. del Castillo, A.; Yebra-Montes, C.; Garibay, M.V.; de Anda, J.; Garcia-Gonzalez, A.; Gradilla-Hernández, M.S. Simple Prediction of an Ecosystem-Specific Water Quality Index and the Water Quality Classification of a Highly Polluted River through Supervised Machine Learning. *Water* **2022**, *14*, 1235. [CrossRef]
144. Ilić, M.; Srdjević, Z.; Srdjević, B. Water quality prediction based on Naïve Bayes algorithm. *Water Sci. Technol.* **2022**, *85*, 1027–1039. [CrossRef]
145. Arora, S.; Keshari, A.K. Dissolved oxygen modelling of the Yamuna River using different ANFIS models. *Water Sci. Technol.* **2021**, *84*, 3359–3371. [CrossRef]
146. Moghadam, S.V.; Sharafati, A.; Feizi, H.; Marjaie, S.M.S.; Asadollah, S.B.H.S.; Motta, D. An efficient strategy for predicting river dissolved oxygen concentration: Application of deep recurrent neural network model. *Environ. Monit. Assess.* **2021**, *193*, 798. [CrossRef]
147. Xu, C.; Chen, X.; Zhang, L. Predicting river dissolved oxygen time series based on stand-alone models and hybrid wavelet-based models. *J. Environ. Manag.* **2021**, *295*, 113085. [CrossRef]
148. Najah, A.; Teo, F.Y.; Chow, M.F.; Huang, Y.F.; Latif, S.D.; Abdullah, S.; Ismail, M.; El-Shafie, A. Surface water quality status and prediction during movement control operation order under COVID-19 pandemic: Case studies in Malaysia. *Int. J. Environ. Sci. Technol.* **2021**, *18*, 1009–1018. [CrossRef]
149. Kim, S.; Maleki, N.; Rezaie-Balf, M.; Singh, V.P.; Alizamir, M.; Kim, N.W.; Lee, J.T.; Kisi, O. Assessment of the total organic carbon employing the different nature-inspired approaches in the Nakdong River, South Korea. *Environ. Monit. Assess.* **2021**, *193*, 445. [CrossRef] [PubMed]
150. Yan, J.; Liu, J.; Yu, Y.; Xu, H. Water Quality Prediction in the Luan River Based on 1-DRCNN and BiGRU Hybrid Neural Network Model. *Water* **2021**, *13*, 1273. [CrossRef]

151. Setshedi, J.; Mutingwende, N.; Ngqwala, N. The Use of Artificial Neural Networks to Predict the Physicochemical Characteristics of Water Quality in Three District Municipalities, Eastern Cape Province, South Africa. *Int. J. Environ. Res. Public Health* **2021**, *18*, 5248. [CrossRef] [PubMed]
152. Abba, S.I.; Abdulkadir, R.A.; Sammen, S.S.; Usman, A.G.; Meshram, S.G.; Malik, A.; Shahid, S. Comparative implementation between neuro-emotional genetic algorithm and novel ensemble computing techniques for modelling dissolved oxygen concentration. *Hydrol. Sci. J.* **2021**, *66*, 1584–1596. [CrossRef]
153. Sha, J.; Li, X.; Zhang, M.; Wang, Z.-L. Comparison of Forecasting Models for Real-Time Monitoring of Water Quality Parameters Based on Hybrid Deep Learning Neural Networks. *Water* **2021**, *13*, 1547. [CrossRef]
154. Zhang, Y.-F.; Fitch, P.; Thorburn, J. Predicting the Trend of Dissolved Oxygen Based on the kPCA-RNN Model. *Water* **2020**, *12*, 585. [CrossRef]
155. Baek, S.-S.; Pyo, J.; Chun, J.A. Prediction of Water Level and Water Quality Using a CNN-LSTM Combined Deep Learning Approach. *Water* **2020**, *12*, 3399. [CrossRef]
156. Jamei, M.; Ahmadianfar, I.; Chu, X.; Yaseen, Z.M. Prediction of surface water total dissolved solids using hybridized wavelet-multigene genetic programming: New approach. *J. Hydrol.* **2020**, *589*, 125335. [CrossRef]
157. Bui, D.T.; Khosravi, K.; Tiefenbacher, J.; Nguyen, H.; Kazakis, N. Improving prediction of water quality indices using novel hybrid machine-learning algorithms. *Sci. Total Environ.* **2020**, *721*, 137612. [CrossRef] [PubMed]
158. Chen, S.; Fang, G.; Huang, X.; Zhang, Y. Water Quality Prediction Model of a Water Diversion Project Based on the Improved Artificial Bee Colony–Backpropagation Neural Network. *Water* **2018**, *10*, 806. [CrossRef]
159. Raheli, B.; Aalami, M.T.; El-Shafie, A.; Ghorbani, M.A.; Deo, R.C. Uncertainty assessment of the multilayer perceptron (MLP) neural network model with implementation of the novel hybrid MLP-FFA method for prediction of biochemical oxygen demand and dissolved oxygen: A case study of Langat River. *Environ. Earth Sci.* **2017**, *76*, 503. [CrossRef]
160. Stoica, C.; Camejo, J.; Banciu, A.; Nita-Lazar, M.; Paun, I.; Cristofor, S.; Pacheco, O.R.; Guevara, M. Water quality of Danube Delta systems: Ecological status and prediction using machine-learning algorithms. *Water Sci. Technol.* **2016**, *73*, 2413–2421. [CrossRef]
161. Zhang, Q.; You, X. Recent Advances in Surface Water Quality Prediction Using Artificial Intelligence Models. *Water Resour. Manag.* **2024**, *38*, 235–250. [CrossRef]
162. Gómez-Escalonilla, V.; Montero-González, E.; Díaz-Alcaide, S.; Martín-Loeches, M.; del Rosario, M.R.; Martínez-Santos, P. A machine learning approach to site groundwater contamination monitoring wells. *Appl. Water Sci.* **2024**, *14*, 250. [CrossRef]
163. Zhang, J.; Xiao, C.; Yang, W.; Liang, X.; Zhang, L.; Wang, X.; Dai, R. Improving prediction of groundwater quality in situations of limited monitoring data based on virtual sample generation and Gaussian process regression. *Water Res.* **2024**, *267*, 122498. [CrossRef]
164. Jeong, H.; Abbas, A.; Kim, H.G.; Hoan, H.V.; Tuan, P.V.; Long, P.T.; Lee, E.; Cho, K.H. Spatial prediction of groundwater salinity in multiple aquifers of the Mekong Delta region using explainable machine learning models. *Water Res.* **2024**, *266*, 122404. [CrossRef]
165. Li, X.; Liang, G.; Wang, L.; Yang, Y.; Li, Y.; Li, Z.; He, B.; Wang, G. Identifying the spatial pattern and driving factors of nitrate in groundwater using a novel framework of interpretable stacking ensemble learning. *Environ. Geochem. Health* **2024**, *46*, 482. [CrossRef] [PubMed]
166. Boufekane, A.; Meddi, M.; Maizi, D.; Busico, G. Performance of artificial intelligence model (LSTM model) for estimating and predicting water quality index for irrigation purposes in order to improve agricultural production. *Environ. Monit. Assess.* **2024**, *196*, 1049. [CrossRef] [PubMed]
167. Lal, A.; Sharan, A.; Sharma, K.; Ram, A.; Roy, D.K.; Datta, B. Scrutinizing different predictive modeling validation methodologies and data-partitioning strategies: New insights using groundwater modeling case study. *Environ. Monit. Assess.* **2024**, *196*, 623. [CrossRef] [PubMed]
168. Khan, I.; Ayaz, M. Sensitivity analysis-driven machine learning approach for groundwater quality prediction: Insights from integrating ENTROPY and CRITIC methods. *Groundw. Sustain. Dev.* **2024**, *26*, 101309. [CrossRef]
169. Cao, W.; Zhang, Z.; Fu, Y.; Zhao, L.; Ren, Y.; Nan, T.; Guo, H. Prediction of arsenic and fluoride in groundwater of the North China Plain using enhanced stacking ensemble learning. *Water Res.* **2024**, *259*, 121848. [CrossRef]
170. Das, R.; Das, S. Coastal groundwater quality prediction using objective-weighted WQI and machine learning approach. *Environ. Sci. Pollut. Res.* **2024**, *31*, 1943–19457. [CrossRef]
171. Chatterjee, T.; Gogoi, U.R.; Samanta, A.; Chatterjee, A.; Singh, M.K.; Pasupuleti, S. Identifying the Most Discriminative Parameter for Water Quality Prediction Using Machine Learning Algorithms. *Water* **2024**, *16*, 481. [CrossRef]
172. Tesoriero, J.; Wherry, S.A.; Dupuy, D.I.; Johnson, T.D. Predicting Redox Conditions in Groundwater at a National Scale Using Random Forest Classification. *Environ. Sci. Technol.* **2024**, *58*, 5079–5092. [CrossRef]

173. Elzain, H.E.; Abdalla, O.; Ahmed, H.A.; Kacimov, A.; Al-Maktoumi, A.; Al-Higgi, K.; Abdallah, M.; Yassin, M.A.; Senapathi, V. An innovative approach for predicting groundwater TDS using optimized ensemble machine learning algorithms at two levels of modeling strategy. *J. Environ. Manag.* **2024**, *351*, 119896. [CrossRef]
174. Iqbal, J.; Su, C.; Ahmad, M.; Baloch, M.Y.J.; Rashid, A.; Ullah, Z.; Abbas, H.; Nigar, A.; Ali, A.; Ullah, A.; et al. Hydrogeochemistry and prediction of arsenic contamination in groundwater of Vehari, Pakistan: Comparison of artificial neural network, random forest and logistic regression models. *Environ. Geochem. Health* **2023**, *46*, 14. [CrossRef]
175. Krishnamoorthy, L.; Lakshmanan, V.R. Groundwater quality assessment using machine learning models: A comprehensive study on the industrial corridor of a semi-arid region. *Environ. Sci. Pollut. Res.* **2024**. [CrossRef] [PubMed]
176. Mahboobi, H.; Shakiba, A.; Mirbagheri, B. Improving groundwater nitrate concentration prediction using local ensemble of machine learning models. *J. Environ. Manag.* **2023**, *345*, 118782. [CrossRef] [PubMed]
177. Sajib, M.; Diganta, M.T.M.; Rahman, A.; Dabrowski, T.; Olbert, A.I.; Uddin, M.G. Developing a novel tool for assessing the groundwater incorporating water quality index and machine learning approach. *Groundw. Sustain. Dev.* **2023**, *23*, 101049. [CrossRef]
178. Masoudi, R.; Mousavi, S.R.; Rahimabadi, D.; Panahi, M.; Rahmani, A. Assessing data mining algorithms to predict the quality of groundwater resources for determining irrigation hazard. *Environ. Monit. Assess.* **2023**, *195*, 319. [CrossRef]
179. Liu, C.; Xu, M.; Liu, Y.; Li, X.; Pang, Z.; Miao, S. Predicting Groundwater Indicator Concentration Based on Long Short-Term Memory Neural Network: A Case Study. *Int. J. Environ. Res. Public Health* **2022**, *19*, 15612. [CrossRef]
180. Huynh, T.-M.-T.; Ni, C.-F.; Su, Y.-S.; Nguyen, V.-C.-N.; Lee, I.-H.; Lin, C.-P.; Nguyen, H.-H. Predicting Heavy Metal Concentrations in Shallow Aquifer Systems Based on Low-Cost Physiochemical Parameters Using Machine Learning Techniques. *Int. J. Environ. Res. Public Health* **2022**, *19*, 12180. [CrossRef]
181. Taşan, M.; Taşan, S.; Demir, Y. Estimation and uncertainty analysis of groundwater quality parameters in a coastal aquifer under seawater intrusion: A comparative study of deep learning and classic machine learning methods. *Environ. Sci. Pollut. Res.* **2023**, *30*, 2866–2890. [CrossRef]
182. Banerjee, K.; Bali, V.; Nawaz, N.; Bali, S.; Mathur, S.; Mishra, R.K.; Rani, S. A Machine-Learning Approach for Prediction of Water Contamination Using Latitude, Longitude, and Elevation. *Water* **2022**, *14*, 728. [CrossRef]
183. Kouadri, S.; Pande, C.B.; Panneerselvam, B.; Moharir, K.N.; Elbeltagi, A. Prediction of irrigation groundwater quality parameters using ANN, LSTM, and MLR models. *Environ. Sci. Pollut. Res.* **2022**, *29*, 21067–21091. [CrossRef]
184. Messier, K.; Wheeler, D.C.; Flory, A.R.; Jones, R.R.; Patel, D.; Nolan, B.T.; Ward, M.H. Modeling groundwater nitrate exposure in private wells of North Carolina for the Agricultural Health Study. *Sci. Total Environ.* **2019**, *655*, 512–519. [CrossRef]
185. Sakizadeh, M. Spatial analysis of total dissolved solids in Dezful Aquifer: Comparison between universal and fixed rank kriging. *J. Contam. Hydrol.* **2019**, *221*, 26–34. [CrossRef] [PubMed]
186. Abbas, F.; Cai, Z.; Shoaib, M.; Iqbal, J.; Ismail, M.; Arifullah; Alrefaei, A.F.; Albeshr, M.F. Machine Learning Models for Water Quality Prediction: A Comprehensive Analysis and Uncertainty Assessment in Mirpurkhas, Sindh, Pakistan. *Water* **2024**, *16*, 941. [CrossRef]
187. Zounemat-Kermani, M.; Batelaan, O.; Fadaee, M.; Hinkelmann, R. Ensemble machine learning paradigms in hydrology: A review. *J. Hydrol.* **2021**, *598*, 126266. [CrossRef]
188. Aldrees, A.; Awan, H.H.; Javed, M.F.; Mohamed, A.M. Prediction of water quality indexes with ensemble learners: Bagging and boosting. *Process Saf. Environ. Prot.* **2022**, *168*, 344–361. [CrossRef]
189. Chen, Y.; Yao, K.; Zhu, B.; Gao, Z.; Xu, J.; Li, Y.; Hu, Y.; Lin, F.; Zhang, X. Water Quality Inversion of a Typical Rural Small River in Southeastern China Based on UAV Multispectral Imagery: A Comparison of Multiple Machine Learning Algorithms. *Water* **2024**, *16*, 553. [CrossRef]
190. Najah, A.; El-Shafie, A.; Karim, O.A.; El-Shafie, A.H. Application of artificial neural networks for water quality prediction. *Neural Comput. Appl.* **2013**, *22*, 187–201. [CrossRef]
191. Gulati, S.; Pal, A. Tuning fuzzy logic controller with SGWO for river water quality modelling. *Mater. Today Proc.* **2022**, *54*, 733–737. [CrossRef]
192. Dilipkumar, J.; Shanmugam, P. Fuzzy-based global water quality assessment and water quality cells identification using satellite data. *Mar. Pollut. Bull.* **2023**, *193*, 115148. [CrossRef]
193. Kalyakulina, A.; Yusipov, I.; Moskalev, A.; Franceschi, C.; Ivanchenko, M. eXplainable Artificial Intelligence (XAI) in aging clock models. *Ageing Res. Rev.* **2024**, *93*, 102144. [CrossRef]
194. Nallakaruppan, K.; Gangadevi, E.; Shri, M.L.; Balusamy, B.; Bhattacharya, S.; Selvarajan, S. Reliable water quality prediction and parametric analysis using explainable AI models. *Sci. Rep.* **2024**, *14*, 7520. [CrossRef]
195. Kundu, S.; Datta, P.; Pal, P.; Ghosh, K.; Das, A.; Das, B.K. Unveiling the hidden connections: Using explainable artificial intelligence to assess water quality criteria in nine giant rivers. *J. Clean. Prod.* **2025**, *492*, 144861. [CrossRef]

196. Nong, X.; Lai, C.; Chen, L.; Wei, J. A novel coupling interpretable machine learning framework for water quality prediction and environmental effect understanding in different flow discharge regulations of hydro-projects. *Sci. Total Environ.* **2024**, *950*, 175281. [CrossRef] [PubMed]
197. Juna, A.; Umer, M.; Sadiq, S.; Karamti, H.; Eshmawi, A.A.; Mohamed, A.; Ashraf, I. Water Quality Prediction Using KNN Imputer and Multilayer Perceptron. *Water* **2022**, *14*, 2592. [CrossRef]
198. Ahmed, U.; Mumtaz, R.; Anwar, H.; Shah, A.A.; Irfan, R.; García-Nieto, J. Efficient Water Quality Prediction Using Supervised Machine Learning. *Water* **2019**, *11*, 2210. [CrossRef]
199. Pany, R.; Rath, A.; Swain, C. Water quality assessment for River Mahanadi of Odisha, India using statistical techniques and Artificial Neural Networks. *J. Clean. Prod.* **2023**, *417*, 137713. [CrossRef]
200. Yang, S.; Luo, D.; Tan, J.; Li, S.; Song, X.; Xiong, R.; Wang, J.; Ma, C.; Xiong, H. Spatial Mapping and Prediction of Groundwater Quality Using Ensemble Learning Models and SHapley Additive exPlanations with Spatial Uncertainty Analysis. *Water* **2024**, *16*, 2375. [CrossRef]
201. Heydari, S.; Nikoo, M.R.; Mohammadi, A.; Barzegar, R. Two-stage meta-ensembling machine learning model for enhanced water quality forecasting. *J. Hydrol.* **2024**, *641*, 131767. [CrossRef]
202. Rodríguez-López, L.; Usta, D.B.; Duran-Llacer, I.; Alvarez, L.B.; Yépez, S.; Bourrel, L.; Frappart, F.; Urrutia, R. Estimation of Water Quality Parameters through a Combination of Deep Learning and Remote Sensing Techniques in a Lake in Southern Chile. *Remote Sens.* **2023**, *15*, 4157. [CrossRef]
203. Zhang, T.; Wu, J.; Chu, H.; Liu, J.; Wang, G. Interpretable Machine Learning Based Quantification of the Impact of Water Quality Indicators on Groundwater Under Multiple Pollution Sources. *Water* **2025**, *17*, 905. [CrossRef]
204. Yao, Z.; Wang, Z.; Huang, J.; Xu, N.; Cui, X.; Wu, T. Interpretable prediction, classification and regulation of water quality: A case study of Poyang Lake, China. *Sci. Total Environ.* **2024**, *951*, 175407. [CrossRef]
205. Liang, Y.; Ding, F.; Liu, L.; Yin, F.; Hao, M.; Kang, T.; Zhao, C.; Wang, Z.; Jiang, D. Monitoring water quality parameters in urban rivers using multi-source data and machine learning approach. *J. Hydrol.* **2025**, *648*, 132394. [CrossRef]
206. Shadkani, S.; Hemmatzadeh, Y.; Saber, A.; Sergini, M.M. Enhanced predictive modeling of dissolved oxygen concentrations in riverine systems using novel hybrid temporal pattern attention deep neural networks. *Environ. Res.* **2024**, *263*, 120015. [CrossRef] [PubMed]
207. Abuzir, S.Y.; Abuzir, Y.S. Machine learning for water quality classification. *Water Qual. Res. J.* **2022**, *57*, 152–164. [CrossRef]
208. Zhang, Q.; Li, Z.; Zhu, L.; Zhang, F.; Sekerinski, E.; Han, J.C.; Zhou, Y. Real-time prediction of river chloride concentration using ensemble learning. *Environ. Pollut.* **2021**, *291*, 118116. [CrossRef]
209. Fertikh, S.; Boutaghane, H.; Boumaaza, M.; Belaadi, A.; Bouslah, S. Assessment and prediction of water quality indices by machine learning-genetic algorithm and response surface methodology. *Model. Earth Syst. Environ.* **2024**, *10*, 5573–5604. [CrossRef]
210. Fooladi, M.; Nikoo, M.R.; Mirghafari, R.; Madramootoo, C.A.; Al-Rawas, G.; Nazari, R. Robust clustering-based hybrid technique enabling reliable reservoir water quality prediction with uncertainty quantification and spatial analysis. *J. Environ. Manag.* **2024**, *362*, 121259. [CrossRef]
211. Ghashghaie, M.; Eslami, H.; Ostad-Ali-Askari, K. Applications of time series analysis to investigate components of Madiyan-rood river water quality. *Appl. Water Sci.* **2022**, *12*, 202. [CrossRef]
212. Bojer, K.; Biru, B.H.; Al-Quraishi, A.M.F.; Debelee, T.G.; Negera, W.G.; Woldesillasie, F.F.; Esubalew, S.Z. Machine learning and remote sensing based time series analysis for drought risk prediction in Borena Zone, Southwest Ethiopia. *J. Arid Environ.* **2024**, *222*, 105160. [CrossRef]
213. Huan, S. A novel interval decomposition correlation particle swarm optimization-extreme learning machine model for short-term and long-term water quality prediction. *J. Hydrol.* **2023**, *625*, 130034. [CrossRef]
214. Yadav, A.; Raj, A.; Yadav, B. Enhancing local-scale groundwater quality predictions using advanced machine learning approaches. *J. Environ. Manag.* **2024**, *370*, 122903. [CrossRef]
215. Hasani, S.S.; Arias, M.E.; Nguyen, H.Q.; Tarabih, O.M.; Welch, Z.; Zhang, Q. Leveraging explainable machine learning for enhanced management of lake water quality. *J. Environ. Manag.* **2024**, *370*, 122890. [CrossRef]
216. Ezzat, D.; Soliman, M.; Ahmed, E.; Hassanien, A.E. An optimized explainable artificial intelligence approach for sustainable clean water. *Environ. Dev. Sustain.* **2024**, *26*, 25899–25919. [CrossRef]
217. Maroufpoor, S.; Jalali, M.; Nikmehr, S.; Shiri, N.; Shiri, J.; Maroufpoor, E. Modeling groundwater quality by using hybrid intelligent and geostatistical methods. *Environ. Sci. Pollut. Res.* **2020**, *27*, 28183–28197. [CrossRef]
218. Shah, M.I.; Javed, M.F.; Alqahtani, A.; Aldrees, A. Environmental assessment based surface water quality prediction using hyper-parameter optimized machine learning models based on consistent big data. *Process Saf. Environ. Prot.* **2021**, *151*, 324–340. [CrossRef]

219. Moayedi, H.; Salari, M.; Ali, S.A.-J.; Dehrashid, A.A.; Azadi, H. Modeling the total hardness (TH) of groundwater in aquifers using novel hybrid soft computing optimizer models. *Environ. Earth Sci.* **2024**, *83*, 392. [CrossRef]
220. Kaya, Y. Slope-aware and self-adaptive forecasting of water levels: A transparent model for the Great Lakes under climate variability. *J. Hydrol.* **2025**, *662*, 133948. [CrossRef]
221. Egbemhenghe, U.; Ojeyemi, T.; Iwuozor, K.O.; Emenike, E.C.; Ogunsanya, T.I.; Anidiobi, S.U.; Adeniyi, A.G. Revolutionizing water treatment, conservation, and management: Harnessing the power of AI-driven ChatGPT solutions. *Environ. Chall.* **2023**, *13*, 100782. [CrossRef]
222. Jiao, J.; Ma, Q.; Huang, S.; Liu, F.; Wan, Z. A hybrid water quality prediction model based on variational mode decomposition and bidirectional gated recursive unit. *Water Sci. Technol.* **2024**, *89*, 2273–2289. [CrossRef]
223. Makumbura, R.K.; Mampitiya, L.; Rathnayake, N.; Meddage, D.P.P.; Henna, S.; Dang, T.L.; Hoshino, Y.; Rathnayake, U. Advancing water quality assessment and prediction using machine learning models, coupled with explainable artificial intelligence (XAI) techniques like shapley additive explanations (SHAP) for interpreting the black-box nature. *Results Eng.* **2024**, *23*, 102831. [CrossRef]
224. Bordbar, M.; Busico, G.; Sirna, M.; Tedesco, D.; Mastrocicco, M. A multi-step approach to evaluate the sustainable use of groundwater resources for human consumption and agriculture. *J. Environ. Manag.* **2023**, *347*, 119041. [CrossRef]
225. Lee, J.M.; Ko, K.-S.; Yoo, K. A machine learning-based approach to predict groundwater nitrate susceptibility using field measurements and hydrogeological variables in the Nonsan Stream Watershed, South Korea. *Appl. Water Sci.* **2023**, *13*, 242. [CrossRef]
226. Zheng, H.; Liu, Y.; Wan, W.; Zhao, J.; Xie, G. Large-scale prediction of stream water quality using an interpretable deep learning approach. *J. Environ. Manag.* **2023**, *331*, 117309. [CrossRef]
227. Zhang, Z.; Huang, J.; Duan, S.; Huang, Y.; Cai, J.; Bian, J. Use of interpretable machine learning to identify the factors influencing the nonlinear linkage between land use and river water quality in the Chesapeake Bay watershed. *Ecol. Indic.* **2022**, *140*, 108977. [CrossRef]

water

MDPI

Article

Groundwater Level Estimation Using Improved Transformer Model: A Case Study of the Yellow River Basin

Tianming Zhou [1,2], Chun Fu [1,2,3,*], Yezhong Liu [1,2] and Libin Xiang [1,2]

[1] School of Engineering Construction, Nanchang University, Nanchang 330031, China
[2] Jiangxi Institute of Regional Economy, Nanchang University, Nanchang 330031, China
[3] School of Public Policy and Administration, Nanchang University, Nanchang 330031, China
* Correspondence: ccfu@ncu.edu.cn

Abstract: Accurate estimation of groundwater levels in river basins is essential for effective water resource planning. Innovations in deep learning and artificial intelligence (AI) have been introduced into this field to enhance the accuracy of long-term groundwater level estimation. This study employs the Transformer deep learning model to estimate groundwater levels, with a benchmark comparison against the long short-term memory (LSTM) model. These models were applied to estimate groundwater levels in the Yellow River Basin, where approximately 1100 monitoring wells are located. Monthly average groundwater level data from the period 2018–2023 were collected from these wells. The two models were used to estimate groundwater levels for the period 2003–2017 by incorporating remote sensing information. The Transformer model was enhanced to simultaneously capture features from both historical temporal data and surrounding spatial data, while automatically enhancing key features, effectively improving estimation accuracy and robustness. At the basin-averaged scale, the enhanced Transformer model outperformed the LSTM model: R^2 increased by approximately 17.5%, while RMSE and MAE decreased by approximately 12.4% and 10.9%, respectively. The proportion of poorly predicted samples decreased by an average of approximately 12.1%. The estimation model established in this study contributes to improving the quantitative analysis capability of long-term groundwater level variations in the Yellow River Basin. This could be helpful for water resource development planning in this densely populated region and likely has broad applicability in other river basins.

Keywords: GRACE; groundwater; deep learning; transformer; estimation; time series prediction; Yellow River Basin

1. Introduction

Groundwater constitutes a critical global water resource, providing drinking water to approximately half of the world's population [1] and supporting around 33% of global irrigation water demand [2]. However, driven by continuous population growth and exacerbated by climate change [3], unsustainable groundwater over-extraction has emerged as a widespread phenomenon. This leads to the rapid depletion of groundwater resources in many regions worldwide, surpassing anticipated rates [4], thereby posing a severe threat to both human societal development and natural ecosystem protection. Consequently, accurately assessing the dynamic trends of groundwater storage changes becomes essential for ensuring water security and achieving regional sustainable development.

Groundwater monitoring has traditionally relied on in situ measurement methods, such as monitoring wells [5]. But these approaches entail substantial labor and equipment costs and exhibit limited spatial and temporal coverage, presenting significant challenges for acquiring continuous, long-term groundwater level data across large regional scales. Within this context, remote sensing technologies offer a critical alternative. Current satellite systems, including GRACE/GRACE-FO [6], GLDAS, SMAP, and NISAR, have demonstrated substantial utility for groundwater level estimation and hold promise for broad application.

Groundwater level estimation primarily relies on two types of methods: physical models and data-driven models. Physical models are typically based on the groundwater-subsidence coupling theory [7], the groundwater–surface water hydrological coupling theory [8], or the Darcy's law–mass conservation law coupling model [9]. These models require a large amount of precise hydrological observation data for parameter calibration to ensure the reliability of the simulation results. However, physical models face significant challenges: the required hydrological parameters are diverse and demand high accuracy, and their calibration process involves complex reasoning and iterative calculations [10]. Additionally, uncertainties inherent in the physical models, such as the choice of parameter types and boundary condition settings, can significantly constrain the interpretability of the estimation results [11].

Machine learning methods, particularly deep learning techniques, have demonstrated significant utility in time series estimation tasks in recent years [12]. Among these, long short-term memory (LSTM) networks utilize gating mechanisms to selectively retain or discard information, enabling effective capture of temporal dependencies within extended sequences. This capability has established LSTM as a widely applied approach for groundwater level estimation; researchers have successfully employed it using actual monitoring data [13]. But LSTM exhibits notable limitations in modeling long-term temporal dependencies: its relatively simple gating architecture is susceptible to challenges such as vanishing or exploding gradients [14]. Furthermore, its performance is constrained in scenarios characterized by severe data sparsity or sequences exhibiting strong nonlinearity.

The Transformer model is a deep learning architecture based on the self-attention mechanism [15], which abandons the traditional Recurrent Neural Network (RNN) and Convolutional Neural Network (CNN) structures, instead adopting a multi-head self-attention mechanism to capture global dependencies between arbitrary elements in the input sequence, particularly excelling at processing long sequence data. This model has been widely applied in the field of Natural Language Processing (NLP) for tasks such as machine translation, text generation, and question answering systems [16]. In recent years, its powerful self-attention mechanism has attracted the attention of time series estimation researchers [17]. Researchers treat time series as sequential data and utilize the Transformer to capture complex long-term patterns and correlations within the sequence. These explorations have demonstrated the potential of the Transformer in handling complex long-sequence time prediction tasks. Moreover, the successful applications of Transformer in hydrology, meteorology, and other earth science fields [18,19] further indicate its applicability to groundwater level estimation tasks. This study aims to leverage the advantages of the Transformer model in complex time series modeling, particularly its strong representation capability of long-sequence temporal features through the self-attention mechanism, with the goal of improving groundwater level estimation accuracy and exploring its new application value beyond traditional fields such as Natural Language Processing.

The Transformer model serves as the foundational architecture for this study. To address specific characteristics of groundwater level estimation, we implement the following enhancements. Groundwater level fluctuations are governed by both temporal variations within the local region and hydrological interactions with surrounding areas [20], but previous studies on groundwater time series estimation have predominantly focused on local temporal dynamics [21], with limited incorporation of spatial interdependencies between regions. This limitation may particularly impact areas characterized by sparse observation points.

To effectively integrate information from surrounding regions into the estimation process, we propose a Dynamic–Static Feature Fusion (DSFF) mechanism tailored for the Transformer architecture. This mechanism utilizes the contextual relationships between input keys to guide the learning of a dynamic attention matrix, thereby capturing intrinsic temporal dependencies within the target region. Concurrently, a 3×3 convolutional layer extracts static spatial–contextual information from adjacent areas. The fused static and dynamic contextual representations subsequently serve as the input for modeling. This enhancement successfully augments the model's information acquisition capability.

Groundwater level variations are governed by multiple driving factors, with dominant influences and their relative contributions exhibiting significant spatial heterogeneity [22]. While prior studies have addressed factor influence, they predominantly employed correlation analyses of input hydrological features (e.g., Pearson's correlation coefficient) [23], providing relatively limited insight into the dynamic weight allocation mechanisms within temporal feature representations. This inherent constraint fundamentally restricts model interpretability during estimation.

Consequently, when applying the Transformer architecture for groundwater level estimation, the dynamic evaluation and utilization of feature channel weights constitutes a critical challenge. To address this, we design a Key Feature Enhancement Architecture (KFEA) tailored for the Transformer model, explicitly modeling inter-channel dependencies and integrating the squeeze-and-excitation (SE) mechanism to generate pixel-specific channel weights. This mechanism adaptively enhances feature channels, conveying information-rich signals while suppressing those containing redundant or less relevant information. This selective emphasis enables the model to concentrate on the most critical temporal patterns during estimation, thereby substantially enhancing the interpretability of the estimation process.

This study addresses three principal challenges in large-scale groundwater time series estimation: the inadequate modeling of long-range temporal dependencies, imprecise identification of critical temporal features, and insufficient incorporation of spatiotemporal information correlation. To overcome these limitations, we propose a groundwater level estimation model founded on an enhanced Transformer architecture. Our core contributions are summarized as follows: (1) We introduce a Transformer encoder framework for time series estimation, which employs sliding windows to extract local temporal features, self-attention mechanisms to capture global dependencies, and positional encoding to preserve temporal integrity. This approach enhances the model's capacity to represent long-range dependencies. (2) The KFEA dynamically calibrates feature channel weights via squeeze-and-excitation (SE) operations, adaptively amplifying information-rich channels and suppressing noise, thus improving sensitivity to critical features and robustness against anomalies. (3) The DSFF mechanism integrates the target pixel's dynamic temporal features with adjacent pixels' static spatial contexts using a 3×3 convolutional layer, addressing spatiotemporal correlation challenges and enhancing spatial consistency of estimation results.

2. Materials and Methods

2.1. Study Area

The study area encompasses the Yellow River Basin (32° N–42° N, 96° E–119° E), spanning approximately 795,000 km^2 (Figure 1). This basin traverses the Qinghai–Tibet Plateau, Loess Plateau, Inner Mongolia Plateau, and North China Plain, exhibiting complex topographical gradients across China's three major geomorphological steps. Climatically, the basin is characterized by a mean annual temperature of 5.8 °C, mean annual precipitation of 466.5 mm, and potential evapotranspiration of 653.6 mm [24], based on long-term observational records. Groundwater constitutes a critical component of the basin's hydrological cycle, regulating regional water resource dynamics. However, comprehensive basin-scale assessment is challenged by spatially heterogeneous groundwater monitoring infrastructure, with significant regions exhibiting sparse or nonexistent well observation points [25]. This constraint necessitates the development of novel data acquisition approaches to support quantitative groundwater resource evaluation.

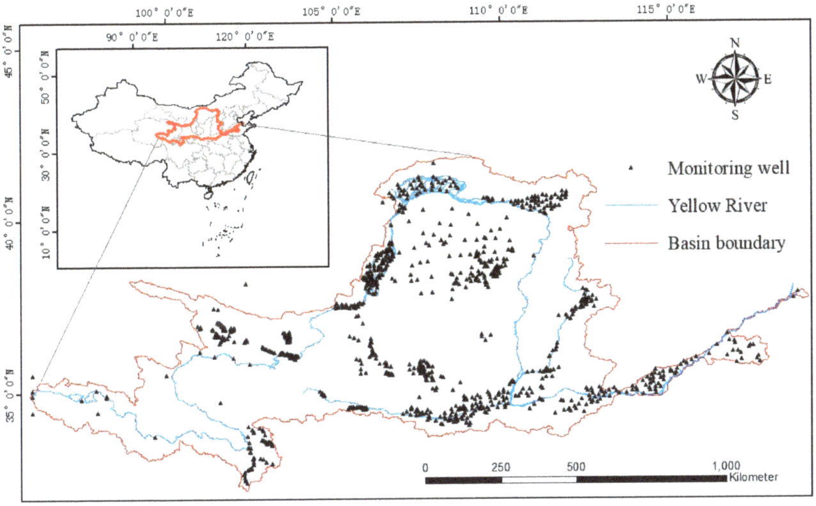

Figure 1. Distribution of groundwater monitoring wells in the Yellow River Basin.

2.2. Data Sources

2.2.1. The Gravity Recovery and Climate Experiment (GRACE)

The GRACE mission (National Aeronautics and Space Administration (NASA), Washington, DC, USA) provides a methodology for monitoring large-scale mass variations within the Earth system. Utilizing a twin-satellite ranging system, GRACE observes the Earth's mass distribution and transport dynamics, enabling estimation of terrestrial water storage changes—comprising snow water equivalent, surface water, soil moisture, and groundwater storage. The derived GRACE dataset quantifies terrestrial water storage anomalies as equivalent water height (EWH). An eleven-month observational gap exists between the conclusion of GRACE operations and the initiation of GRACE Follow-On (GRACE-FO). To meet the needs of remote sensing data for groundwater level in the model training parts and inversion estimation parts must ensure data continuity, missing monthly EWH values within this gap were reconstructed using cubic spline interpolation, resulting in a continuous terrestrial water storage time series from April 2003 to October 2023 [26].

2.2.2. Global Land Data Assimilation System (GLDAS)

The GLDAS soil moisture dataset, collaboratively developed by NASA and the National Oceanic and Atmospheric Administration (NOAA), leverages land surface modeling frameworks and data assimilation methodologies. This integration merges satellite remote sensing observations with in situ ground-based measurements to enable precise estimations of land surface flux dynamics and state variable distributions. For the present study, we utilize the 2 m depth soil moisture outputs from the GLDAS Noah-V2.1 model to characterize the spatiotemporal patterns of soil water storage within the Yellow River Basin.

2.2.3. Measured Groundwater Level

The groundwater level data used in this study are provided by the China Geological Environmental Monitoring Institute (Geological Cloud 3.0, cgs.gov.cn), which are obtained from observation wells (approximately 1100) in certain regions of the Yellow River Basin. The data covers the period from October 2018 to October 2023 at monthly intervals, with high precision. The well observation points are mainly distributed in plains, river valleys, and other areas with high human activity (Figure 1).

2.3. *Methods*

2.3.1. Calculation Method of Groundwater Level

Observing and quantifying groundwater level variations presents significant challenges, especially in large aquifer systems [27]. Direct measurements of groundwater levels depend on a network of observation wells with sparse spatial distribution, which often lacks the spatial coverage required to resolve aquifer heterogeneity. Consequently, large-scale groundwater level monitoring necessitates the integration of remote sensing data with in situ measurements to enable the estimation of groundwater level over extensive areas.

In the estimation of groundwater level from remote sensing data, the conversion of GRACE terrestrial water storage to groundwater storage is essential. Current methodologies primarily include the mass balance equation approach [28–31], lithology method [32], and filtering technique [33]. The mass balance equation, the most widely adopted approach, quantifies total mass or water volume changes within time-varying gravitational fields by integrating complementary hydrological datasets to derive target variables [34]. Its clear physical basis has enabled frequent applications in groundwater storage estimation and assessments of extreme climatic events (e.g., droughts and floods) [35]. This study employs the mass balance equation method for analytical purposes, leveraging its rigorous framework to resolve groundwater storage from GRACE-derived terrestrial water storage signals, and the groundwater storage and groundwater level well observation values are modeled for time series estimation.

The mass balance equation represents the influence of factors such as soil water, snowmelt, and surface water [36].

$$\Delta GWS = \Delta TWS - \Delta SWS - \Delta SMS \tag{1}$$

In the equation, ΔGWS, ΔTWS, ΔSWS, and ΔSMS denote the anomalies of groundwater storage change, total water storage change, surface water storage change, and soil moisture storage change, respectively, during the study period (cm). GRACE satellite data are typically employed to characterize the total water storage change (ΔTWS), whereas GLDAS soil moisture data represent the soil moisture storage change (ΔSMS). The surface water storage change (ΔSWS) is often omitted in groundwater storage estimation, as groundwater constitutes over 90% of the Earth's total liquid freshwater resources [37,38]. Groundwater storage quantifies the vertical water change, while surface water storage reflects horizontal water movement. The vertical component of surface water storage change is generally negligible compared to its horizontal counterpart [39].

2.3.2. Spatial Interpolation Methods

The highest-resolution remote sensing data used in this study is the 0.25° GLDAS-2 dataset. Therefore, other datasets in this research underwent spatial interpolation or resampling to uniformly align with the 0.25° × 0.25° grid of GLDAS, covering a total of 1277 pixels across the Yellow River Basin.

1. Inverse Distance Weighting

Inverse distance weighting (IDW) represents a deterministic interpolation approach rooted in distance-weighting principles. Its core premise is that proximal data points exert a greater influence on the predicted value, with the impact of distant points diminishing monotonically with increasing spatial separation. The IDW formulation employs a weighted average framework, where the assigned weights are inversely proportional to the distance raised to a power (commonly set to ($p = 2$)). This method eschews reliance on statistical modeling, featuring a computationally parsimonious architecture that facilitates straightforward implementation, thereby rendering it suitable for rapid generation of smooth spatial distributions. For the present study, a K-nearest neighbor (K-NN) variant of IDW is adopted to enhance computational efficiency and mitigate the influence of remote data points. This is achieved by imposing a constraint on the number of neighboring points included in the interpolation calculation. Through systematic accuracy comparisons, the optimal K value is determined to be 10, striking a balance between computational tractability and spatial interpolation fidelity.

2. Ordinary Kriging

Kriging interpolation is a statistical-based spatial interpolation method assuming geographic data have spatial autocorrelation, which is quantified by a variogram. This method provides not only predicted values but also prediction error estimates, making it suitable for variables with spatial continuity. The values at unknown points are calculated by computing the experimental variogram, fitting a theoretical variogram model (spherical model in this study), and using the Linear Unbiased Estimation (BLUE) method. Compared with simpler interpolation methods, Kriging interpolation handles uneven data distributions more effectively. However, it has higher computational complexity and is sensitive to the choice of variogram model.

2.3.3. Machine Learning Methods

In this study, we utilized machine learning methods to integrate ground well data with remote sensing data for estimating groundwater levels. Ground well data provide direct measurements of actual groundwater levels but suffer from short temporal coverage and uneven spatial distribution, while remote sensing data capture groundwater storage anomalies with longer temporal records and broad spatial coverage. A physical relationship exists between these two datasets, and this relationship function remains relatively stable over time at any given pixel location. Therefore, we employed machine learning methods to analyze the relationship between these datasets at each pixel and subsequently used this to estimate groundwater levels for years lacking actual well measurements.

Our study period was divided into a 'Model Training Period' and a 'Retrieval Estimation Period': the Model Training Period spanned from October 2018 to October 2023.

During this period, both ground well data and remote sensing data were available. Various machine learning models, as will be introduced later, were trained 200 times on the data for each pixel within this period to obtain the optimal training results. The Retrieval Estimation Period covered April 2003 to September 2018. For this period, only remote sensing data were available. The optimal relationships derived during the Model Training Period were applied to the remote sensing data from the Retrieval Estimation Period to estimate the actual groundwater levels.

1. Transformer model

The main structure of the time series forecasting Transformer model consists of a linear embedding layer, a positional encoding layer, Transformer encoder layers, and a fully connected layer (Figure 2).

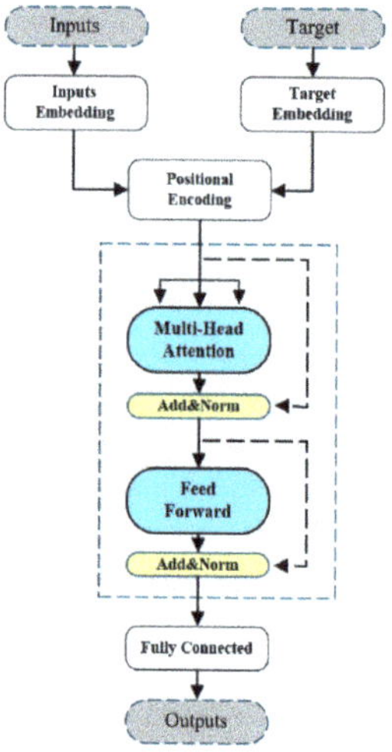

Figure 2. Time series prediction Transformer model structure.

The Transformer encoder layer serves as the core of the entire model, undertaking the critical tasks of time series feature extraction and feature representation learning. Composed of multiple identical encoder sub-layers, each sub-layer consists of a multi-head self-attention layer and a feedforward neural network layer. The multi-head self-attention layer enables the model to focus on information from different positions within the input sequence during processing; by utilizing input query, key, and value matrices to compute attention scores, the attention results from multiple heads are concatenated and linearly transformed to generate the final output (Figure 3). This mechanism allows the model to capture input sequence information from different representation subspaces, thereby enhancing both its expressive and generalization capabilities.

$$\text{Multi-Head}\ (Q, K, V) = \text{Concat}\ (\ \text{head}_{\ 1}, \text{head}_{\ 2}, \ldots, \text{head}_{\ h})W^{0} \quad (2)$$

$$\text{head}_{\ i} = \text{Attention}\left(QW_i^Q, KW_i^K, VW_i^V\right) \quad (3)$$

In the self-attention mechanism, Q (query), K (key), and V (value) are essential components. The query represents the information request from each input position, allowing the model to search for information from other positions. The key identifies the information associated with each input position, assisting in matching with the query to find relevant positions. The value contains the actual information associated with the key, which, once a match is found, is used for computation and output generation.

The feedforward network (FFN) performs nonlinear transformations and feature extraction on the output of the multi-head self-attention mechanism, followed by residual connections and normalization operations.

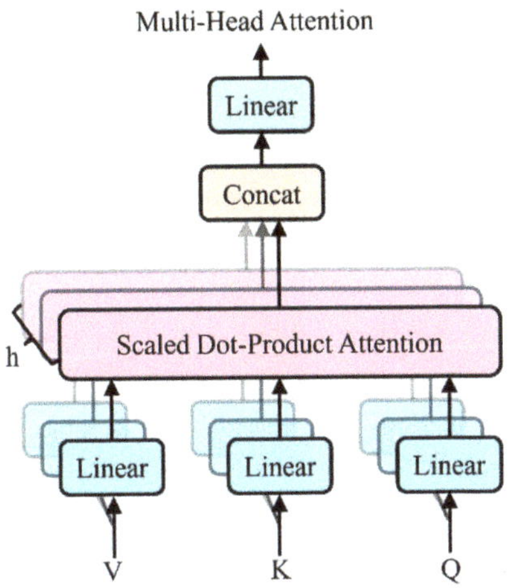

Figure 3. Multi-head attention parallel layers.

The linear embedding layer maps the input data from the original low-dimensional space to a higher-dimensional feature space, thereby providing a richer feature representation.

The positional encoding layer generates a unique encoding vector for each position in the input sequence and adds this vector to the input embedding vector at that position, enabling the Transformer model to distinguish elements at different positions when processing the input data. The positional encoding vectors are generated by a combination of sine and cosine functions, with each dimension corresponding to a different frequency. The positional encoding model is computed as follows:

$$PE_{(pos,\ 2i)} = \sin\left(\frac{pos}{10,000^{\frac{2i}{d_{model}}}}\right) \tag{4}$$

$$PE_{(pos,\ 2i+1)} = \cos\left(\frac{pos}{10,000^{\frac{2i}{d_{model}}}}\right) \tag{5}$$

Here, PE denotes the positional encoding matrix obtained through the positional encoding operation, where (i) represents the dimension index, (pos) represents the position index, and d_{model} represents the input dimensionality.

The time series prediction process of Transformer model can be roughly described as follows, as shown in Figure 4a, predicting the future time point dataset with the past time series data (estimation in this paper). Different combinations of solid lines represent different sample learning strategies. The red dashed line on the right in the figure represents the predicted object for this step, and self-represents the input for this step.

To grasp the relationship between long-term dependence and short-term dependence on data, different "layers" pay different attention to different time series patterns. Figure 5 shows the attention patterns of several layers. For example, some layers focus on the most recent month and give the highest weight to these months (Figure 5: red line); some layers focus on long-term periodicity and give higher weight to the similar months of each year (Figure 5: green line); while some layers have no obvious tendency and give uniform weight to the data of each month (Figure 5: yellow line); and some layers attend from near to far, taking into account the short-term and long-term characteristics (Figure 5: blue line). The set meaning is shown in Figure 4b–d, where the purple input is the high-weight object. By stacking multiple self-attention layers and applying appropriate methods to determine the importance of different layers later, we obtain the final estimation result.

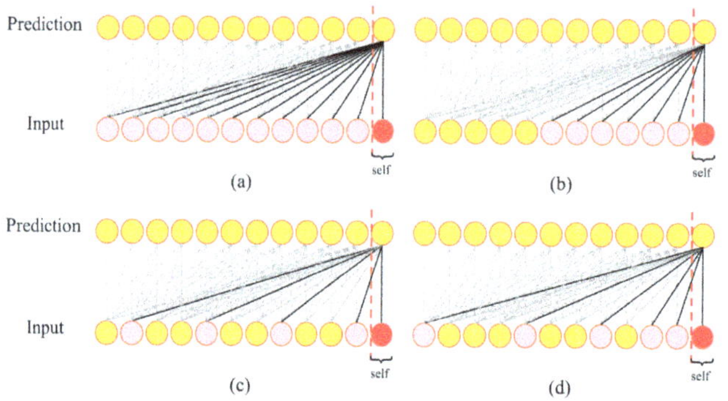

Figure 4. (**a**) Self-attention estimation, (**b**) tend to learn recent months, (**c**) tend to learn similar months (schematic only), and (**d**) tend to learn from near to far.

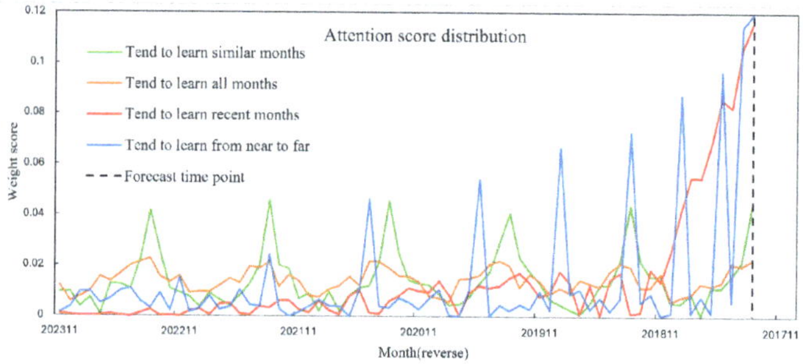

Figure 5. Distribution of attention scores at different layers.

2. Key Feature Enhancement Architecture

In large-scale groundwater estimation, abnormal time series induced by extreme hydrological events yield fluctuations with characteristics inconsistent with actual well observations. To address this, a Key Feature Enhancement Architecture (KFEA) based on SE-Net (squeeze-and-excitation networks) is designed in this paper. This module enables the model to assign preferential attention to specific layers by learning the weights of each channel in the feature vector, thereby improving the network's adaptability and sensitivity to the importance of different channel features and enhancing estimation accuracy.

This module consists of three main components: the squeeze operation, which extracts channel features from the feature map (x); the excitation operation, which computes the weights for these extracted features; and the scaling operation, which applies the computed weights to the input feature map (x). These components can be described by the following equations [40].

Squeeze:

$$z_c = F_{sq} = \frac{1}{HW}\sum_{t=1}^{H}\sum_{j=1}^{W} u_c(i, j) \quad (6)$$

In the equation, (u_c) represents the feature vector of a specific channel in the feature map (x), while (H) and (W) denote the height and width of the feature map, respectively.

Excitation:

$$S = F_{ex}(z_c) = \sigma(W_{s2}\delta(W_{s1}z_c)) \quad (7)$$

In the equation, W_{s1} and W_{s2} are the weights of the fully connected layers, the activation function (σ) is ReLU, (δ) is sigmoid function.

Reweighting:

$$\tilde{x} = F_{scale}(S, x) = Sx \quad (8)$$

In the entire time series prediction model, this module processes the features outputted by the Transformer encoder, analyzing different layers, enhancing the model's adaptive ability to feature importance, improving the prediction accuracy of the model.

3. Dynamic–Static Feature Fusion

In regional groundwater change analysis, dynamics are governed by both temporal dynamics and spatial contextual dependencies from surrounding regions. To address this, a Dynamic–Static Feature Fusion (DSFF) is designed in this paper based on the COT Attention framework, integrating contextual mining and self-attention learning into a unified architecture. This approach leverages static key–value relationships while exploring dynamic feature interactions, enabling the model to capture both local temporal patterns and global spatial dependencies in time series data.

The module first extracts static contextual relationships between keys via a 3 × 3 convolution layer, followed by two successive 1 × 1 convolutional operations for self-attention computation based on queries and context keys to generate dynamic context. The static and dynamic contexts are then fused to produce the final output (Figure 6). In time series prediction tasks, the simultaneous capture of local and global sequence features is critical. As a novel attention mechanism, this module enhances the Transformer model's feature extraction capability for time series by integrating convolutional operations with dynamic attention weight calculations.

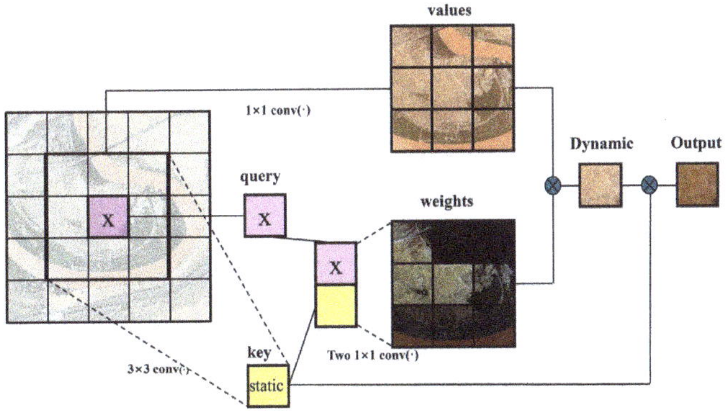

Figure 6. Schematic diagram of COTAttention.

4. Integrated Transformer model

In time series tasks, SEnet exhibits specific adaptability for capturing key information in sequential data. Through its squeeze-and-excitation mechanism, SEnet effectively enhances the weights of informative features while suppressing irrelevant ones. Conversely, the COT framework's output pattern integrating dynamic and static contexts enables comprehensive acquisition of local and global information. Therefore, in addition to independent computations using the Transformer model, this paper performs comparative analyses on combinations of DSFF, KFEA structures, and the Transformer model. The comparative analysis framework evaluates the synergistic effects of DSFF and KFEA within the Transformer architecture. The structures of the three types of variants are shown in Figure 7.

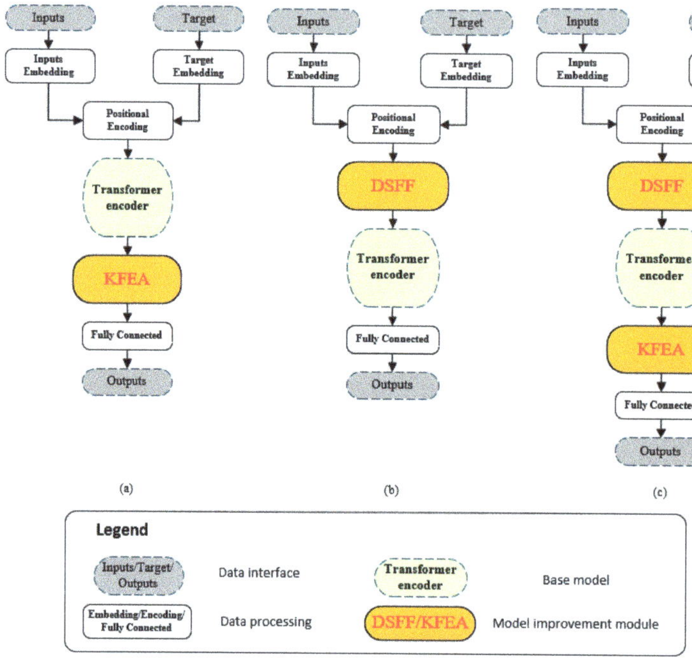

Figure 7. Transformer model variants: (**a**) Transformer-KFEA (**b**) DSFF-Transformer (**c**) DSFF-Transformer-KFEA.

The improved models are referred to as Transformer-KFEA (a), DSFF-Transformer (b), and DSFF-Transformer-KFEA (c). In the Transformer-KFEA model, the SEnet structure is introduced at the end of the Transformer encoder, allowing the model to learn the importance weights of each channel through global average pooling and fully connected layers. This enhances the features of important channels while suppressing those of less important ones. This mechanism helps the model focus on feature dimensions with more informative content, thereby improving time series estimation accuracy. In the DSFF-Transformer model, the DSFF structure is integrated into the Transformer encoder, capturing local static context information through convolution operations and dynamically generating attention key–value pairs based on input features. This aids in better capturing both global and local information in time series estimation. The DSFF-Transformer- KFEA model combines both methods to investigate whether their combined use can jointly improve model performance.

5. LSTM

The LSTM model, a specialized variant of recurrent neural networks (RNNs), is widely employed in sequence data analysis owing to its capability to capture long-term temporal dependencies, demonstrating notable achievements in diverse time series prediction tasks. By leveraging memory cells and gating mechanisms, LSTM exhibits significant efficacy in simulating complex temporal patterns and resolving dynamic groundwater variations over time. However, when processing extremely long sequences, LSTM may encounter challenges associated with vanishing or exploding gradient issues, which arise from the inherent limitations in propagating error signals through extended temporal steps. The gating mechanism of LSTM comprises three primary components: the forget gate, which regulates information retention; the input gate, which controls new information integration; and the output gate, which modulates state information exposure.

$$f_t = \sigma(W_f \cdot [h_{t-1}, x_t] + b_f) \tag{9}$$

$$i_t = \sigma(W_i \cdot [h_{t-1}, x_t] + b_i) \tag{10}$$

$$C_t = f_t \odot C_{t-1} + i_t \odot \tanh(W_C \cdot [h_{t-1}, x_t] + b_C) \tag{11}$$

$$o_t = \sigma(W_o \cdot [h_{t-1}, x_t] + b_o) \tag{12}$$

$$h_t = o_t \odot \tanh(C_t) \tag{13}$$

In this context, f_t represents the activation value of the forget gate, i_t denotes the activation value of the input gate, C_t refers to the cell state, o_t indicates the activation value of the output gate, and h_t represents the hidden state.

As a model that also focuses on long-range dependencies and positional information, this study attempts to utilize LSTM to perform the same task and outputs the results for comparative analysis with the Transformer model and its variants.

2.3.4. Core Metrics and Hyperparameter Configuration

This study employed three core metrics—Root Mean Square Error (RMSE), Mean Absolute Error (MAE), and the Coefficient of Determination (R^2). During the model training period, the information contained within each pixel was divided into an 80% training set and a 20% test set, thereby obtaining these metrics. RMSE is a metric for quantifying prediction error, calculated by averaging the squared differences between predicted and actual values, followed by taking the square root of the result. RMSE emphasizes larger prediction errors, making it more sensitive to outliers, with smaller RMSE values indicating predictions that are closer to the true values. MAE is another quantitative measure of prediction error, computed as the average of the absolute differences between predicted values and actual observations. MAE directly reflects the average magnitude of prediction errors, with smaller values indicating more accurate predictions. Unlike RMSE, MAE penalizes large errors less severely, making it more robust to outliers. R^2 represents the proportion of variance in the observed values that is explained by the model, with values ranging from 0 to 1. The closer R^2 is to 1, the higher the goodness of fit of the model.

$$RMSE = \sqrt{\frac{1}{n}\sum_{i=1}^{n}\left(y_i - \hat{y}_i\right)^2} \tag{14}$$

$$MAE = \frac{1}{n}\sum_{i=1}^{n}\left|y_i - \hat{y}_i\right| \tag{15}$$

$$R^2 = 1 - \frac{\sum_{i=1}^{n}\left(y_i - \hat{y}_i\right)^2}{\sum_{i=1}^{n}\left(y_i - \bar{y}\right)^2} \tag{16}$$

In this context, y_i represents the true value of the i -the observation, $\hat{y}_i$ denotes the predicted value of the i -the observation, and n refers to the number of samples.

Hyperparameter configuration (see Table 1) significantly impacts the performance and prediction accuracy of the algorithm during the model training period. For instance, insufficient iterations may lead to the results not converging to the optimal value, while excessive iterations may cause overfitting and degrade performance, thereby affecting prediction accuracy. To prevent overfitting while ensuring fitting accuracy, an early stopping mechanism is implemented in the model. Training will be terminated when the validation loss (val-loss) does not improve for a specified number of epochs, as determined by the patience parameter. By analyzing the performance of R^2 achieved under different iteration counts during the model training period, the training performance for each iteration configuration was determined. Consequently, the maximum number of iterations was set to 200, with a patience value of 75. Other hyperparameters were determined in a similar manner. Potential effective hyperparameter options were identified by preliminary experimental results and reviewing relevant studies [41,42]. For the Transformer and its variants, potential optimal selections were obtained for parameters such as learning rate, dimension, self-attention heads, and encoder layers, and the optimal combination with the highest R^2 was determined through grid search testing on the test set (A partial illustration of the Transformer model prototype is shown in Table 2). For model variants incorporating the DSFF module, with other parameters fixed, combinations of COT Convolution Kernel sizes of 3×3 and 5×5 were tested, respectively. The 5×5 kernel showed significantly poorer R^2 performance than the 3×3 kernel across almost all pixels in the test set. For model variants incorporating the KFEA module, the SENet Channel Compression Ratio was tested in a similar manner, with 16 being selected as it yielded the best R^2 performance among the tested values of 8, 16, and 32. The specific hyperparameter configurations are as follows.

Table 1. Hyperparameter configuration.

Model Name	Epochs/ Patience	Learning Rate (Adam)	Dim	Self-Attention Heads	Encoder Layers	COT Convolution Kernel	SEnet Channel Compression Ratio
Transformer	200/75	0.0001	128	4	4	-	-
Transformer-KFEA	200/75	0.00005	128	4	4	-	16
DSFF-Transformer	200/75	0.0001	128	4	4	3×3	-
DSFF-Transformer-KFEA	200/75	0.0001	128	4	4	3×3	16
LSTM	200/75	0.002	-	-	-	-	-

Table 2. Results of hyperparameter optimization via grid search (evaluated by test set R^2).

Model Name	Epochs/ Patience	Learning Rate (Adam)	Dim	Self-Attention Heads	Encoder Layers	Test Set R^2
Transformer	200/75	0.0001	32	4	4	0.59
Transformer	200/75	0.00005	32	4	4	0.52
Transformer	200/75	0.00001	32	4	4	0.5
Transformer	200/75	0.0001	64	4	4	0.62
Transformer	200/75	0.00005	64	4	4	0.54
Transformer	200/75	0.00001	64	4	4	0.53
Transformer	200/75	0.0001	128	4	4	0.68
...	...	...	...	...	...	...

3. Results

3.1. Comparison of Estimation Performance of Various Machine Learning Models

This study employed various machine learning models to estimate groundwater levels in the Yellow River Basin. During the model training period, well groundwater level data were converted into areal datasets using inverse distance weighting (IDW) and Kriging interpolation methods. These datasets were then resampled to match 0.25° × 0.25° resolution remote sensing pixels, thereby obtaining paired data of well groundwater levels and remote sensing-derived groundwater storage. Subsequently, time series estimations were performed for groundwater levels across all 1277 pixels.

3.1.1. Influence of Spatial Interpolation Method on Model Estimation

The two interpolation schemes exhibit differential impacts on the model architectures: negligible influence on the baseline LSTM and Transformer models, moderate effect on the KFEA-integrated variant, and more pronounced impact on the DSFF module-incorporated architecture. Under the IDW scheme, the DSFF-Transformer model demonstrates a 10.7% reduction in R^2 variance compared to the baseline Transformer, whereas the DSFF-Transformer-KFEA model shows only a 0.5% variance decrease.

This phenomenon can be attributed to the DSFF model's reliance on surrounding data for static information acquisition, where different interpolation schemes cause variations in peripheral data to exert influence. The DSFF model's behavior of collecting relatively large-scale data and employing time series estimation introduces additional information into the estimation process. This approach helps reduce sensitivity to outliers but may simultaneously introduce new noise. Consequently, the number of samples decreases in high and low R^2 zones while increasing in intermediate R^2 zones, as shown in Figure 8.

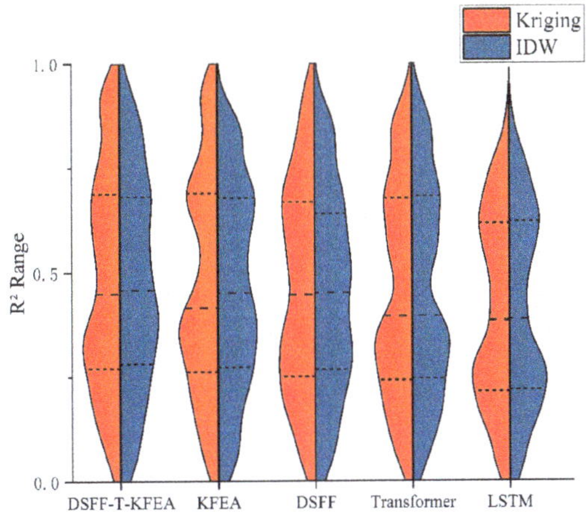

Figure 8. R^2 distribution of each model.

The RMSE values for the five models under the dual interpolation scheme across the entire region are 0.292, 0.299, 0.311, 0.317, and 0.331, respectively, with corresponding MAE values of 0.227, 0.227, 0.241, 0.240, and 0.250 (Table 3). The results show that Kriging method yields lower error metrics compared to the IDW method. The IDW method assigns weights solely based on distance, which may lead to the "bull's-eye effect" where excessive weight is placed on nearby data points, causing concentric variations in the interpolation surface. In regions with uneven distribution of monitoring points (such as dense urban areas) or spatial correlation features (such as directionality or anisotropy), the distance-based weight distribution of IDW may result in non-uniform interpolation outcomes.

Table 3. R^2, RMSE, and MAE of each model.

Index	Interpolation Scheme	DSFF-Transformer-KFEA	Transformer-KFEA	DSFF-Transformer	Transformer	LSTM
R^2	Kriging	0.478	0.468	0.461	0.450	0.404
	IDW	0.476	0.464	0.451	0.453	0.408
RMSE	Kriging	0.467	0.479	0.494	0.494	0.534
	IDW	0.759	0.778	0.805	0.811	0.865
MAE	Kriging	0.362	0.369	0.376	0.380	0.408
	IDW	0.589	0.596	0.617	0.620	0.658

3.1.2. Influence of Machine Learning Model on Estimation

Significant differences in predictive performance exist across the 1277 sample regions for the same model structure, with the percentage of samples for which the models fail to produce valid predictions ($R^2 < 0$) shown in the table. The results indicate that the use of different spatial interpolation schemes has a certain impact on data availability.

Under the Transformer model's core architecture, the IDW method yields fewer unusable points. Among the model variants, the Transformer-KFEA demonstrates the highest data adaptability, with the lowest percentages of unusable points under both spatial interpolation schemes (5.40% and 5.87%, respectively). In contrast, the DSFF-Transformer model exhibits the lowest data adaptability (12.37% and 11.12%), as shown in Table 4, a phenomenon likely attributed to the DSFF module's reliance on structured data inputs during the 3 × 3 convolution step. This dependency may limit its predictive efficacy in regions with pronounced anomalies, as the module requires coherent spatial context for optimal feature extraction. The KFEA module's channel-wise weighting mechanism appears to mitigate sensitivity to data irregularities, contributing to the Transformer-KFEA model's enhanced robustness across interpolation schemes.

Table 4. Proportion of invalid samples ($R^2 < 0$) in each model.

Model	DSFF-Transformer-KFEA	Transformer-KFEA	DSFF-Transformer	Transformer	LSTM
Kriging	9.08%	5.40%	12.37%	8.61%	9.40%
IDW	8.07%	5.87%	11.12%	8.93%	10.18%

After excluding samples with ineffective predictions, Figure 8 (the dashed line represents the quartiles and median) presents the distribution of R^2 values for samples obtained using different machine learning models through five sets of paired violin plots, where each pixel is treated as a sample. Within each set, the left violin plot corresponds to Kriging interpolation results, while the right represents IDW results. The R^2 values of the five models primarily cluster within the range of 0.3 to 0.7, illustrating that time series forecasting models are substantially influenced by the volatility of real-world data. In regions with distinct data characteristics, the models effectively capture trend patterns, whereas in areas with obscure data features or anomalous fluctuations (such as extreme hydrological events), the models' learning capacity is notably compromised, leading to a sharp decline in R^2 values. This results in a distribution roughly resembling a "dumbbell" shape.

After excluding a small subset of non-representative samples, the table presents the R^2, RMSE, and MAE metrics for the five models under two spatial interpolation schemes. The average RMSE values across the entire region for the dual interpolation scheme are

0.613, 0.629, 0.650, 0.653, and 0.699 for the five models, with corresponding MAE values of 0.476, 0.483, 0.497, 0.500, and 0.533. The results demonstrate that the Transformer model exhibits reduced prediction volatility and higher accuracy compared to the LSTM model.

Since our model primarily improves upon existing frameworks, the traditional Diebold–Mariano (DM) test tends to be conservative—often failing to reject the null hypothesis that new and old models exhibit no significant difference. Therefore, this study employs the Clark–West (CW) test for significance testing, a method designed for comparing nested models. In CW tests of the enhanced models, the DSFF-Transformer-KFEA demonstrated statistically significant improvements over both the Transformer prototype and LSTM model in approximately 56.2% and 57.4% of the 1277 pixel-based samples, respectively.

Furthermore, the accuracy improvement conferred by the DSFF module is found to be limited, potentially due to the module introducing additional information that propagates the influence of short-term significant fluctuations in local data caused by extreme hydrological events in the original dataset. In contrast, the KFEA structure effectively filters the new information introduced by the DSFF module, assigning preferential weights to informative channels. As a result, the model integrating both KFEA and DSFF modules demonstrates a notable accuracy enhancement relative to the baseline model and single-module configurations. These findings highlight the synergistic effect of channel-wise feature weighting (via KFEA) and contextual–temporal attention (via DSFF) in mitigating interpolation-induced noise and improving predictive robustness.

The regional average of the deviation from long-term mean groundwater level between model estimates and well groundwater level across the Yellow River Basin is visualized in Figure 9, demonstrating the propagation of model estimates from more recent years toward older years, depicting the prediction performance across the entire study area. The correlation coefficients between the estimated groundwater levels obtained by the Transformer model and its variants and the actual well groundwater level from ground-based wells are 0.829, 0.822, 0.847, and 0.838, respectively. During periods of relatively stable hydrological changes, all models effectively capture and fit the trends of groundwater variations. Analysis of well measured data and GRACE-derived signals indicates that the models can accurately resolve annual periodic fluctuations and long-term trends in groundwater storage.

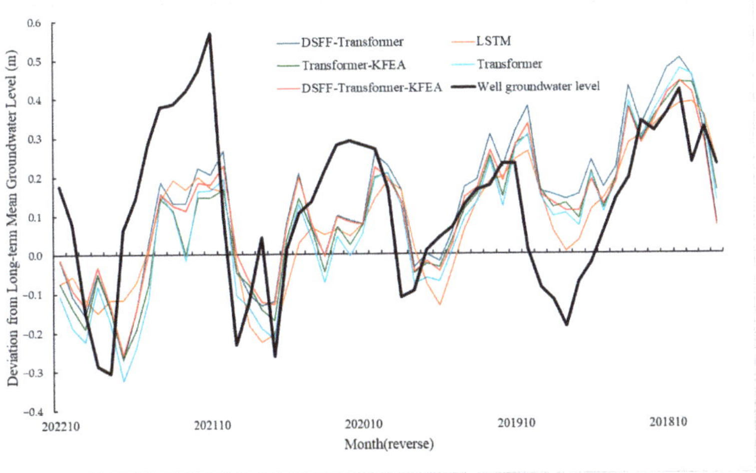

Figure 9. Comparison between the well groundwater level and models.

Notably, the Transformer model exhibits certain limitations in learning local extrema, leading to systematic overestimation of low values and underestimation of high values. Among the model variants, the Transformer-KFEA demonstrates superior trend simulation, achieving the highest correlation coefficient. This performance advantage is likely related to the KFEA module's capability to prioritize critical components in long-term time series information, enabling enhanced trend fitting across the Yellow River Basin. The module's channel-wise weighting mechanism may enhance the model's sensitivity to dominant hydrological signals, thereby mitigating the impact of noise in extreme value regions.

3.2. *Spatial Representation of Estimation Results*

The R^2 values of the five models across the upper, middle, and lower reaches of the Yellow River are tabulated (Table 5). Under Kriging interpolation, the average R^2 values for the five models in the upper, middle, and lower reaches are 0.426, 0.466, and 0.793, respectively, whereas IDW yields corresponding values of 0.461, 0.426, and 0.763. Both interpolation methods show the highest estimation accuracy in the lower reach, primarily due to the significantly higher density of monitoring wells in this region. This dense monitoring network enables comprehensive spatial coverage of groundwater conditions. Conversely, the upper reach exhibits lower accuracy owing to sparse well distribution and extremely limited monitoring points.

Table 5. Models R^2 for different basins.

Basin	Interpolation Scheme	DSFF-Transformer -KFEA	Transformer -KFEA	DSFF-Transformer	Transformer	LSTM
Upstream	Kriging	0.446	0.430	0.425	0.441	0.387
	IDW	0.496	0.445	0.471	0.489	0.405
midstream	Kriging	0.495	0.498	0.490	0.435	0.413
	IDW	0.445	0.461	0.419	0.410	0.396
downstream	Kriging	0.880	0.875	0.783	0.796	0.637
	IDW	0.785	0.777	0.800	0.804	0.650

Notably, Kriging outperforms IDW in the relatively data-rich middle reach, while IDW demonstrates slightly better performance in the data-sparse upper reach. This discrepancy may be linked to Kriging's underlying assumptions of stationarity and isotropy, which are more susceptible to violation in areas with sparse sampling. In contrast, IDW's simpler distance-based weighting mechanism does not rely on such strict statistical assumptions, potentially conferring greater robustness in regions with limited well observational data.

The results demonstrate that the DSFF module's incorporation of peripheral data provides fundamental support for upstream areas lacking well-measured data. Simultaneously, in the middle reaches affected by soil erosion, broader data collection mitigates adverse impacts from systematic anomalies in gravity satellite observations. However, introducing additional data may propagate deficiencies from other regions to adjacent pixels. Consequently, standalone application of the DSFF module yields limited accuracy improvements. After implementing the KFEA module's screening mechanism—which filters incorporated data—this limitation is effectively overcome. The integrated dual-module model consequently achieves the highest accuracy in both data-sparse upstream regions and data-deficient middle reach areas.

The overall performance of different models in the Yellow River Basin is visualized in Figure 10, showing that the high and low accuracy areas of each model exhibit similar spatial distributions: lower accuracy in the central and western regions, and higher accuracy in the northeastern regions. This pattern is likely associated with the fact that a significant portion of the central and western regions are in mountainous areas and the Loess Plateau, which are characterized by complex geological structures, substantial fluctuations in surface water flow, and the impact of soil erosion. The heterogeneous geological structures in these regions may diminish the representativeness of monitoring points in reflecting the actual groundwater conditions of their surrounding areas. Long-term soil erosion can also lead to surface mass migration, thereby affecting the gravity readings of GRACE satellites and indirectly impacting on estimation accuracy. The accuracy in the source region of the Yellow River is also relatively limited. While geological structural changes in this area may exert some influence on gravity satellite readings, the associated mass variations are not significant when compared to the systematic mass transfer from soil erosion in the middle reaches of the Yellow River. Therefore, the primary cause of the poorer accuracy performance in the source region does not lie in this factor, but rather in the lower density of monitoring wells, as shown in Figure 1.

In the eastern part of the Yellow River (i.e., the middle and lower reaches), regions with intensive human activities typically feature a denser network of monitoring wells, which enhances the ability of monitoring point data to reflect the groundwater conditions of adjacent areas, contributing to improved estimation accuracy in these regions. The results highlight the combined influence of geological heterogeneity and anthropogenic factors on the spatial distribution of model performance in hydrological estimation.

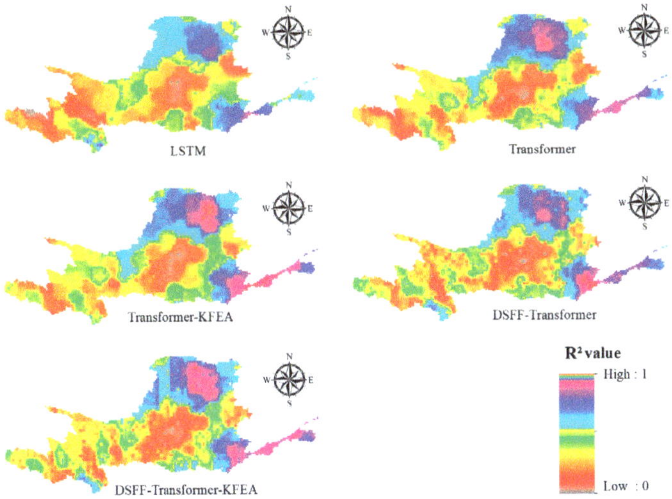

Figure 10. R^2 distribution under Kriging method.

The spatial distribution of high and low accuracy regions under data distribution constraints exhibits similar patterns across all models, though the proportions of regions with $R^2 > 0.7$ differ significantly: 30.2%, 25.8%, 30.0%, 26.2%, and 19.5% for the five models, respectively. The DSFF-Transformer-KFEA and DSFF-Transformer models show the highest proportions of high-value regions, while the LSTM model exhibits the lowest. This indicates that the DSFF module's capability to capture both dynamic and static features enhances the Transformer model's regional estimation stability, effectively expanding the area with relatively higher accuracy. In contrast, the LSTM model lacks such a mechanism, leading to more limited estimation accuracy.

Notably, the Transformer-KFEA model displays a lower proportion of high-value regions than the baseline Transformer model, suggesting that applying squeeze-and-excitation operations to sequence features may, in certain contexts, impact the model's fitting performance. The results highlight the trade-off between channel-wise feature weighting (via KFEA) and contextual feature integration (via DSFF) in modulating spatial prediction accuracy, underscoring the importance of architectural design in aligning with data distribution characteristics.

The annual groundwater level change rates derived from different estimation models are −2.65, −2.15, −2.19, −1.90, and −2.26 mm/month, exhibiting a consistent west-high, east-low spatial distribution trend, as show in Figure 11. The western basin shows relatively minor groundwater level declines, whereas several northeastern areas exhibit more pronounced groundwater level declines. This pattern is primarily attributed to the western region serving as the Yellow River's headwater area, characterized by sparse population density and limited anthropogenic water consumption. Additionally, upstream glacial meltwater has increased slightly due to global warming, contributing to groundwater recharge. In contrast, the northeastern basin—marked by higher population density and flat topography—experiences intensive groundwater extraction for agricultural and industrial activities, leading to sustained declines in groundwater levels.

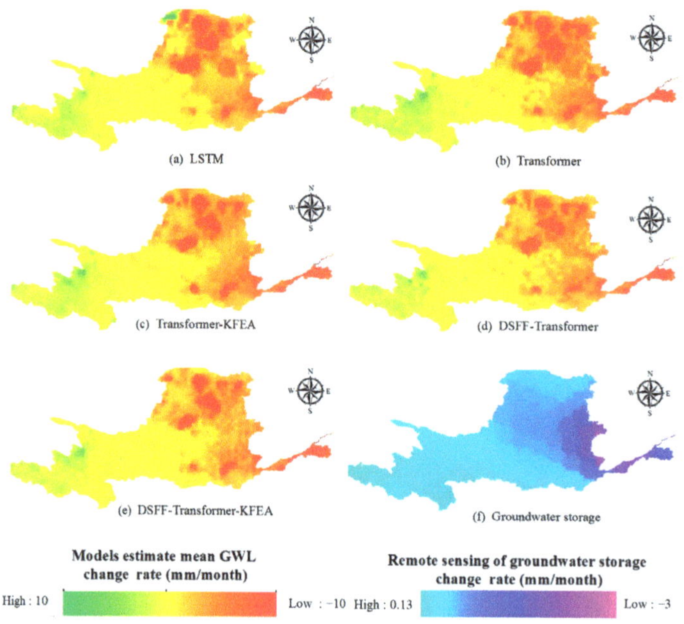

Figure 11. (**a–e**) Groundwater level change rate under different models (**f**) Remote sensing of groundwater storage change rate.

Subplot (f) of Figure 11 displays the average change rate of remote sensing-derived groundwater storage over the 20-year period. This reflects groundwater trends obtained solely from remote sensing data, exhibiting a general pattern of higher values in the west and lower values in the east. This spatial distribution remains broadly consistent with results derived from machine learning models incorporating well groundwater data. Comparison with the other five that while remote sensing data alone can provide an approximate understanding of regional groundwater changes during the study period, they fail to reflect differences in groundwater level responses when identical storage changes occur at different locations. Concurrently, Transformer models and their variants not only consider current monthly groundwater storage changes but also incorporate historical information and spatial context. This approach facilitates more accurate conversion of groundwater storage data into groundwater level estimates without requiring high-precision specific yield information across the entire region.

4. Discussion

In this study, four Transformer model variants and an LSTM model were employed for groundwater level estimation. Experimental results indicate that the Transformer model integrating a dynamic-static fusion perception mechanism and KFEA outperforms the traditional LSTM model, with R^2 improvements ranging from 16.7% to 18.3%. This performance advantage stems from the model's ability to mitigate critical limitations of conventional time series estimation methods in groundwater prediction: unclear long-term dependency modeling, insufficient spatiotemporal information integration, and suboptimal key feature weighting.

4.1. Application of Self-Attention Mechanism in Estimation

The Transformer model, by virtue of its unique self-attention mechanism, enables direct modeling of temporal dependencies over arbitrary distances. Through this global dependency modeling framework, it significantly mitigates the issue of ambiguous long-term dependencies in traditional LSTM/GRU models, which arises from gradient vanishing in long time series sequences. This advantage is corroborated by numerous studies. For instance, Azad Deihim [43] applied the Transformer encoder to capture latent spatiotemporal dependencies in multivariate time series for Beijing PM2.5 prediction and metro interstate traffic volume forecasting, demonstrating a 20.6% reduction in RMSE compared to LSTM, which aligns with our findings. Siying Zhu [44] leveraged the Transformer's attention mechanism to resolve long-term trends in meteorological datasets, incorporating a time series clustering framework for unsupervised learning. In meteorological applications, this approach yielded a 13.6% reduction in MAE relative to DLinear, further validating the Transformer's superiority in capturing temporal dynamics. These results highlight the potential of Transformer models in resolving long-term temporal variations in groundwater levels, offering promising prospects for enhancing estimation accuracy and interpretability in hydrological modeling.

4.2. Application of Spatial Information

By integrating the DSFF module into the Transformer framework, this study presents a hybrid modeling approach that aligns closely with hydrological realities for groundwater level estimation while preserving the model's inherent characteristics. This is achieved through the introduction of a DSFF mechanism and a combined "spatial convolution + temporal attention" design, which significantly mitigates the issue of insufficient spatiotemporal information integration. These findings are corroborated by complementary research efforts. For instance, Lingtong Min [45] demonstrated in remote sensing small target detection that mixed modeling of spatial context and temporal channel information substantially reduces the likelihood of valuable information being obfuscated or overlooked. By introducing COT into the decoupling head, their model achieved a 0.4% increase in mAP@0.5:0.95. Similarly, Deli Zhu [46] employed a DSFF module in post-fire forest restoration studies to enhance the model's capacity for exploring spatial context in complex environments, resulting in a 3.26% improvement in mean intersection over union (mIoU) relative to ResNet. These results highlight that groundwater level prediction requires the consideration of both temporal features within pixels and spatial features between pixels. The introduction of the DSFF module effectively expands feature acquisition pathways, which is conducive to improving the accuracy of large-scale groundwater temporal estimation using remote sensing data. This approach underscores the importance of integrating contextual–temporal attention mechanisms in hydrological modeling to better resolve the spatiotemporal complexity of groundwater systems.

4.3. Application of Feature Filtering

The Transformer variant with the KFEA, by introducing a channel feature adaptive weighting mechanism while preserving the original model characteristics, coupled with a compression–excitation operation after the encoder output, is capable of automatically assigning weights to different channel features. This improves the issue of inaccurate key element weight allocation. Numerous research findings support this perspective. Chen Liu's research [47] in the domain of multi-dimensional time series classification, extracted features across the time, variance, and channel dimensions of multi-dimensional time series data, guiding the features into the channel interactions within a squeeze-and-excitation network. This successfully enhanced the model's representation ability. Compared to the baseline model ResNet, the model achieved a 3.31% improvement in weighted average accuracy, further validating our viewpoint.

4.4. Application of Improved Model

The dual-module coupling model integrates the two modules, fully leveraging the advantages of both. The DSFF module introduces additional information at the encoder's front end, while the KFEA processes this information via adaptive weighting at the encoder output stage. This design ensures a reasonable weight distribution between the introduced static information and the dynamic information captured by the Transformer model through the temporal attention mechanism. Experimental results show that the coupled model achieves an R^2 improvement of approximately 5.1% to 6.2% compared to the Transformer prototype.

This analysis reveals that prior studies utilizing GRACE terrestrial water storage for groundwater estimation predominantly focused on temporal sequence variations at specific locations, with limited consideration of multi-source information weighting and long-term temporal dependencies. Our findings indicate that incorporating surrounding contextual information, rationally assigning channel weights to input data, and accounting for long-term correlations during single-pixel estimation processes can facilitate a clearer and more accurate characterization of long-term groundwater variation patterns. This approach addresses the limitations of conventional methods by integrating spatial contextual cues and temporal dependency modeling, thereby enhancing the interpretability of groundwater dynamics at both local and regional scales.

4.5. Deficiencies and Prospects

In the upper reaches of the Yellow River, sparse well observational data points hinder robust estimation, while in the middle reaches' loss plateau region, long-term soil–water erosion impacts GRACE-derived gravity measurements. This study demonstrates that integrating surrounding pixel information during groundwater estimation alleviates these effects. However, systematic supplementation of in situ observations and quantification of erosion-induced mass transfer lie beyond the current scope but represent promising directions for future research.

5. Conclusions

This study investigates groundwater level time series estimation in the Yellow River Basin using five deep learning algorithms: Transformer, Transformer-KFEA, DSFF-Transformer, DSFF-Transformer-KFEA, and LSTM. The key conclusions are as follows:

(1) The Transformer-based estimation model effectively captures long-range temporal dependencies in groundwater time series, demonstrating significantly higher accuracy than the LSTM model. (2) The dynamic–static fusion perception mechanism and KFEA address critical limitations of conventional estimation methods, including neglect of spatial contextual information and ambiguous feature weighting. These innovations enhance estimation accuracy without substantial increases in computational complexity. (3) IDW exhibits poorer performance than Kriging in regions with sparse data, indicating that distance-based interpolation schemes lack flexibility for low-quality datasets. (4) Groundwater levels in the Yellow River Basin show a fluctuating downward trend during our research period (2003–2024), which downward trend increases from upstream to downstream. The self-attention estimation model proposed herein demonstrates potential for application in other basins. It is noted that this research focuses on satellite pixel-scale analysis ($0.25^\circ \times 0.25^\circ$), where topographic, geomorphic, and geological heterogeneities within pixels may compromise estimation precision. Future studies will investigate finer-resolution groundwater distributions and incorporate additional hydrological parameters to enhance model fidelity.

Author Contributions: Data curation, T.Z.; Methodology, T.Z.; Project administration, C.F.; Software, T.Z.; Supervision, C.F.; Validation, Y.L. and L.X.; Writing—original draft, T.Z.; Writing—review and editing, Y.L. and L.X. All authors have read and agreed to the published version of the manuscript.

Funding: This research was supported by the 2024 Project of the Key Research Base of Philosophy and Social Sciences of Jiangxi Province (24ZXSKJD01).

Data Availability Statement: The data used in this study are available upon request from the corresponding author.

Conflicts of Interest: The authors declare no conflicts of interest.

References

1. van der Gun, J. *Groundwater and Global Change*; United Nations Educational, Scientific and Cultural Organization (UNESCO): Paris, France, 2012; pp. 1–42+44.
2. Siebert, S.; Burke, J.; Faures, J.M.; Frenken, K.; Hoogeveen, J.; Döll, P.; Portmann, F.T. Groundwater use for irrigation—A global inventory. *Hydrol. Earth Syst. Sci.* **2010**, *14*, 1863–1880. [CrossRef]
3. Vorosmarty, C.J.; Green, P.; Salisbury, J.E.; Lammers, R.B. Global Water Resources:Vulnerability from Climate Change and Population Growth. *Science* **2000**, *289*, 284–288. [CrossRef]
4. Jasechko, S.; Perrone, D. Global groundwater wells at risk of running dry. *Science* **2021**, *372*, 418–421. [CrossRef]
5. Shin, Y.S. On the "Groundwater". *Logging Landslides* **1976**, *13*, 16–21. [CrossRef]
6. Ding, K.; Zhao, X.; Cheng, J.; Yu, Y.; Couchot, J.; Zheng, K.; Lin, Y.; Wang, Y. GRACE/ML-based analysis of the spatiotemporal variations of groundwater storage in Africa. *J. Hydrol.* **2025**, *647*, 132336. [CrossRef]
7. Peng, M.; Lu, Z.; Zhao, C.; Motagh, M.; Bai, L.; Conway, B.D.; Chen, H. Mapping land subsidence and aquifer system properties of the Willcox Basin, Arizona, from InSAR observations and independent component analysis. *Remote Sens. Environ.* **2022**, *271*, 112894. [CrossRef]
8. Ma, R.; Chen, K.; Andrews, C.B. Methods for Quantifying Interactions Between Groundwater and Surface Water. *Annu. Rev. Environ. Resour.* **2024**, *49*, 623–653. [CrossRef]
9. Noorduijn, S.L.; Harrington, G.A.; Cook, P.G. The representative stream length for estimating surface water–groundwater exchange using Darcy's Law. *J. Hydrol.* **2014**, *513*, 353–361. [CrossRef]
10. Thanh, N.N.; Thunyawatcharakul, P.; Ngu, N.H.; Chotpantarat, S. Global review of groundwater potential models in the last decade: Parameters, model techniques, and validation. *J. Hydrol.* **2022**, *614*, 128501. [CrossRef]
11. Yang, M.; Liu, H.; Meng, W. An analytical solution of the tide-induced groundwater table overheight under a three-dimensional kinematic boundary condition. *J. Hydrol.* **2021**, *595*, 125986. [CrossRef]
12. Islam, Z.; Abdel-Aty, M.; Mahmoud, N. Using CNN-LSTM to predict signal phasing and timing aided by High-Resolution detector data. *Transp. Res. Part C Emerg. Technol.* **2022**, *141*, 103742. [CrossRef]

13. Alabdulkreem, E.; Alruwais, N.; Mahgoub, H.; Dutta, A.K.; Khalid, M.; Marzouk, R.; Motwakel, A.; Drar, S. Sustainable groundwater management using stacked LSTM with deep neural network. *Urban Clim.* **2023**, *49*, 101469. [CrossRef]
14. Turkoglu, M.O.; Stefano, A.; Wegner, J.D.; Schindler, K. Gating Revisited: Deep Multi-layer RNNs That Can Be Trained. *IEEE Trans. Pattern Anal. Mach. Intell.* **2021**, *44*, 4081–4092. [CrossRef]
15. Choudhary, A.; Arora, A. Assessment of bidirectional transformer encoder model and attention based bidirectional LSTM language models for fake news detection. *J. Retail. Consum. Serv.* **2024**, *76*, 103545. [CrossRef]
16. Zhang, H.; Shafiq, M.O. Survey of transformers and towards ensemble learning using transformers for natural language processing. *J. Big Data* **2024**, *11*, 25. [CrossRef]
17. Su, L.; Zuo, X.; Li, R.; Wang, X.; Zhao, H.; Huang, B. A systematic review for transformer-based long-term series forecasting. *Artif. Intell. Rev.* **2025**, *58*, 80. [CrossRef]
18. Subhadarsini, S.; Kumar, D.N.; Govindaraju, R.S. Enhancing Hydro-climatic and land parameter forecasting using Transformer networks. *J. Hydrol.* **2025**, *655*, 132906. [CrossRef]
19. Xiong, Z.; Zhang, Z.; Gui, H. A Meteorology-Driven Transformer Network to Predict Soil Moisture for Agriculture Drought Forecasting. *IEEE Trans. Geosci. Remote Sens.* **2025**, *63*, 4405818. [CrossRef]
20. Zhang, D.; Madsen, H.; Ridler, M.E.; Kidmose, J.; Jensen, K.H.; Refsgaard, J.C. Multivariate hydrological data assimilation of soil moisture and groundwaterhead. *Hydrol. Earth Syst. Sci.* **2016**, *20*, 4341–4357. [CrossRef]
21. Boo, K.B.W.; El-Shafie, A.; Othman, F.; Khan, M.H.; Birima, A.H.; Ahmed, A.N. Groundwater level forecasting with machine learning models: A review. *Water Res.* **2024**, *252*, 121249. [CrossRef]
22. Kostić, S.; Stojković, M.; Guranov, I.; Vasović, N. Revealing the background of groundwater level dynamics: Contributing factors, complex modeling and engineering applications. *Chaos Solitons Fractals* **2019**, *127*, 408–421. [CrossRef]
23. Sabzehee, F.; Amiri-Simkooei, A.; Iran-Pour, S.; Vishwakarma, B.; Kerachian, R. Enhancing spatial resolution of GRACE-derived groundwater storage anomalies in Urmia catchment using machine learning downscaling methods. *J. Environ. Manag.* **2023**, *330*, 117180. [CrossRef]
24. Wang, Y.; Sun, L.; Li, H. Monthly 8-km Gridded Meteorological Dataset for the Upper and Middle Reaches of the Yellow River Basin (1980–2015). *J. Glob. Change Data Discov.* **2022**, *6*, 25–36, 184–195.
25. Akl, M.; Thomas, B.F. Challenges in applying water budget framework for estimating groundwater storage changes from GRACE observations. *J. Hydrol.* **2024**, *639*, 131600. [CrossRef]
26. Shi, Z.; Zhu, X.; Tang, Y. Analysis of Terrestrial Water Storage Changes and Influencing Factors in China Based on GRACE Satellite Data. *Arid. Land Geogr.* **2023**, *46*, 1397–1406.
27. Castellazzi, P.; Ransley, T.; McPherson, A.; Slatter, E.; Frost, A.; Shokri, A.; Wallace, L.; Crosbie, R.; Janardhanan, S.; Kilgour, P.; et al. Assessing Groundwater Storage Change in the Great Artesian Basin Using GRACE and Groundwater Budgets. *Water Resour. Res.* **2024**, *60*, e2024WR037334. [CrossRef]
28. Zhang, L.; Ke, P.; Zhang, L. Study on Spatiotemporal Variation of Groundwater Storage in China Based on GRACE Data. *J. Hydroecol. Ecol.* **2024**, 9–17.
29. Gong, H.; Pan, Y.; Zheng, L.; Li, X.; Zhu, L.; Zhang, C.; Huang, Z.; Li, Z.; Wang, H.; Zhou, C. Long-term groundwater storage changes and land subsidence development in the North China Plain (1971–2015). *Hydrogeol. J.* **2018**, *26*, 1417–1427. [CrossRef]
30. Castellazzi, P.; Martel, R.; Rivera, A.; Huang, J.; Pavlic, G.; Calderhead, A.I.; Chaussard, E.; Garfias, J.; Salas, J. Groundwater depletion in Central Mexico: Use of GRACE and InSAR to support water resources management. *Water Resour. Res.* **2016**, *52*, 5985–6003. [CrossRef]
31. Diego-Alejandro, S.-A.; Alexandra, S.; Luiz-Carlos, F. Characterization of groundwater storage changes in the Amazon River Basin based on downscaling of GRACE/GRACE-FO data with machine learning models. *Sci. Total Environ.* **2024**, *912*, 168958.
32. Hellwig, J.; de Graaf, I.E.M.; Weiler, M.; Stahl, K. Large-scale assessment of delayed groundwater responses to drought. *Water Resour. Res.* **2020**, *56*, e2019WR025441. [CrossRef]
33. Lu, F.; You, W.; Fan, D. Inversion of Water Storage and Ocean Mass Changes in Mainland China over the Last Decade from GRACE RL05 Data. *Acta Geod. Cartogr. Sin.* **2015**, *44*, 160–167.
34. Frappart, F.; Ramillien, G. Monitoring Groundwater Storage Changes Using the Gravity Recovery and Climate Experiment (GRACE) Satellite Mission: A Review. *Remote Sens.* **2018**, *10*, 829. [CrossRef]
35. Chen, J.L.; Wilson, C.R.; Tapley, B.D.; Yang, Z.L.; Niu, G.Y. 2005 drought event in the Amazon River basin as measured by GRACE and estimated by climate models. *J. Geophys. Res.* **2009**, *114*, B05404. [CrossRef]
36. Long, D.; Yang, W.; Sun, Z. Satellite Gravimetry Inversion and Basin Water Balance for Groundwater Storage Changes in the Haihe Plain. *J. Hydraul. Eng.* **2023**, *54*, 255–267.
37. Kang, X.; Li, L.; Sun, C. Sustainability Study of Groundwater in the Gansu Section of the Yellow River Basin Based on GRACE and GLDAS Data. *J. Desert Res.* **2024**, *44*, 196–206.

38. UNESCO. The United Nations World Water Development Report 2022: Groundwater: Making the Invisible Visible. 2022. Available online: https://www.unwater.org/publications/un-world-water-development-report-2022 (accessed on 24 July 2024).
39. Cao, J.; Xiao, Y.; Long, D. Monitoring Groundwater Storage Changes in the North China Plain by Combining Satellite Gravimetry and Well Data. *Geomat. Inf. Sci. Wuhan Univ.* **2024**, *5*, 805–818.
40. Zhu, X.-H.; Li, K.-R.; Deng, Y.-J.; Long, C.-F.; Wang, W.-Y.; Tan, S.-Q. Center-Highlighted Multiscale CNN for Classifcation of Hyperspectral Images. *Remote Sens.* **2024**, *16*, 4055. [CrossRef]
41. Huang, J.; Yan, H.; Chen, Q.; Liu, Y. Multi-Granularity Temporal Embedding Transformer Network for Traffc Flow Forecasting. *Sensors* **2024**, *24*, 8106. [CrossRef]
42. Vasan, V.; Sridharan, N.V.; Vaithiyanathan, S.; Aghaei, M. Detection and classification of surface de-fects on hot-rolled steel using vision transformers. *Heliyon* **2024**, *10*, e38498. [CrossRef]
43. Deihim, A.; Alonso, E.; Apostolopoulou, D. STTRE: A Spatio-Temporal Transformer with Relative Embeddings for multivariate time series forecasting. *Neural Netw.* **2023**, *168*, 549–559. [CrossRef] [PubMed]
44. Wang, Y.; Wu, X.; Zhang, J.; Wang, W.; Zheng, L.; Shang, J. Series clustering and dynamic periodic patching-based transformer for multivariate time series forecasting. *Appl. Soft Comput.* **2025**, *174*, 112980. [CrossRef]
45. Min, L.; Fan, Z.; Lv, Q.; Reda, M.; Shen, L.; Wang, B. YOLO-DCTI: Small Object Detection in Remote Sensing Base on Contextual Transformer Enhancement. *Remote Sens.* **2023**, *15*, 3970. [CrossRef]
46. Zhu, D.; Yang, P. Study on the Evolutionary Characteristics of Post-Fire Forest Recovery Using Unmanned Aerial Vehicle Imagery and Deep Learning: A Case Study of Jinyun Mountain in Chongqing, China. *Sustainability* **2024**, *16*, 9717. [CrossRef]
47. Liu, C.; Wei, Z.; Zhou, L.; Shao, Y. Multidimensional time series classification with multiple attention mechanism. *Complex Intell. Syst.* **2025**, *11*, 14. [CrossRef]

water

MDPI

Article

Daily Reservoir Evaporation Estimation Using MLP and ANFIS: A Comparative Study for Sustainable Water Management

Funda Dökmen [1], Çiğdem Coşkun Dilcan [2] and Yeşim Ahi [2,*]

[1] Vocational School of İzmit, University of Kocaeli, 41285 Kocaeli, Türkiye; funda.dokmen@kocaeli.edu.tr
[2] Water Management Institute, Ankara University, 06135 Ankara, Türkiye; coskunc@ankara.edu.tr
* Correspondence: ysmahi@ankara.edu.tr; Tel.: +90-3126000162

Abstract: Reservoir evaporation is a vital component of the hydrological cycle and presents considerable challenges for sustainable water management, especially in arid and semi-arid regions. This study assesses the effectiveness of two Artificial Intelligence (AI) methods: Multilayer Perceptron (MLP) and Adaptive Neuro-Fuzzy Inference System (ANFIS), a combination ANN with fuzzy logic, in estimating daily evaporation from a large reservoir in a semi-arid region. Using eight years of hydrometeorological data from a nearby station, the study employed the ReliefF algorithm as a feature selection method for relevant input variables. The dataset was divided into training, validation, and testing subsets with 5% and 10% validation ratios, using four train–test splits of 70:30, 75:25, 80:20, and 85:15. Various training algorithms (e.g., Levenberg–Marquardt) and membership functions (e.g., generalized bell-shaped functions) were tested for both models. MLP consistently outperformed ANFIS on the test sets, showing higher R^2 and lower RMSE values. In the best-performing 70:30 split, MLP achieved an R^2 of 0.8069 and RMSE of 0.0923, compared to ANFIS with an R^2 of 0.3192 and RMSE of 0.2254. The findings highlight the AI-based approaches' potential to support improved evaporation forecasting and integration into decision support tools for water resource planning amid changing climatic conditions.

Keywords: evaporation prediction; meteorological data; machine learning algorithms; modeling

1. Introduction

In recent decades, growing concerns over water scarcity and climate variability have intensified the need for precise estimation of hydrological processes, with evaporation emerging as one of the most challenging yet crucial components to model. However, accurately estimating evaporation poses significant challenges due to the complex and nonlinear relationships among meteorological variables [1–3]. Traditional empirical formulas often fall short in capturing these intricate interactions, resulting in limited estimation reliability. These shortcomings have led to the extensive application of Artificial Intelligence (AI)-based methods in evaporation modeling processes in recent years.

Recent studies highlight the value of Artificial Intelligence (AI) models in evaporation prediction, revealing that these shortcomings have led to the widespread application of AI-based methods in evaporation modeling processes. For instance, Jayasinghe et al. (2022) [4] employed deep learning models such as LSTM for pan evaporation forecasting and found that recurrent architectures significantly improved temporal accuracy over traditional methods. Architectures such as LSTM and GRU consistently outperform traditional models in hydrological time series prediction, including runoff and water level forecasting [5–8]. In a related study, Sharma et al. (2020) [9] showed that Random Forest effectively predicts pan evaporation and outperforms artificial neural networks over long-term datasets, while Bilali et al. (2022) [10] demonstrated that XGBoost achieves high accuracy and interpretability using tools like SHAP and LIME to identify key climatic variables influencing evaporation. Furthermore, many researchers emphasized the role of AI-integrated remote sensing approaches in predicting evapotranspiration under climate variability scenarios, underscoring the need for models that can generalize across diverse environmental settings [11–13]. These advancements confirm that AI offers distinct advantages in modeling complex hydrological processes; however, the relative performance of AI models under different data conditions remains an open question. This motivates a comparative approach to evaluate and optimize AI techniques such as ANNs and ANFIS for operational evaporation forecasting.

Artificial Neural Networks (ANNs), Adaptive Neuro-Fuzzy Inference System (ANFIS), Support Vector Regression (SVR), deep learning models, and various hybrid Machine Learning (ML) techniques are widely employed in this context. Owing to their ability to model nonlinear relationships, these methods demonstrate superior modeling capability over conventional techniques, particularly under uncertain or incomplete data conditions [1–3,14].

The literature highlights that ANNs and ANFIS exhibit notably strong modeling potential, especially in scenarios where climatic data are limited [15,16]. Nevertheless, the superiority of one model over another is largely dependent on the characteristics of the dataset, prevailing climate conditions, and the selected input variables. Hybrid and subset-based modeling strategies have proven effective in enhancing ANFIS, in some cases achieving better results than standard models [1,17].

Comparative analyses using ANN and ANFIS models demonstrate that both methods can effectively estimate daily reservoir evaporation [1–3,15]. Allawi et al. [18] stated that ANFIS outperformed ANNs in long-term forecasts in evaporation prediction. Accordingly, both are considered powerful modeling tools, and the choice between them may vary depending on the application purpose, data structure, and available computational resources [1–3,14–17,19–21]. ANN models are generally distinguished by faster learning processes, better generalization capabilities, and lower computational costs. In contrast, ANFIS models tend to yield enhanced precision, particularly when limited datasets and optimized input variables are used [1,2,20]. In this context, powerful feature selection algorithms like ReliefF significantly enhance the efficiency and accuracy of both ANN and ANFIS models [22,23]. ReliefF identifies the most relevant input features associated with the target variable, eliminating irrelevant and noisy variables and ensuring that the models are trained with only meaningful data. This leads to faster learning and reduces computational demands in ANN models, while in ANFIS models, it reduces the number of fuzzy rules, thereby improving interpretability and generalization ability. Moreover, in cases with small or limited datasets, the rule-based structure of ANFIS becomes much more effective when combined with carefully selected key features [24,25]. Consequently, the proper integration of preprocessing steps such as ReliefF contributes to making these models more reliable and applicable under real-world conditions, enhancing both reliability and operational efficiency [26].

Located in Turkey's Southeastern Anatolia region, the Atatürk Dam plays a strategic role in regional water management through large-scale irrigation and hydroelectric power generation; however, the use of decision support systems such as machine learning, which could contribute to sustainable water management, remains limited. This study aims to analyze and compare the estimation performance and generalization ability of Artificial Neural Networks (ANNs) in the form of Multilayer Perceptron (MLP) and Adaptive Neuro-Fuzzy Inference System (ANFIS) models for estimating daily evaporation at the Atatürk Dam using long-term hydrometeorological data. Feature selection is performed using the ReliefF algorithm to identify relevant input variables, and the models' estimation success is evaluated based on statistical metrics under different data partitioning strategies.

2. Materials and Methods

2.1. Study Area and Data Collection

This study was conducted using data from the Atatürk Dam Reservoir, which is situated on the Euphrates River between the provinces of Adıyaman and Şanlıurfa in southeastern Turkey. As one of the central components of the Southeastern Anatolia Project, the dam serves as a key infrastructure for hydroelectric energy production and irrigation. Completed in 1992, the Atatürk Dam has a height of 169 m and forms one of Turkey's largest reservoirs, with a surface area of approximately 817 km^2 and a total storage capacity of 48.7 billion cubic meters [27].

Daily hydrometeorological data were obtained from the Turkish State Meteorological Service (TSMS) for the period between 2004 and 2011 to develop predictive evaporation models [28]. The selected meteorological station is near the reservoir and provides consistent, long-term observational records. The dataset spans eight years and includes key variables influencing evaporation such as air temperature (°C), relative humidity (%), wind speed (m/s), sunshine duration (hours), and atmospheric pressure (hPa). Daily pan evaporation measurements obtained from the same station were used as target output data for model training and evaluation.

2.2. Data Preprocessing

Descriptive statistics, including measures such as the mean, median, and standard deviation, were employed to summarize the fundamental characteristics of the dataset and are presented in Table 1.

Table 1. Descriptive statistics of the dataset.

Parameters		Statistical Data							
		Min	Mean	Max	SD	Sample Variance	Kurtosis	Skewness	Units
Input	Min. Air Temperature	−4.30	15.71	30.20	5.95	35.44	−0.62	−0.27	°C
	Mean Air Temperature	4.70	23.95	36.50	6.52	42.55	−0.75	−0.39	°C
	Max. Air Temperature	9.30	31.59	44.60	7.06	49.83	−0.50	−0.52	°C
	Total Precipitation	0.00	0.58	40.60	2.91	8.47	74.49	7.89	mm
	Sunshine Hours	0.00	10.41	14.10	3.24	10.47	1.70	−1.48	hours
	Max Humidity	15.00	61.58	100.00	17.51	306.56	−0.69	0.11	%
	Min Humidity	5.00	27.39	86.00	12.28	150.71	3.42	1.39	%
	Mean Humidity	11.90	42.97	92.30	14.10	198.84	0.31	0.67	%
Output	Total Evaporation	0.00	9.27	19.40	4.29	18.42	−1.02	−0.15	mm

The ReliefF algorithm, a statistical feature selection method, was utilized to enhance model performance by identifying and selecting input variables with the strongest relationship to the target variable [29]. The ReliefF feature selection algorithm was applied to rank the importance of input variables and confirm their relevance in evaporation prediction. The selected features were then normalized using min–max scaling to bring all variables into the [0, 1] range (Equation (1)), thus improving the convergence rate of the learning algorithms, reduce inconsistencies, standardize feature ranges, and improve computational stability [30].

$$Y = \frac{X_i - X_{min}}{X_{max} - X_{min}} \quad (1)$$

Here, X_i represents the raw data, X_{min} and X_{max} denote the minimum and maximum values of X, respectively, and Y is the normalized data value. Normalization is a vital preprocessing step in machine learning that transforms data into a consistent scale, making it more suitable for analysis and modeling.

Following feature selection and normalization, a 1712 × 7 matrix was formed, representing the final dataset to be used in model training and evaluation. Each of the 1712 rows corresponds to a sample, and each of the 7 columns to a selected input feature. Prior to model training, all input variables were normalized to ensure uniform scaling, facilitating improved convergence and model stability. The normalized dataset was then divided into training, validation, and testing subsets. Two fixed validation proportions, 5% and 10% of the total dataset, were examined to evaluate the effect of validation size on MLP and ANFIS model performances. This dataset was not used during training but only for validation purposes. For each validation level, the remaining data were partitioned into training and testing sets using various ratios: specifically, training sets comprising 70%, 75%, 80%, and 85% of the nonvalidation data, with the remainder allocated to testing. For ANFIS modeling, the "checking data" option of the MATLAB R2017b (9.3.0.713579) ANFIS toolbox was used to assign validation subsets during training. This feature enabled performance monitoring and validation-based adjustment of the fuzzy inference system. Consequently, both MLP and ANFIS models were trained and validated under comparable conditions using the selected data splits. This flexible partitioning approach enabled a robust comparison of model behavior under different data availability scenarios, supporting a more comprehensive evaluation of model accuracy, generalization ability, and sensitivity to training data volume.

2.3. Model Development

2.3.1. Artificial Neural Network (ANN)

Artificial Neural Networks (ANNs) are widely recognized as robust computational models for knowledge extraction, offering the capacity to infer complex, and often nonlinear, causal relationships between input and output variables. In addition, ANNs enable the identification of hidden neurons and facilitate the analysis of their internal dynamics, thereby enhancing the interpretability of classification processes [31]. As a supervised learning method, ANNs are particularly advantageous due to their adaptability and predictive accuracy in time series forecasting.

In the present study, separate ANN models were constructed for each scenario derived through the ReliefF feature selection algorithm. A feedforward Multilayer Perceptron (MLP) architecture was utilized via MATLAB's (R2017b (9.3.0.713579)) neural network fitting application. To evaluate model performance, three distinct training algorithms, Bayesian Regularization (BR), Levenberg–Marquardt (L-M), and Scaled Conjugate Gradient (SCG), were applied across various data partitioning schemes as mentioned above. Model performance was iteratively evaluated on the validation set to avoid overfitting. The overall structure of the MLP model used in this study, including input variables, hidden layer configuration, and output layer, is illustrated in Figure 1. Among these, the scenario yielding the most optimal performance determined based on the coefficient of determination (R^2) and Root Mean Square Error (RMSE) was selected for final model verification as mentioned in Section 2.4.

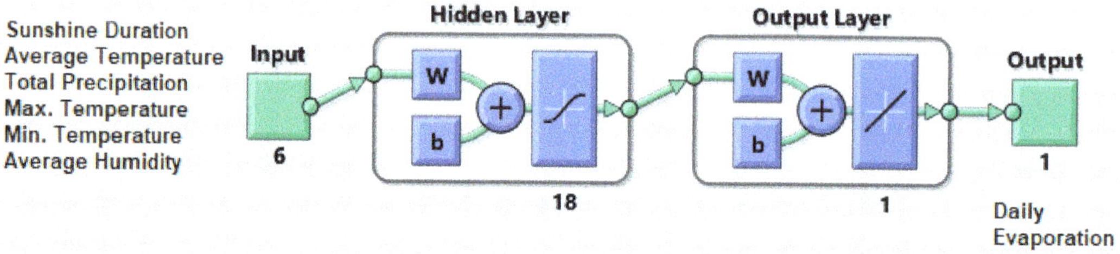

Figure 1. Architecture of the MLP model used for daily evaporation estimation, showing six input features selected by the ReliefF algorithm and a single output node representing daily reservoir evaporation.

2.3.2. Adaptive Neuro-Fuzzy Inference System (ANFIS)

The Adaptive Neuro-Fuzzy Inference System (ANFIS), developed by Jang et al. [32], combines the learning capabilities of Artificial Neural Networks (ANNs) with the reasoning structure of Fuzzy Inference Systems (FISs). This integration enables automatic tuning of fuzzy rules through adaptive networks, improving training efficiency and convergence speed compared to traditional ANNs.

ANFIS modeling proceeds in three main stages. Initially, system parameters, including inputs, outputs, antecedents, and consequents, are defined using the Takagi–Sugeno approach. Fuzzy rules follow the "if–then" structure, where antecedents involve nonlinear membership functions and consequents are expressed as linear functions of inputs or constants. In the second stage, these rules are mapped onto a multilayer adaptive network architecture.

In parallel, an ANFIS model was constructed to incorporate human-like reasoning through fuzzy logic combined with the adaptive learning capacity of neural networks. Separate ANFIS models were constructed for each scenario derived through the ReliefF feature selection algorithm. The scenario demonstrating the highest performance was selected for final model verification, as outlined in Section 2.4.

2.4. Model Evaluation and Comparison

The trained MLP and ANFIS models were evaluated on the test dataset using two standard performance metrics: coefficient of determination (R^2) and Root Mean Square Error (RMSE).

RMSE is one of the most widely used indicators for assessing a model's prediction accuracy. In the formulas below [33], X_i denotes the simulated values, Y_i the observed values, and m the total number of data points.

$$\text{RMSE} = \sqrt{\frac{1}{n}\sum_{i=1}^{n}(X_i - Y_i)^2} \tag{2}$$

$$R^2 = 1 - \frac{\sum_{i=1}^{m}(X_i - Y_i)^2}{\sum_{i=1}^{m}\left(\bar{Y} - Y_i\right)^2} \tag{3}$$

R-squared, which ranges from 0 to 1, represents the squared correlation coefficient (R^2) between the observed and predicted datasets. It indicates the proportion of variance in the dependent variable that is explained by the independent variables, effectively reflecting the model's goodness of fit. These metrics allowed for both quantitative comparison and error structure analysis.

2.5. Software and Implementation

All model simulations and performance analyses were conducted in MATLAB R2017b (9.3.0.713579). ReliefF and MLP/ANFIS implementations were carried out using MATLAB's (R2017b (9.3.0.713579)) built-in Neural Network and Fuzzy Logic Toolboxes. Additional validation routines and statistical analysis were implemented using custom scripts.

3. Results

Input Selection and Model Development

As described in Section 2.2, the ReliefF algorithm implemented via MATLAB R2017b (9.3.0.713579) was utilized to identify the most influential meteorological variables affecting evaporation, based on data obtained from the Turkish State Meteorological Service. The output of the algorithm, presented in Table 2, includes weight scores assigned to each parameter, where values closer to +1 indicate a stronger positive contribution to the target variable. Variables with the highest weights were selected as inputs for subsequent MLP and ANFIS modeling processes.

Table 2. ReliefF algorithm results.

Parameter Name	ReliefF Ordering	Weights
Sunshine duration (hour)	5	0.0034
Average temperature (°C)	2	0.0016
Total precipitation (mm)	4	0.0014
Maximum temperature (°C)	3	0.0012
Minimum temperature (°C)	1	0.0012
Average humidity (%)	8	0.0005
Min humidity (%)	7	0.0000
Max humidity (%)	6	−0.0002

Among the evaluated parameters, maximum humidity (%) exhibited a negative weight, signifying an inverse association with evaporation. Furthermore, minimum humidity (%) displayed a similar pattern and was determined to have negligible predictive influence. Based on these findings, the final set of input variables included sunshine duration (hours), average temperature (°C), total precipitation (mm), maximum temperature (°C), minimum temperature (°C), and average humidity (%).

Following the feature selection process using the ReliefF algorithm, all input variables were subjected to normalization to ensure uniform scaling, thereby enhancing model stability and facilitating faster convergence during training. The normalized dataset was subsequently partitioned into training, validation, and testing subsets. To assess the impact of validation set size on model performance, two fixed validation proportions, 5% and 10% of the total dataset, were employed. For each validation level, the remaining data were further split into training and testing sets using varying ratios, wherein 70%, 75%, 80%, and 85% of the nonvalidation data were allocated for training, with the remainder reserved for testing.

This flexible data partitioning strategy enabled a robust assessment of model performance under different data availability conditions, allowing for a comprehensive evaluation of prediction accuracy, generalization capacity, and sensitivity to training data volume. Among the developed MLP and ANFIS models, the optimal configuration selected based on the lowest Root Mean Square Error (RMSE) and highest coefficient of determination (R^2) was identified for final model validation and deployment. The detailed model development parameters corresponding to the best-performing MLP and ANFIS configurations are summarized in Table 3.

Table 3. ANFIS model development parameters.

Model Name	Parameters	Value
MLP	Training Method	L-M
	Number of Epochs	15
	Number of Neurons in Hidden Layer	18
ANFIS	Input Membership Function (MF) Type	gbellmf
	Number of Input MFs	[2 2 2 2 2 2]
	Output MF Type	Linear
	FIS Generation Method	Grid Partitioning
	Learning Algorithm	Hybrid
	Number of Epochs	150

The best-performing MLP model was obtained using the Levenberg–Marquardt (L-M) training with a data split of 70% for training, 5% for validation, and 25% for testing. For ANFIS modeling, the best configuration employed the hybrid learning algorithm with gbell membership functions, using a 75% training, 10% validation, and 15% testing split.

The estimation results, showcasing the highest prediction performance for the training and testing, are visually presented in Figure 2. The RMSE for MLP modeling, utilizing 70% training data, is approximately 0.0995 and it is 0.0923 for the testing data. And the R^2 was 0.8069. It is important to note that, although the test performance curve is displayed alongside the training and validation curves in Figure 2, the test data was not used during training or model selection. In the Neural Network Fitting App of MATLAB R2017b (9.3.0.713579), the model is trained using only the training set, while the validation set is used to prevent overfitting via early stopping. The test set is evaluated only once after training is completed, and the resulting performance metrics and plots are generated for visualization purposes only. Therefore, the independence of the test set was strictly preserved throughout the modeling process.

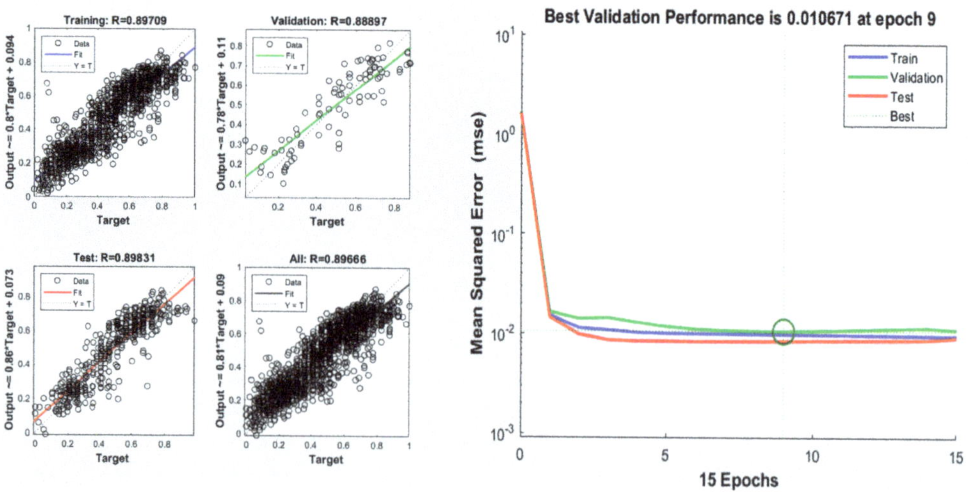

Figure 2. Best-fitting MLP Model results.

The estimation results, showcasing the highest prediction performance for the training and testing, are visually presented in Figure 3. For ANFIS modeling with 75% training data, the RMSE was approximately 0.0830 for the training set and it was around 0.1633 for the testing set. The R^2 values varied across datasets, with about 0.859 for training, 0.4474 for validation, and 0.5053 for testing.

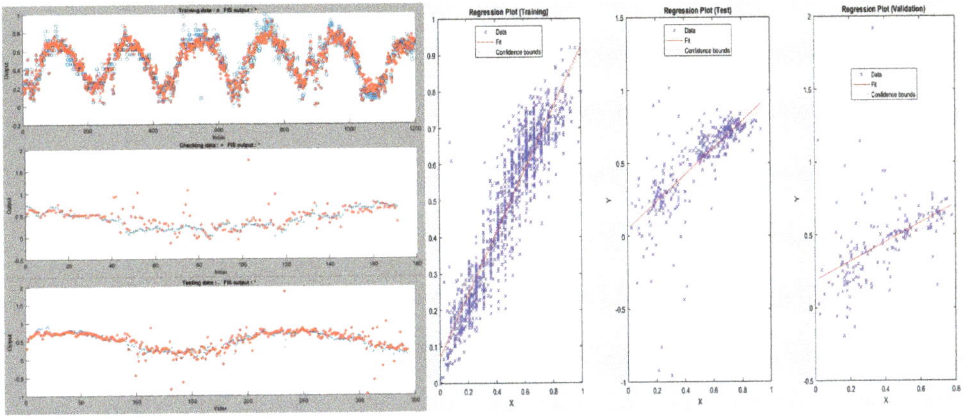

Figure 3. Best-fitting ANFIS Model results.

The comparative evaluation of MLP and ANFIS models across different data split configurations is presented in Table 4, revealing distinct performance patterns. During the training phase, ANFIS achieved marginally lower RMSE and higher R^2 values than MLP, suggesting strong fitting capacity. However, MLP demonstrated superior generalization in the testing phase, with lower RMSE and higher R^2 values compared to ANFIS. For instance, in the 70:5:25 split, MLP yielded an RMSE of 0.0923 and R^2 of 0.8069 on the test set, whereas ANFIS showed a notable decline with an RMSE of 0.1633 and R^2 of 0.5053 in the 75:10:15 split. Similar patterns were observed across other testing ratios, confirming that, although ANFIS is effective in capturing training patterns, MLP provides more robust and reliable predictions on unseen data, making it a more suitable model for operational evaporation forecasting.

Table 4. Comparison of MLP and ANFIS performance based on RMSE and R^2 values computed on the test datasets for different training–validation–testing splits.

Data Splitting Percentage	MLP		ANFIS	
	RMSE	R^2	RMSE	R^2
70% Train	0.0995	0.8048	0.0835	0.8565
5% Validation	0.1033	0.7903	0.2026	0.4796
25% Test	0.0923	0.8069	0.2254	0.3192
70% Train	0.0959	0.8117	0.0835	0.8565
10% Validation	0.0968	0.7916	0.1887	0.5492
20% Test	0.0999	0.8041	0.2386	0.2604
75% Train	0.0933	0.8241	0.0830	0.8588
5% Validation	0.1186	0.7093	0.2027	0.5100
20% Test	0.1137	0.7306	0.2198	0.2791
75% Train	0.0999	0.7959	0.0830	0.8588
10% Validation	0.1026	0.7892	0.2304	0.4474
15% Test	0.1059	0.7634	0.1633	0.5053

Table 4. *Cont.*

Data Splitting Percentage	MLP		ANFIS	
	RMSE	R^2	RMSE	R^2
80% Train	0.0955	0.8186	0.0852	0.8499
5% Validation	0.1000	0.7732	0.2513	0.4167
15% Test	0.1037	0.7725	0.0702	0.6388
80% Train	0.0971	0.8067	0.0852	0.8499
10% Validation	0.1051	0.7840	0.2440	0.4460
10% Test	0.1048	0.7780	0.1936	0.5555
85% Train	0.0997	0.7947	0.0843	0.8552
5% Validation	0.0932	0.8247	0.2539	0.4391
10% Test	0.1129	0.7695	0.2659	0.1021
85% Train	0.1015	0.7931	0.0843	0.8552
10% Validation	0.0967	0.7953	0.2864	0.3790
5% Test	0.1026	0.7505	0.2406	0.4517

4. Discussion

This study assessed and compared the performance of Artificial Neural Networks (ANNs) in the form of a Multilayer Perceptron (MLP) and Adaptive Neuro-Fuzzy Inference System (ANFIS) for daily reservoir evaporation estimation using long-term hydrometeorological data from the Atatürk Dam.

The findings regarding the performance of Multilayer Perceptron (MLP) and Adaptive Neuro-Fuzzy Inference System (ANFIS) models in predicting daily reservoir evaporation across varying training–test splits demonstrate that both models possess considerable predictive capabilities. Nonetheless, MLP consistently yields more robust and generalizable outcomes during the testing phase, whereas ANFIS exhibits a pronounced tendency toward overfitting.

The most balanced configuration for the MLP model was achieved with a data split of 70% for training, 5% for validation, and 25% for testing, yielding R^2 values of nearly 0.80 in both the training and testing phases. In contrast, while ANFIS exhibited its highest training performance at a 75% for training, 10% for validation, and 15% for testing ratio (R^2 = 0.8588), the corresponding test R^2 declined to 0.5053, highlighting the model's sensitivity to data partitioning. During the training phase, the Adaptive Neuro-Fuzzy Inference System (ANFIS) consistently outperformed the Multilayer Perceptron (MLP) in terms of lower Root Mean Square Error (RMSE) and higher coefficient of determination (R^2) values. For example, with a 75% training split, ANFIS achieved an RMSE of 0.08300 and an R^2 of 0.85876, whereas MLP, under identical conditions, recorded an RMSE of 0.09993 and an R^2 of 0.79592. These results indicate that ANFIS is highly effective at identifying localized patterns within the training data. However, this strength also reflects a strong tendency toward overfitting, limiting its ability to generalize to unseen data. Adnan et al. [34] revealed that various ANFIS variants outperformed ANNs, while the studies by Salih et al. [35] and Allawi et al. [36] confirmed the high predictive capacity of ANFIS. However, this proficiency also entails a heightened susceptibility to overfitting. Although ANFIS attained high predictive accuracy across all training configurations, a marked deterioration in test performance was consistently observed. This trend has been corroborated by recent studies (e.g., [15,37,38], which highlight ANFIS's propensity to overlearn noise and local fluctuations in the training data, ultimately compromising its generalization performance on unseen datasets. Zhou et al. [38] also confirmed that ANNs with ensemble feature selection yielded superior evapotranspiration predictions across heterogeneous climatic zones.

MLP models demonstrated more balanced and stable performance across both training and test datasets, exhibiting high accuracy even under lower training ratios. For instance, with a 70% training split, MLP achieved an RMSE of 0.0923 and an R^2 of 0.8069 in testing, outperforming ANFIS, which under the same conditions yielded an RMSE of 0.2254 and an R^2 of 0.3192. Furthermore, MLP models can be efficiently trained using algorithms such as Levenberg–Marquardt. These findings are consistent with those of Arya-Azar et al. [17], who attribute the superior generalization ability of ANN architectures to their layered nonlinear transformation capacity and flexible weight optimization mechanisms.

The ability of ANNs to deliver consistent performance, particularly during the testing phase, is of critical importance for water management systems, where reliable predictions under variable meteorological conditions are essential. Evaporation processes are highly sensitive to variations in temperature, humidity, solar radiation, and wind; therefore, models such as Artificial Neural Networks (ANNs), which can effectively adapt to such fluctuations, are becoming increasingly critical for long-term water resource planning. Bouramtane et al. [39] likewise emphasized that models capable of maintaining stability amid fluctuating input parameters are indispensable for operational forecasting systems.

Recent advances in hydrology and water resource modeling have increasingly highlighted the potential of Artificial Neural Networks (ANN) and the Adaptive Neuro-Fuzzy Inference System (ANFIS) in evaporation prediction. However, it is widely acknowledged that the success of these models is strongly dependent on the choice of input features and model configuration. Khosravi et al. [40] demonstrated this by comparing the predictive accuracy of eight Machine Learning (ML) and Deep Learning (DL) algorithms using 30 years of meteorological data in Iran. Their results showed that hybrid approaches consistently outperformed individual models.

The importance of input feature selection and model structure is also well-documented. For example, Marouane et al. [19] reported that including water temperature improved the performance of ANFIS-HHO models in Algeria. This is consistent with our ReliefF-based variable selection, which identified key predictors and significantly enhanced model performance. Likewise, Adnan Ikram et al. [41] integrated ANFIS with the Whale Optimization Algorithm (WOA) for monthly pan evaporation estimation in the Dongting Lake basin. Their findings emphasized the dominant role of maximum temperature and the added value of periodicity inputs, even under data-scarce conditions.

In mountainous regions such as the Indian Himalayas, Malik et al. [42] demonstrated that a hybrid neuro-fuzzy system (CANFIS) outperformed conventional ANN models for monthly evaporation simulation. These results collectively underscore that integrating neuro-fuzzy logic with optimization algorithms offers a powerful and adaptive framework for modeling complex hydrological processes.

In parallel, ensemble models have gained attention for their enhanced robustness and generalizability. Ehteram et al. [43] introduced a composite ANN framework that consistently outperformed individual hybrid models across different stations, reinforcing the value of ensemble learning. Similarly, our ANN-based model maintained stable performance across all data splits, supporting its reliability for operational forecasting.

Despite these advances, model transferability remains a key limitation. Alsumaiei et al. [44] found that ANN models trained with limited inputs (e.g., temperature and wind) performed adequately at local scales but failed to capture extreme evaporation events in arid environments. Our findings corroborate this concern, suggesting that AI-based models require regional calibration to preserve accuracy under diverse climatic regimes.

In addition to the comparison between ANN and ANFIS, it is essential to contextualize the findings against other widely used evaporation prediction models. Recent studies have shown that ensemble learning and deep learning techniques have emerged as powerful alternatives to traditional AI approaches in evaporation modeling. For instance, Abed et al. (2022) [45] reported that Random Forest (RF) outperformed ANNs in long-term pan evaporation prediction, primarily due to its inherent capability to rank variable importance and handle complex interactions without overfitting. Likewise, Bilali et al. (2022) [10] demonstrated that Extreme Gradient Boosting (XGBoost) yielded high predictive accuracy (NSE from 0.76–0.83) and interpretability through integration with SHAP and LIME tools, which facilitated identification of dominant climatic drivers such as solar radiation and air temperature.

Furthermore, Mehra et al. (2025) [46] applied Long Short-Term Memory (LSTM) and Gated Recurrent Unit (GRU) networks for evaporation modeling and showed that these deep learning architectures outperformed both classical machine learning methods and traditional ANNs, especially in capturing temporal dependencies in time series data. In a recent study, deep LSTM architectures alone have shown improved accuracy over traditional ANN and empirical models for daily pan evaporation estimation, especially with limited input features. Similarly, CNN models have demonstrated high accuracy in capturing the nonlinearities of evaporation processes, often outperforming other machine learning frameworks [45]. Additionally, hybrid and optimization-enhanced frameworks such as ANFIS coupled with Particle Swarm Optimization (PSO) or the Whale Optimization Algorithm (WOA) have demonstrated superior predictive capabilities across various hydroclimatic regions [19,41].

Compared to these models, our findings suggest that, while Artificial Neural Networks (ANNs) in the form of a Multilayer Perceptron (MLP) exhibited superior generalization performance over ANFIS in this study, the core architecture of many recent high-performing models still builds upon MLP or ANFIS structures—whether through optimization, hybridization, or deep expansion. This reinforces the practical value of our direct comparative analysis and highlights that enhancing and tuning these foundational architectures continues to be a central strategy in evaporation prediction modeling. Future work could extend the current comparative framework by including RF, XGBoost, and other emerging models such as deep learning architectures (e.g., LSTM), which have also shown promise in evaporation and evapotranspiration modeling under variable climate conditions.

Taken together, the findings from Table 4 and recent literature indicate that the MLP model provides a more stable and generalizable approach for evaporation prediction, maintaining its effectiveness especially in large-scale and real-time applications. In contrast, although ANFIS models can achieve high accuracy on training data, they require additional support from optimization algorithms (e.g., PSO, GA, SFLA) or pruning techniques to improve their generalization performance and prevent overfitting. Ultimately, model selection should be guided by data volume, preprocessing level, and application objectives, with MLP being the preferred choice for general-purpose and operational use.

5. Conclusions

The findings highlight the strategic potential of AI-based models in supporting more adaptive and climate-resilient water resource management. Combining the interpretability of ANFIS with the adaptability of MLP or enriching these models with evolutionary algorithms and feature selection techniques such as ReliefF can yield high-performance and flexible systems for evaporation prediction. When integrated with decision support tools, such systems can contribute to mitigating the impacts of drought, managing reservoir operations, optimizing water distribution plans, formulating climate-responsive policies, and ultimately supporting inclusive and sustainable water resource management.

Moreover, the implementation of these models within early warning and monitoring systems can enhance preparedness against hydrological extremes. They also enable real-time updates of operational rules for reservoirs, allowing for more responsive and efficient water allocation.

To further strengthen the applicability of such models, future research should explore regional calibration, integration with climate projections, and real-time data assimilation techniques.

Author Contributions: The research was jointly developed by F.D., Ç.C.D. and Y.A., who contributed collectively to the conceptualization, methodology, data analysis, model implementation, and manuscript writing. All authors have read and agreed to the published version of the manuscript.

Funding: This research received no external funding.

Data Availability Statement: The hydrometeorological dataset used in this study was obtained from the Turkish State Meteorological Service (TSMS). Due to institutional restrictions, the data are not publicly available.

Acknowledgments: The authors would like to thank the Turkish State Meteorological Service (TSMS) for providing access to the hydrometeorological data used in this study.

Conflicts of Interest: The authors declare no conflicts of interest.

Abbreviations

AI	Artificial Intelligence
ANFIS	Adaptive Neuro-Fuzzy Inference System
ANN	Artificial Neural Network
BR	Bayesian Regularization
GA	Genetic Algorithm
L-M	Levenberg–Marquardt
MATLAB	Matrix Laboratory
MF	Membership Function
ML	Machine Learning
MLP	Multilayer Perceptron
PSO	Particle Swarm Optimization
RMSE	Root Mean Square Error
SCG	Scaled Conjugate Gradient
SD	Standard Deviation
SFLA	Shuffled Frog-Leaping Algorithm
SVR	Support Vector Regression
TSMS	Turkish State Meteorological Service

References

1. Shiri, J.; Dierickx, W.; Baba, A.; Neamati, S.; Ghorbani, M. Estimating daily pan evaporation from climatic data of the State of Illinois, USA using adaptive neuro-fuzzy inference system (ANFIS) and artificial neural network (ANN). *Hydrol. Res.* **2011**, *42*, 491–502. [CrossRef]
2. Moghaddamnia, A.; Gousheh, M.; Piri, J.; Amin, S.; Han, D. Evaporation Estimation Using Artificial Neural Networks and Adaptive Neuro-Fuzzy Inference System Techniques. *Adv. Water Resour.* **2009**, *32*, 88–97. [CrossRef]
3. Kumar, P.; Kumar, D.; Jaipaul; Tiwari, A. Evaporation Estimation Using Artificial Neural Networks and Adaptive Neuro-Fuzzy Inference System Techniques. *Pak. J. Meteorol.* **2012**, *8*, 81–88.

4. Jayasinghe, W.; Deo, R.; Ghahramani, A.; Ghimire, S.; Raj, N. Development and evaluation of hybrid deep learning Long Short-Term Memory Network model for pan evaporation estimation trained with satellite and ground-based data. *J. Hydrol.* **2022**, *607*, 127534. [CrossRef]
5. Gao, S.; Huang, Y.; Han, J.; Wang, G.; Zhang, M.; Lin, Q. Short-term runoff prediction with GRU and LSTM networks without requiring time step optimization during sample generation. *J. Hydrol.* **2020**, *589*, 125188. [CrossRef]
6. Park, K.; Jung, Y.; Seong, Y.; Lee, S. Development of Deep Learning Models to Improve the Accuracy of Water Levels Time Series Prediction through Multivariate Hydrological Data. *Water* **2022**, *14*, 469. [CrossRef]
7. Wei, X.; Wang, G.; Schmalz, B.; Hagan, D.; Duan, Z. Evaluate Transformer model and Self-Attention mechanism in the Yangtze River basin runoff prediction. *J. Hydrol. Reg. Stud.* **2023**, *47*, 101438. [CrossRef]
8. Xie, H.; Randall, M.; Chau, K. Green Roof Hydrological Modelling With GRU and LSTM Networks. *Water Resour. Manag.* **2021**, *36*, 1107–1122. [CrossRef]
9. Sharma, R.; Singh, A.; Mittal, M.; Kumar, A. Predictive modeling of Pan Evaporation using Random Forest Algorithm along with Features Selection. In Proceedings of the 2020 10th International Conference on Cloud Computing, Data Science & Engineering (Confluence), Noida, India, 29–31 January 2020; pp. 380–384. [CrossRef]
10. Bilali, A.; Abdeslam, T.; Ayoub, N.; Lamane, H.; Ezzaouini, M.; Elbeltagi, A. An interpretable machine learning approach based on DNN, SVR, Extra Tree, and XGBoost models for predicting daily pan evaporation. *J. Environ. Manag.* **2022**, *327*, 116890. [CrossRef]
11. Sun, D.; Zhang, H.; Qi, Y.; Ren, Y.; Zhang, Z.; Li, X.; Lv, Y.; Cheng, M. A Comparative Analysis of Different Algorithms for Estimating Evapotranspiration with Limited Observation Variables: A Case Study in Beijing, China. *Remote Sens.* **2025**, *17*, 636. [CrossRef]
12. Hadadi, F.; Moazenzadeh, R.; Mohammadi, B. Estimation of actual evapotranspiration: A novel hybrid method based on remote sensing and artificial intelligence. *J. Hydrol.* **2022**, *609*, 127774. [CrossRef]
13. Elbeltagi, A.; Kushwaha, N.; Rajput, J.; Vishwakarma, D.; Kulimushi, L.; Kumar, M.; Zhang, J.; Pande, C.; Choudhari, P.; Meshram, S.; et al. Modelling daily reference evapotranspiration based on stacking hybridization of ANN with meta-heuristic algorithms under diverse agro-climatic conditions. *Stoch. Environ. Res. Risk Assess.* **2022**, *36*, 3311–3334. [CrossRef]
14. Goyal, M.; Bharti, B.; Quilty, J.; Adamowski, J.; Pandey, A. Modeling of daily pan evaporation in sub tropical climates using ANN, LS-SVR, Fuzzy Logic, and ANFIS. *Expert. Syst. Appl.* **2014**, *41*, 5267–5276. [CrossRef]
15. Piri, J.; Mohammadi, K.; Shamshirband, S.; Akib, S. Assessing the suitability of hybridizing the Cuckoo optimization algorithm with ANN and ANFIS techniques to predict daily evaporation. *Environ. Earth Sci.* **2016**, *75*, 246. [CrossRef]
16. Allawi, M.; El-Shafie, A. Utilizing RBF-NN and ANFIS Methods for Multi-Lead ahead Prediction Model of Evaporation from Reservoir. *Water Resour. Manag.* **2016**, *30*, 4773–4788. [CrossRef]
17. Arya-Azar, N.; Kardan, N.; Ghordoyee-Milan, S. Developing the artificial neural network–evolutionary algorithms hybrid models (ANN–EA) to predict the daily evaporation from dam reservoirs. *Eng. Comput.* **2021**, *39*, 1375–1393. [CrossRef]
18. Allawi, M.F.; Binti Othman, F.; Afan, H.A.; Ahmed, A.N.; Hossain, M.S.; Fai, C.M.; El-Shafie, A. Reservoir Evaporation Prediction Modeling Based on Artificial Intelligence Methods. *Water* **2019**, *11*, 1226. [CrossRef]
19. Marouane, B.; Mu'azu, M.; Petroselli, A. Prediction of reservoir evaporation considering water temperature and using ANFIS hybridized with metaheuristic algorithms. *Earth Sci. Inform.* **2024**, *17*, 1779–1798. [CrossRef]
20. Patil, A.; Deka, P. Performance evaluation of hybrid Wavelet-ANN and Wavelet-ANFIS models for estimating evapotranspiration in arid regions of India. *Neural Comput. Appl.* **2015**, *28*, 275–285. [CrossRef]
21. Keshtegar, B.; Kisi, O.; Arab, H.; Zounemat-Kermani, M. Subset Modeling Basis ANFIS for Prediction of the Reference Evapotranspiration. *Water Resour. Manag.* **2018**, *32*, 1101–1116. [CrossRef]
22. Palma-Mendoza, R.; Rodríguez, D.; De-Marcos, L. Distributed ReliefF-based feature selection in Spark. *Knowl. Inf. Syst.* **2018**, *57*, 1–20. [CrossRef]
23. Xue, Y.; Zhu, H.; Neri, F. A feature selection approach based on NSGA-II with ReliefF. *Appl. Soft Comput.* **2023**, *134*, 109987. [CrossRef]
24. Ouadfel, S.; Elaziz, M. Efficient high-dimension feature selection based on enhanced equilibrium optimizer. *Expert Syst. Appl.* **2022**, *187*, 115882. [CrossRef]
25. Zhang, M.; Wang, J.; Hou, J.; Song, H.; Li, X.; Guo, F. RG-NBEO: A ReliefF guided novel binary equilibrium optimizer with opposition-based S-shaped and V-shaped transfer functions for feature selection. *Artif. Intell. Rev.* **2022**, *56*, 6509–6556. [CrossRef]
26. Eiras-Franco, C.; Guijarro-Berdiñas, B.; Alonso-Betanzos, A.; Bahamonde, A. Scalable feature selection using ReliefF aided by locality-sensitive hashing. *Int. J. Intell. Syst.* **2021**, *36*, 6161–6179. [CrossRef]
27. State Hydraulic Works General Directorate. Massive Economic Contribution from the 32-Year-Old Atatürk Dam. 2025. Available online: https://www.dsi.gov.tr/Haber/Detay/13076 (accessed on 25 May 2025).

28. TSMS Turkish State Meteorological Service. Sanlıurfa Provincial Meteorological Data Records. 2024. Available online: https://bulut.mgm.gov.tr/share/preview/vY2TdgRzVwq9tNPg (accessed on 17 July 2024).
29. Robnik-Šikonja, M.; Kononenko, I. Theoretical and empirical analysis of ReliefF and RReliefF. *Mach. Learn.* **2003**, *53*, 23–69. [CrossRef]
30. Sinsomboonthong, S. Performance Comparison of New Adjusted Min-Max with Decimal Scaling and Statistical Column Normalization Methods for Artificial Neural Network Classification. *Int. J. Math. Math. Sci.* **2022**, *1*, 3584406. [CrossRef]
31. Boger, Z.; Guterman, H. Knowledge extraction from artificial neural network models. In Proceedings of the IEEE International Conference on Systems, Man, and Cybernetics. Computational Cybernetics and Simulation, Orlando, FL, USA, 12–15 October 1997; Volume 4, pp. 3030–3035.
32. Jang, J.S.R.; Sun, C.T.; Mizutani, E. Neuro Fuzzy and Soft Computing A Computational Approach to Learning and Machine Intelligence. *Prentice Hall.* **1997**, *42*, 1482–1484. [CrossRef]
33. Wright, S. Correlation and causation. *J. Agric. Res.* **1921**, *20*, 557–585.
34. Adnan, R.M.; Malik, A.; Kumar, A.; Parmar, K.S.; Kişi, O. Pan evaporation modeling by three different neuro-fuzzy intelligent systems using climatic inputs. *Arab. J. Geosci.* **2019**, *12*, 606. [CrossRef]
35. Salih, S.Q.; Allawi, M.F.; Yousif, A.A.; Armanuos, A.M.; Saggi, M.K.; Ali, M.; Shahid, S.; Al-Ansari, N.; Yaseen, Z.M.; Chau, K.W. Viability of the advanced adaptive neuro-fuzzy inference system model on reservoir evaporation process simulation: Case study of Nasser Lake in Egypt. *Eng. Appl. Comput. Fluid Mech.* **2019**, *13*, 878–891. [CrossRef]
36. Allawi, M.F.; Ahmed, M.L.; Aidan, I.A.; Deo, R.C.; El-Shafie, A. Developing reservoir evaporation predictive model for successful dam management. *Stoch. Environ. Res. Risk Assess.* **2021**, *35*, 499–514. [CrossRef]
37. Mohammed, S.J.; Zubaidi, S.L.; Ortega-Martorell, S.; Al-Ansari, N.; Ethaib, S.; Hashim, K. Application of hybrid machine learning models and data pre-processing to predict water level of watersheds: Recent trends and future perspective. *Cogent Eng.* **2022**, *9*, 2143051. [CrossRef]
38. Zhou, H.; Ma, L.; Niu, X.; Xiang, Y.; Chen, J.; Su, Y.; Li, J.; Lu, S.; Chen, C.; Wu, Q. A Novel Hybrid Model Combined with Ensemble Embedded Feature Selection Method for Estimating Reference Evapotranspiration in the North China Plain. *Agric. Water Manag.* **2024**, *296*, 108807. [CrossRef]
39. Bouramtane, T.; Leblanc, M.; Kacimi, I.; Ouatiki, H.; Boudhar, A. The contribution of remote sensing and input feature selection for groundwater level prediction using LSTM neural networks in the Oum Er-Rbia Basin, Morocco. *Front. Water* **2023**, *5*, 1241451. [CrossRef]
40. Khosravi, K.; Farooque, A.A.; Naghibi, A.; Heddam, S.; Sharafati, A.; Hatamiafkoueieh, J.; Abolfathi, S. Enhancing Pan evaporation predictions: Accuracy and uncertainty in hybrid machine learning models. *Ecol. Inform.* **2025**, *85*, 102933. [CrossRef]
41. Adnan Ikram, R.M.; Jaafari, A.; Milan, S.G.; Kisi, O.; Heddam, S.; Zounemat-Kermani, M. Hybridized Adaptive Neuro-Fuzzy Inference System with Metaheuristic Algorithms for Modeling Monthly Pan Evaporation. *Water* **2022**, *14*, 3549. [CrossRef]
42. Malik, A.; Kumar, A.; Kisi, O. Monthly pan-evaporation estimation in Indian central Himalayas using different heuristic approaches and climate based models. *Comput. Electron. Agric.* **2017**, *143*, 302–313. [CrossRef]
43. Ehteram, M.; Panahi, F.; Ahmed, A.N.; Mosavi, A.H.; El-Shafie, A. Inclusive Multiple Model Using Hybrid Artificial Neural Networks for Predicting Evaporation. *Front. Environ. Sci.* **2022**, *9*, 789995. [CrossRef]
44. Alsumaiei, A.A. Utility of Artificial Neural Networks in Modeling Pan Evaporation in Hyper-Arid Climates. *Water* **2020**, *12*, 1508. [CrossRef]
45. Abed, M.; Imteaz, M.; Ahmed, A.; Huang, Y. Modelling monthly pan evaporation utilising Random Forest and deep learning algorithms. *Sci. Rep.* **2022**, *12*, 13132. [CrossRef]
46. Mehra, V.; Nakiya, A.; Kumar, J. Estimation of Evaporation Using Long Short Term Memory and Gated Recurrent Unit Based Neural Network. *Adv. Res.* **2025**, *26*, 489–497. [CrossRef]

water

MDPI

Article

Improved Reinforcement Learning for Multi-Objective Optimization Operation of Cascade Reservoir System Based on Monotonic Property

Xiang Li [1,2], Haoyu Ma [1], Sitong Chen [3], Yang Xu [1] and Xiang Zeng [3,*]

[1] Hubei Key Laboratory of Intelligent Yangtze and Hydroelectric Science, China Yangtze Power Co., Ltd., Yichang 443000, China; amber.x.li@whu.edu.cn (X.L.)
[2] State Key Laboratory of Water Resources Engineering and Management, Wuhan University, Wuhan 430072, China
[3] School of Resource and Environmental Sciences, Wuhan University, Wuhan 430079, China
* Correspondence: zengxiang@whu.edu.cn

Abstract: In this paper, improved reinforcement learning (IRL) is designed for the multi-objective optimization operation of a cascade reservoir system. The primary improvement of IRL is searching within limited solution space, based on the derived monotonic property: the first-order derivative relationship between individual reservoir water release decisions for mainstream use (i.e., hydropower generation) as well as tributary use (i.e., regional water supply) and the cascade system's or a particular reservoir's water availability, along with the synchronicity and substitutability assumption of storage distribution in the cascade system. The improved algorithm is then applied to a real-world cascade reservoir system in the Yangtze River of China. The results demonstrate the high computational efficiency and reasonable interpretability of IRL.

Keywords: improved reinforcement learning; monotonic property; cascade reservoir system

1. Introduction

The operation of multi-reservoir systems is essential for regulating the spatial and temporal distribution of water resources to satisfy multiple water use needs, such as hydropower generation and regional water demand [1,2]. However, the optimization of the coordinated operation of a complex multi-reservoir system, particular for reservoirs in series, is always challenged by the high-dimensional decision-making process, nonlinear system dynamics, and stochasticity of runoff [1–3]. Various optimization techniques are proposed to address the above challenges, as can be found from the state-of-the-art reviews of Yeh [1], Wurbs [4], Labadie [5], Rani and Moreira [6], Dobson [7] and Giuliani et al [8].

According to Dobson et al. [7], optimization methods for multi-reservoir system operation can be categorized into release sequences, operating policy, and real-time optimization argument. Among these three categories, only the value function estimation based on Markov Decision Processes (MDP) can explicitly capture the dynamic nature of the optimization problem as well as the stochasticity of inflow by decomposing the complex optimization issues into a sequence of sub-problems. As one of the most famous optimization algorithms based on MDP, stochastic dynamic programming (SDP) has been widely used to solve reservoir optimization operation problems, since it can handle the nonlinear characteristics of the objective and constraints [1,5,9]. However, the applicability of SDP is restricted to only a limited number of individual reservoirs, due to the so-called curse of dimensionality, i.e., the exponential increase in computing time and memory requirements as the number of individual reservoirs increases. Various improvements to SDP have been proposed to achieve controlled computational efficiency [10,11]. For example, Tilmant and Kelman [12] introduced Successive Approximation Dynamic Programming (DPSA), which decomposes multi-reservoir operation problems into a series of single reservoir operation problems; Howson and Sancho [13] developed Progressive Optimization Algorithm (POA), transforming the multi-stage operation problem into multiple one-stage operation problems; Turgeon [14] and Turgeon and Charbonneau [15] proposed aggregation–disaggregation methods to reduce the high dimensionality of state variables. Although the computational complexity of SDP can be significantly reduced, a severe loss of solution accuracy might result from those improvements [1,16–19].

As the other optimization method based on MDP, reinforcement learning (RL) integrates the trial-and-error learning mechanisms with Bellman's principle of optimality [20–23] to approximate the optimal solutions. As one of the most popular RL methods for solving Markov decision problems, the Q-learning method [11] employs a depth-first search strategy and incrementally optimizes the visited state variables [24], which is different from enumeration over all discrete combinations of state variables to develop a value function and derive the optimal decisions in SDP. Based on this high efficiency computational advantage, Q-learning is widely adopted to solve single-reservoir and multi-reservoir optimization operation problems [24–26]. For example, Xu et al. [26] suggested a stochastic optimization method with the integration of Deep Q-Networks (DQN) with aggregation–disaggregation methods for cascade reservoir system optimization operation. However, the limitation of converging into a local optimal solution, as well as a relatively low interpretability of the decision-making process, needs to be further addressed for Q-learning applications in complex reservoir systems [27,28].

Recently, the use of mathematic characteristics of a reservoir system, e.g., the relationships between the decision variables and state variables [17,29,30], has been proposed to improve the MDP by achieving a high level of computational efficiency and effectiveness. For example, Zhao et al. [29–31] proposed improved SDP with a monotonic relationship between the decision variable and state variable for a single reservoir water supply, hydropower generation and flood control operation. Using the spatially distributed characteristics of individual reservoir storage, Zeng et al. [17] proposed an improved SDP algorithm based on a derived monotonic relationship between individual reservoir carryover storage and parallel reservoir system water availability. The high computational efficiency and effectiveness with controlled solution accuracy have been proven in a real case. However, the application of the monotonic decision property to overcome the dimensionality problem of cascade reservoir optimization operation has received little attention.

Following the work of Zeng et al. [17], the objective of this paper is to reduce the computational complexity of RL with a derived monotonic relationship between individual reservoir water release decisions and the water availability of the cascade system or a specific individual reservoir, which provides the theoretical basis for the improved RL algorithm design. According to our knowledge, this study can be the first to apply the monotonic property to address the curse of dimensionality for cascade reservoir optimization operation.

The rest of this paper is organized with a step-by-step structure as follows: Section 2 presents the formulation of the stochastic dynamic optimization model for the operation of reservoirs in series and specifies the optimality condition for the model without binding inequality constraints. On the basis of the specified optimality condition, Section 3 provides the first-order derivative relationship between individual reservoir water release decisions and the water availability of the cascade system or a specific individual reservoir. The monotonic property with reduced searching region of the improved RL method is described in Section 4. A cascade reservoir system in the Yangtze River of China is illustrated as a case study in Section 5. Finally, the conclusions of this paper are given in Section 6.

2. Optimality Condition for Multi-Objective Operation of Cascade Reservoir System

This section first describes a long-term optimization problem for the multi-objective operation of the cascade reservoir system, then develops a two-stage optimization operation model based on the stochastic dynamic programming and reinforcement learning methods. Following that, the optimality condition for the multi-objective operation of the cascade reservoir system is derived.

2.1. Stochastic Optimization Operation Model for Cascade Reservoir System

Considering the stochasticity of hydrologic inflow, the long-term operation of the cascade reservoir system illustrated in Figure 1 requires balancing water release for hydropower generation in the mainstream and water release for regional use in tributaries at each period. The objective is to maximize the expected value of multi-regional basin-wide benefits, integrating diverse water use purposes, such as ecology and hydropower production. In addition, the decision of water release at each period needs to obey the physical and engineering constraints of the cascade reservoir system, which includes the mass balance equation, release, and carryover storage constraints. The mathematical expression of the long-term multi-objective optimization operation for the cascade reservoir system can be written as follows:

$$\max_{IR,OR} \underset{I}{E}\left\langle \sum_{t=1}^{T} \left\{ \sum_{i=1}^{n} \left[B^i(IR_{i,t}) + B_i(OR_{i,t}) \right] \right\} \right\rangle \tag{1a}$$

$$s.t. \begin{cases} WA_{1,t} = S_{1,t-1} + I_{1,t} - E_{1,t} = S_{1,t} + IR_{1,t} + OR_{1,t} & \forall t \\ WA_{i,t} = S_{i,t-1} + I_{i,t} - E_{i,t} = S_{i,t} + IR_{i,t} + OR_{i,t} - IR_{i-1,t} & \forall i \in [2,n], t \\ 0 \le IR_{i,t} \le ID_{i,t} & \forall i \in [1,n], t \\ 0 \le OR_{i,t} \le OD_{i,t} & \forall i \in [1,n], t \\ S_{i,t}^{min} \le S_{i,t} \le S_{i,t}^{max} & \forall i \in [1,n], t \end{cases} \tag{1b}$$

where $E(\,)$ is the expectation function; $B^i(\,)$ is the benefit function of in-stream water release from i reservoir to meet the hydropower generation or ecological water use [32–35]; $B_i(\,)$ is the benefit function of out-stream water release from i reservoir to meet the regional consumptive water demand; particularly, both $B^i(\,)$ and $B_i(\,)$ are assumed to be concave [16,32–34,36]; T is the planning horizon of the cascade reservoir system; n is the number of reservoirs in the cascade system; $WA_{i,t}$ is the water availability of i reservoir at time t, which is defined as the sum of initial water storage plus current inflow and minus evaporation loss during this time period for i reservoir; $S_{i,t-1}$ is the initial storage of the reservoir i at the beginning of period t; $I_{i,t}$ is the inflow of i reservoir at period t; $E_{i,t}$ is the evaporation loss of i reservoir at period t; $IR_{i,t}$ is the water release for hydropower generation from i reservoir into mainstream at period t; $ID_{i,t}$ is the water demand from mainstream at period t; $OR_{i,t}$ is the water supply for tributary water user at period t; $OD_{i,t}$ is the tributary water demand of i reservoir at period t; $S_{i,t}^{min}$ and $S_{i,t}^{max}$ are the minimum and maximum storage of i reservoir, respectively.

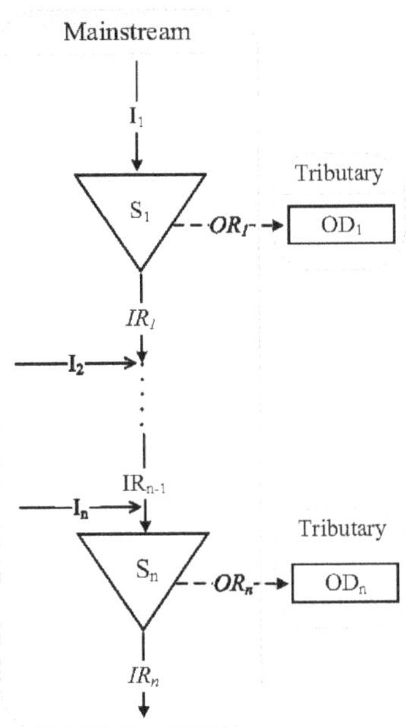

Figure 1. Cascade reservoir system with hydropower generation and tributary water supply objectives.

2.2. Stochastic Dynamic Programming for System Optimization Operation

The aforementioned long-term optimization model encompasses a substantial number of water release decisions for each reservoir, and deriving optimal solutions for such an extensive nonlinear optimization model poses significant challenges [1,2,5]. Based on the optimality principle of dynamic programming, the long-term optimization problem can be decomposed into a series of two-stage stochastic optimization problems. If the runoff of each reservoir can be described as a first-order Markov process, then the randomness of the inflow at the subsequent time $t + 1$ can be characterized using a prior inflow transition probability $P(I_{t+1}|I_t)$ given any specific $I_{i,t}$. Under this assumption, the long-term stochastic optimization model can be expressed using the following backward recursive equation:

$$CV_{t-1}(S_{t-1}, I_t) = \max_{IR_t, OR_t} \left\{ B_t(IR_t, OR_t) + \sum\nolimits_{I_{t+1}} [P(I_{t+1}|I_t) \times CV_t(S_t, I_{t+1})] \right\} \tag{2a}$$

$$B_t(IR_t, OR_t) = \sum\nolimits_{i=1}^{n} \left[B^i(IR_{i,t}) + B_i(OR_{i,t}) \right] \tag{2b}$$

$$S_{t-1} = \begin{bmatrix} S_{1,\,t-1} \\ \vdots \\ S_{n,\,t-1} \end{bmatrix};\ I_t = \begin{bmatrix} I_{1,t} \\ \vdots \\ I_{n,t} \end{bmatrix};\ IR_t = \begin{bmatrix} IR_{1,t} \\ \vdots \\ IR_{n,t} \end{bmatrix};\ OR_t = \begin{bmatrix} OR_{1,t} \\ \vdots \\ OR_{n,t} \end{bmatrix} \tag{2c}$$

where S_{t-1}, I_t, IR_t and OR_t represent the vector of the initial storage, inflow, and water release in the mainstream and tributary of the cascade reservoir system at period t, respectively; $CV(\,)$ is the value function of the cascade reservoir system at period t, representing the maximum benefit that can be achieved from period t to the end of the planning horizon, conditioned on the inflow and storage state combination. Previous studies [35,37] have shown that because the value function represents the cumulative utilities from period t to T, it is considered to share similar mathematical characteristics with the benefit function [29,31]. In other words, the value function $CV(\ )$ is concave.

To maintain consistency with the mathematical expression of the benefit function used in the reinforcement learning method, Equation (2a) is rewritten with the Q-value function as follows:

$$Q_{t-1}(S_{t-1}, I_t, IR_t, OR_t) = B_t(IR_t, OR_t) + \sum_{I_{t+1}} \left[P(I_{t+1}|I_t) \times \max_{IR_{t+1}, OR_{t+1}} [Q_t(S_t, I_{t+1}, IR_{t+1}, OR_{t+1})] \right] \quad (3a)$$

$$CV_{t-1}(S_{t-1}, I_t) = \max_{IR_t, OR_t} [Q_{t-1}(S_{t-1}, I_t, IR_t, OR_t)] \quad (3b)$$

where $Q(\ )$ is the approximate function of $CV(\ )$ in the reinforcement learning method.

2.3. Reinforcement Learning for System Optimization Operation

In order to construct the value function of the cascade reservoir system, the dynamic programming method based on the Bellman equation requires perfect information of the underlying stochastic hydrologic processes throughout the entire planning horizon. The well-known "curse of dimensionality" problem is caused by this assumption, even though it can capture the best solutions for long-term operation. To make efficient use of the searching process and reduce computational complexity, the Q-learning method in reinforcement learning adopts the concept of random sampling to gradually approximate the value function through incremental temporal difference methods. The multi-objective optimization model of the cascade reservoir system in the Q-learning approach can be written as follows:

$$\begin{aligned} Q_{t-1}(S_{t-1}, I_t, IR_t, OR_t) \leftarrow Q_{t-1}(S_{t-1}, I_t, IR_t, OR_t) + \alpha\{B_t(IR_t, OR_t) + \\ \max_{IR_{t+1}, OR_{t+1}} [Q_t(S_t, I_{t+1}, IR_{t+1}, OR_{t+1})] - Q_{t-1}(S_{t-1}, I_t, IR_t, OR_t)\Big\} \end{aligned} \quad (4)$$

where α is the learning rate parameter controlling the balance given to the immediate water use benefit and future value return.

Notably, the Q-learning approach only carries out incremental optimization for the sample state combination, as opposed to exhaustively listing every discrete combination of carryover storage at each period of SDP. This paper specifically assumes that the immediate water use benefit ($B^i(\ )$ and $B_i(\ )$) is provided at each period. Therefore, for any state combination, the convergence condition of the Q-value function is:

$$Q_{t-1}(S_{t-1}, I_t, IR_t, OR_t) \leftarrow B_t(IR_t, OR_t) + \max_{IR_{t+1}, OR_{t+1}} [Q_t(S_t, I_{t+1}, IR_{t+1}, OR_{t+1})] \quad (5)$$

Combining the above equation with Equation (3b), we have:

$$CV_{t-1}(S_{t-1}, I_t) = \max_{IR_t, OR_t} \left\{ B_t(IR_t, OR_t) + \max_{IR_{t+1}, OR_{t+1}} [Q_t(S_t, I_{t+1}, IR_{t+1}, OR_{t+1})] \right\} \quad (6)$$

According to the equation above, "Q-Learning converges to the optimum action values with probability 1 so long as all actions are repeatedly sampled in all states and the action values are represented discretely" [21] (p. 279). Following this, the optimality condition of Q-learning and dynamic programming methods is equivalent, and the computational complexity of Q-learning can be significantly lower.

2.4. The Optimality Principle of Equal Marginal Utility

If boundary constraints on release and carryover storage (i.e., Equation (1b)) are non-binding, the Lagrange multiplier approach [36] can be used to specify the optimality condition of the two-stage optimization model with the mass balance equation mentioned above, which leads to:

$$\begin{cases} \frac{dB^i}{dIR^*_{i,t}} + \frac{\partial Q_t}{\partial S^*_{i,t}} \times \frac{\partial S^*_{i,t}}{\partial IR^*_{i,t}} + \frac{\partial Q_t}{\partial S^*_{i+1,t}} \times \frac{\partial S^*_{i+1,t}}{\partial IR^*_{i,t}} = 0 \\ \frac{dB_i}{dOR^*_{i,t}} + \frac{\partial Q_t}{\partial S^*_{i,t}} \times \frac{\partial S^*_{i,t}}{\partial OR^*_{i,t}} = 0 \end{cases} \rightarrow \begin{cases} \frac{dB^i}{dIR^*_{i,t}} = \frac{\partial Q_t}{\partial S^*_{i,t}} - \frac{\partial Q_t}{\partial S^*_{i+1,t}} \\ \frac{dB_i}{dOR^*_{i,t}} = \frac{\partial Q_t}{\partial S^*_{i,t}} \end{cases} \forall\, i \neq n \tag{7a}$$

$$\begin{cases} \frac{dB^n}{dIR^*_{n,t}} + \frac{\partial Q_t}{\partial S^*_{n,t}} \times \frac{\partial S^*_{n,t}}{\partial IR^*_{n,t}} = 0 \\ \frac{dB_n}{dOR^*_{n,t}} + \frac{\partial Q_t}{\partial S^*_{n,t}} \times \frac{\partial S^*_{n,t}}{\partial OR^*_{n,t}} = 0 \end{cases} \rightarrow \begin{cases} \frac{dB^n}{dIR^*_{n,t}} = \frac{\partial Q_t}{\partial S^*_{n,t}} \\ \frac{dB_n}{dOR^*_{n,t}} = \frac{\partial Q_t}{\partial S^*_{n,t}} \end{cases} \forall\, i = n \tag{7b}$$

The aforementioned formulas illustrate the equal marginal utility optimality principle for the multi-objective optimization operation for the cascade reservoir system. That is, (1) for the individual reservoir not at the lowest elevation $(i \neq n)$, the marginal benefit of releasing water for the tributary use should be equal to the marginal benefit of carryover storage, and the marginal benefit of releasing water in the mainstream for hydropower generation should be equal to the deviation between this reservoir and the downstream reservoir's marginal benefits of carryover storage; (2) for the individual reservoir at the lowest elevation $(i = n)$, the marginal benefits of releasing water for both mainstream and tributary use should be equal to the marginal benefit of carryover storage.

3. Monotonic Property for Multi-Objective Optimization Operation of Cascade Reservoir Systems

This section derives the monotonic relationship between the optimal water release decision of an individual reservoir and the (total) water availability of the cascade system or a specific individual reservoir.

3.1. Monotonic Property of Water Release for the Tributary Use

For further mathematical derivation, two assumptions regarding the synchronicity and substitutability are used to characterize the spatial distribution of individual reservoir storage.

Assumption 1. *The carryover storage of each reservoir is a non-decreasing function of the cascade system's total storage, i.e., synchronic property. The mathematical expression of this assumption is*

$$\frac{\partial S^*_{i,t}}{\partial S^{Sum}_t} \geq 0 \tag{8a}$$

$$S^{Sum}_t = \sum_{i=1}^{n} S^*_{i,t} \tag{8b}$$

In the equations, S^{Sum}_t *is the total carryover storage of the cascade system at time t. The assumption of storage state synchronicity (Equation (8a)) indicates that the reservoir system should allocate the increased storage according to the prior optimal storage distribution. Theoretical research and empirical findings [14–17] have demonstrated that reservoir operation rules based on this assumption can successfully lower the system's invalid water spillage and improve the coordination of decisions regarding water supply and storage across individual reservoirs [16,38,39]. Therefore, this property is widely adopted for the operation of a complex reservoir system, such as the storage target curves [40,41] and the parametric rule [42].*

Assumption 2. *The stored water in each reservoir is substitutable among each reservoir. The mathematical expression of resource substitutability is defined as follows [17]:*

$$\frac{\partial^2 Q_t}{\partial S_{i,t}^* \partial\left(S_t^{Sum} - S_{i,t}^*\right)} \leq 0 \tag{9}$$

The above equation indicates that if the total storage of other reservoirs increases, the marginal value of carryover storage for each reservoir either decreases or stays the same because there is more water available in the system for both mainstream and tributary demand, particularly for the successive drought year [17].

Based on the above two assumptions, the monotonic relationship between the optimal water release for tributary use and the total system water availability at period t can be derived.

For any two optimal cascade systems, carryover storage $S_t^{Sum,1}$ and $S_t^{Sum,2}$ at period t, if $S_t^{Sum,1} \leq S_t^{Sum,2}$, then according to the Assumption 1, we have $S_{i,t}^1 \leq S_{i,t}^2$. Using the contradiction method, it can be deduced that $\left(S_t^{Sum,1} - S_{i,t}^1\right) \leq \left(S_t^{Sum,2} - S_{i,t}^2\right)$, otherwise, it would exist the carryover storage of an individual reservoir: $S_{i,t}^1 \geq S_{i,t}^2$, which is in violation of Assumption 1.

Subsequently, according to Assumption 2, we have

$$\left.\frac{\partial Q_t}{\partial S_{i,t}^1}\right|_{(S_{1,t}^1,\cdots,S_{m,t}^1)} \geq \left.\frac{\partial Q_t}{\partial S_{i,t}^1}\right|_{(S_{1,t}^2,\cdots,S_{m,t}^2)} \tag{10}$$

According to the concavity of the value function (decreasing marginal benefit), i.e., conditioned on the unchanged carryover storage of other reservoirs in the system, a higher storage will lead to a lower marginal benefit:

$$\left.\frac{\partial Q_t}{\partial S_{i,t}^1}\right|_{(S_{1,t}^2,\cdots,S_{m,t}^2)} \geq \left.\frac{\partial Q_t}{\partial S_{i,t}^2}\right|_{(S_{1,t}^2,\cdots,S_{m,t}^2)} \tag{11}$$

Combining Equations (10) and (11), we can obtain:

$$\left.\frac{\partial Q_t}{\partial S_{i,t}^1}\right|_{(S_{1,t}^1,\cdots,S_{m,t}^1)} \geq \left.\frac{\partial Q_t}{\partial S_{i,t}^2}\right|_{(S_{1,t}^2,\cdots,S_{m,t}^2)} \tag{12}$$

Based on the marginal benefit relationship in Equations (7a) and (7b), it can be derived that $\frac{dB_i}{dOR_{i,t}^1} \geq \frac{dB_i}{dOR_{i,t}^2}$. According to the decreasing marginal benefit property of water release, the optimal water release for tributary use must satisfy the following relationship: $OR_{i,t}^1 \leq OR_{i,t}^2$. Similarly, it can be deduced that the optimal water release for hydropower generation from the reservoir at the lowest elevation $(i = n)$ must satisfy $IR_{n,t}^1 \leq IR_{n,t}^2$. The cascade system's total water balance equation can be used to obtain:

$$WA_t^{Sum,1} = S_t^{Sum,1} + \sum_{i=1}^{n} OR_{i,t}^1 + IR_{n,t}^1 \leq S_t^{Sum,2} + \sum_{i=1}^{n} OR_{i,t}^2 + IR_{n,t}^2 = WA_t^{Sum,2} \tag{13a}$$

$$WA_t^{Sum} = \sum_{i=1}^{n} WA_{i,t} \tag{13b}$$

That is, if the total carryover storage of the cascade reservoir system increases, the carryover storage of each reservoir, water release for the tributary use, and the total available water of the cascade system should increase or remain unchanged, and vice versa. As a result, the non-decreasing monotonicity between a cascade reservoir system's total water availability, water release for tributary use, and optimal carryover storage is determined.

3.2. Monotonic Property of Water Release for the Mainstream Use

Assume that only the water availability of reservoir k increases and the water availability of other reservoirs remains the same, i.e., for any two available water combinations, $WA^1_{k,t} \leq WA^2_{k,t}$ and $WA^1_{i,t} = WA^2_{i,t}(i \neq k)$. According to the mass balance equation in Equation (1b), the optimal water release from the upstream and downstream reservoirs can be expressed as:

$$IR_{k-1,t} = \sum_{i=1}^{k-1} \left(WA_{i,t} - S_{i,t} - OR_{i,t} \right) \tag{14a}$$

$$IR_{k,t} = \sum_{i=k+1}^{n} \left(S_{i,t} + OR_{i,t} - WA_{i,t} \right) + IR_{n,t} \tag{14b}$$

Water release for each reservoir's tributary satisfies the monotonically non-decreasing property due to carryover storage. Therefore, according to Equation (14a), the water release of the upstream reservoir has the following relationship: $IR^1_{k-1,t} \geq IR^2_{k-1,t}$, provided that the amount of water available in the reservoir remains constant.

Similarly, from Equation (14b), it can be deduced that the water release of reservoir k and its downstream reservoirs should satisfy $IR^1_{k,t} \leq IR^2_{k,t}$.

The above derivation implies that as the available water of a specific individual reservoir increases, the water release for the mainstream demand from this reservoir and the corresponding downstream reservoirs should not decrease, while the upstream reservoir should not increase. That is, the non-decreasing monotonicity between the water release for the mainstream use from a specific particular reservoir or any downstream reservoir and the water availability of such individual reservoir is derived, whereas the non-increasing monotonicity between the upstream water release and the water availability is specified.

3.3. Quantitative Analysis Based on the Monotonic Property

Based on the monotonicity of the water storage and release decisions of the cascade reservoir system, the following quantitative analysis can be further conducted:

If the cascade system's total available water increases with an equal interval length (i.e., *IL*), the sum increase in the total water storage and the water release for tributaries would not exceed *IL*, according to the system's total water balance equation (Equation (13a)). Following the non-decreasing monotonical relationship between the optimal decision of individual reservoir and the total available water of the system, it can deduce with the contradiction method that the carryover storage and the water release of individual reservoir will not exceed *IL*.

Similarly, if reservoir k's water availability increases with an interval length (i.e., *IL*) while other reservoirs' available water stays constant, then both the decrease amount of water release from upstream reservoir and the increase amount of release from downstream will not exceed IL, since the total increment or decrement does not exceed IL. Following the chain rule of derivation, the monotonic property would be maintained in the subsequent operation periods.

In summary, the monotonic relationship between different water release and the water availability in the cascade system is shown in Figure 2.

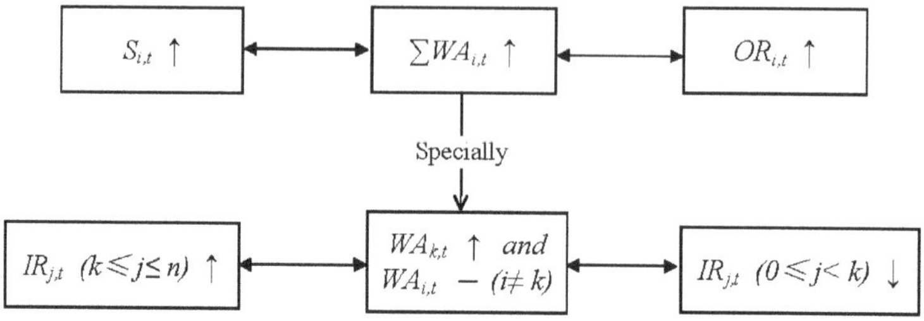

Figure 2. Monotonic relationship between optimization decisions and water availability; where, ↑ denotes non-decreasing monotonic relationship; ↓ denotes non-increasing monotonic relationship; – denotes without any change.

4. Improved Reinforcement Learning Method for Optimization Operation of Cascade Reservoir System

Reinforcement learning for solving the joint optimization operation of the cascade reservoir system would not utilize the prior knowledge of water mass balance of each reservoir, as well as the hydrologic connection equation between the upper and lower reservoirs to specify the monotonic decision-making property. Based on the monotonic relationship between the optimal release decisions and the water availability, the efficiency of the reinforcement learning method can be effectively enhanced. The main principle of the improved optimization method can be summarized as two steps: initial state optimization, and searching space reduction, as shown in Figure 3.

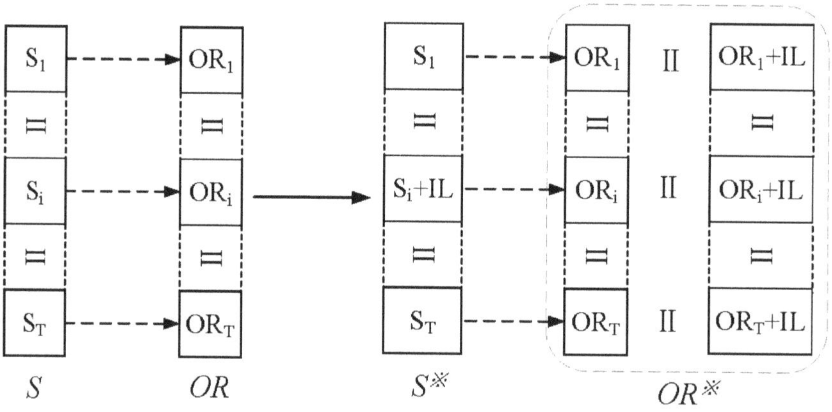

Figure 3. Schematic diagram of the improved reinforcement learning method.

(1) Initial state optimization: First, the storage volume of each reservoir is discretized with an equal interval length (i.e., *IL*) for each reservoir at each period, i.e., the discretization interval of each reservoir is set as IL, and the number of discretization of reservoir i is NS_I. After determining the feasible space of the water release decision for each initial state combination based on the constraints of water supply and storage, Newton's method is applied to specify the optimal release for the mainstream and tributary from the reservoir's water balance equation (Equation (1b)) and the marginal benefit equalization equation (Equations (7a) and (7b)). The feasibility of the decision is checked according to the water release and carryover storage inequality constraints, where, if the decision variables that do not meet these constraints, according the suggested procedure [43], the boundary of

the inequality constraint is set as the optimal value, and the corresponding water release is adjusted according to the mass balance equation. In this process, the reinforcement learning method employs incremental sampling to identify the set of solutions for each reservoir's optimal water release over the course of the scheduling period.

(2) Searching space reduction: Assume that the optimal water release and carry-over storage for the entire periods corresponding to the j sample are $\left(IR_1^{\#},\cdots,IR_T^{\#}\right)$; $\left(OR_1^{\#},\cdots,OR_T^{\#}\right)$ and $(S_1^{\#},\cdots,S_T^{\#})$, respectively. If the initial water storage state of j sample increases with IL, then the water release for the tributary can be defined as the decision variable in Newton's method, and the release for the mainstream can be specified using Equation (14a). According to the monotonicity of the water release decision, the optimal release would be identified within the range between $\left(OR_1^{\#},\cdots,OR_T^{\#}\right)$ and $\left(OR_1^{\#}+IL,\cdots,OR_T^{\#}+IL\right)$. As a result, the improved method only searches in the feasible interval with a variation of IL, which leads to a higher computational efficiency than searching the entire feasible region in the traditional reinforcement learning method.

5. Case Study

5.1. System Description

As illustrated in Figure 4, the WB-Xll-Xjb-Sx reservoir system is a giant cascade reservoir system on the mainstream of the upper reaches of the Yangtze River in China. Among them, the storage capacity of WB reservoir is 13.46 billion m^3, and the annual average runoff is 241.34 billion m^3, with the annual demand of 120 billion m^3 in the mainstream and 3.20 billion m^3 in the tributary; the storage capacity of Xll reservoir is 6.46 billion m^3, and the annual average runoff is 20.64 billion m^3 with the annual demand of 65 billion m^3 in the mainstream and 0.5 billion m^3 in the tributary; the storage capacity of Xjb is 0.90 billion m^3, and the annual average runoff is 7.28 billion m^3 with the annual demand of 15 billion m^3 in the mainstream and 1.8 billion m^3 in the tributary; the storage capacity of Sx is 22.15 billion m^3, and the annual average runoff is 280.18 billion m^3 with the annual demand of 200 billion m^3 in the mainstream and 4.5 billion m^3 in the tributary.

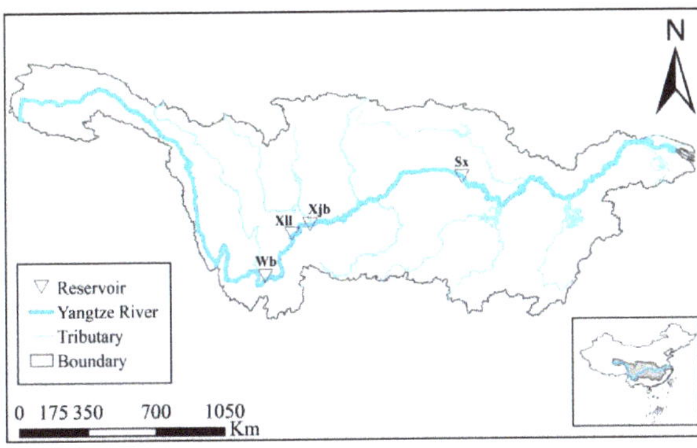

Figure 4. The layout of WB-Xll-Xjb-Sx cascade reservoir system.

In this paper, the commonly used benefit function [44] is set as the objective function to evaluate the system operation performance, as below:

$$B^i(IR_{i,t}) = \begin{cases} \left(\frac{E_{i,t}}{TE_{i,t}}\right)^{\alpha} = \left(\frac{\gamma\times IR_{i,t}}{\gamma\times ID_{i,t}}\right)^{\alpha} = \left(\frac{IR_{i,t}}{ID_{i,t}}\right)^{\alpha}; & IR_{i,t}\le ID_{i,t} \\ 1 \quad ; & IR_{i,t} > ID_{i,t} \end{cases} \quad (15a)$$

$$B_i(OR_{i,t}) = \begin{cases} \left(\frac{OR_{i,t}}{OD_{i,t}}\right)^{\alpha}; & OR_{i,t} \le OD_{i,t} \\ 1 & ; OR_{i,t} > OD_{i,t} \end{cases} \tag{15b}$$

where $E_{i,t}$ and $IE_{i,t}$ are the hydropower generation of the release and maximum release from reservoir i to the mainstream at period t, respectively; γ is the transition coefficient from water into hydropower, which is given from an experimental test; α is the exponential parameter of the benefit function generated by the water release. $\alpha < 1$ is defined for a concave function. According to the literature [35] and sensitivity parameter analysis, the value of α is set to 0.8 in this paper. It is assumed that the weighting factors for various water utilities are equal to 1 for the sake of the subsequent comparison analysis.

5.2. Comparison Analysis of Optimization Results

The WB-Xll-Xjb-Sx cascade reservoir system's comparison analysis of multi-objective optimization techniques divides a hydrological year into 12 periods, with each period's operation time step of a month. Historical inflow runoff from 1959 to 2023 is collected from each reservoir to generate the corresponding runoff state transition probability function. The inflow is discretized as four for each reservoir, and six storage discretization levels (i.e., interval lengths), ranging from 0.2 to 0.4 billion m^3, are employed for the algorithm implementation. Stochastic dynamic programming (SDP), reinforcement learning (RL), and improved reinforcement learning (IRL) methods with a learning rate of 0.9 are executed with GUN Fortran to solve the optimization operation of the cascade reservoir system on a ThinkPad T15g workstation (i7-10750H/16 GB), provided by Lenovo, Beijing, China. The optimization results under different storage discretization levels are shown in Figures 5–7.

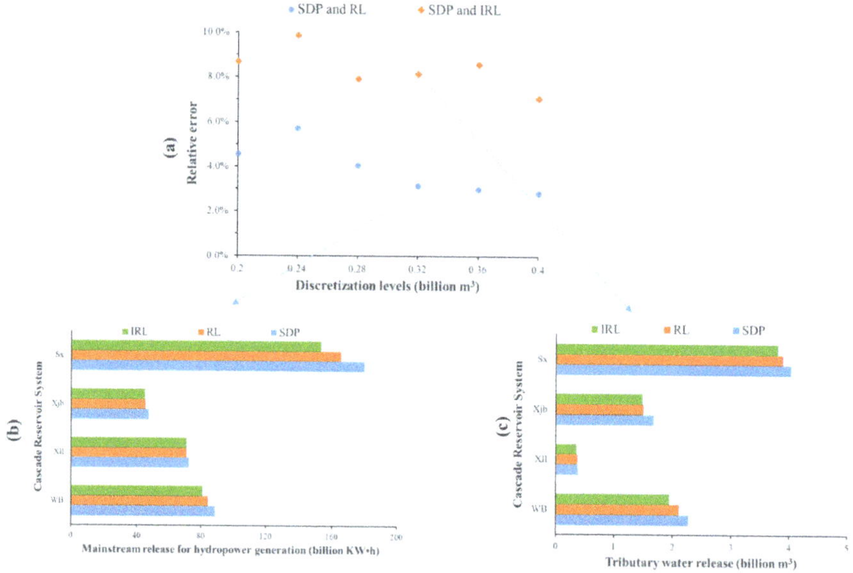

Figure 5. Differences between SDP and RL or SDP and IRL (**a**) relative error under different combinations of storage discretization levels; (**b**) hydropower generation under the discretization level of 0.24 billion m^3; (**c**) tributary release under the discretization level of 0.24 billion m^3.

To compare the results from different methods, the optimization results of SDP are set as a benchmark, and the differences between SDP and RL or SDP and IRL are measured by relative error in Figure 5. As can be seen, there is no significant difference in the optimization results obtained by the three optimization algorithms under six different discretization intervals, with relative errors all kept within 10%. Only when the reservoir capacity discretization interval is set to 0.24 billion m^3, the relative errors between SDP and RL, as well as the error between SDP and IRL, both reach the maximum of 5.73% and 9.87%, respectively. As further illustrated in Figure 5b,c, the maximum relative error mainly results from the optimization operation of the Sx reservoir, which has a relatively large storage capacity and water demand.

The computing time of SDP, RL, and IRL with different discretization levels is illustrated in Figure 6a,b. The computation time of IRL is consistently less than 0.10% of that from SDP and 5% of that from RL for all storage discretization levels. For example, the time of SDP is 4109.47 s and RL of 52.83 s, while that of RIDP is 2.10 s under the discretization level of 0.32 billion m^3 for running the models over a one-year time horizon. This conclusion indicates the high computational efficiency of IRL.

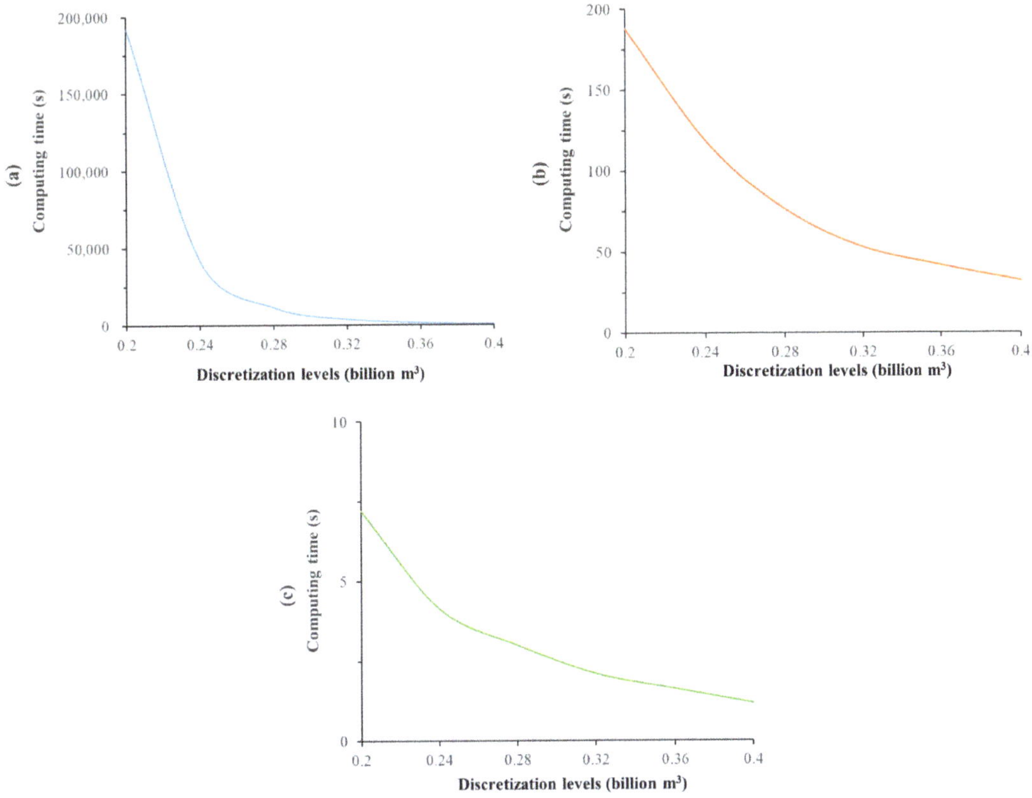

Figure 6. Computing time under the discretization level of 0.32 billion m^3 for running the models over a one-year time horizon (**a**) SDP; (**b**) RL; (**c**) IRL.

The carryover storage distribution of the cascade reservoir system in September is given in Figure 7. As can be seen from Figure 7a,b, the carryover storage of the Xll reservoir obtained from SDP or RL would not always increase with the total storage of the cascade reservoir system increases, particularly for a high system storage level. In contrast, the carryover storage of all individual reservoirs derived from IRL would increase with the total storage of the cascade system in Figure 7c, which follows the synchronic storage distribution property. Since September is often defined as the refill period in the Yangtze River, a relatively high storage of the system would seldom result in a relatively low storage of the individual reservoir, which is not consistent with the previous numerical study [45] or the commonly used operating rules [32–34]. These distribution results imply the relatively high interpretability of the solutions from IRL.

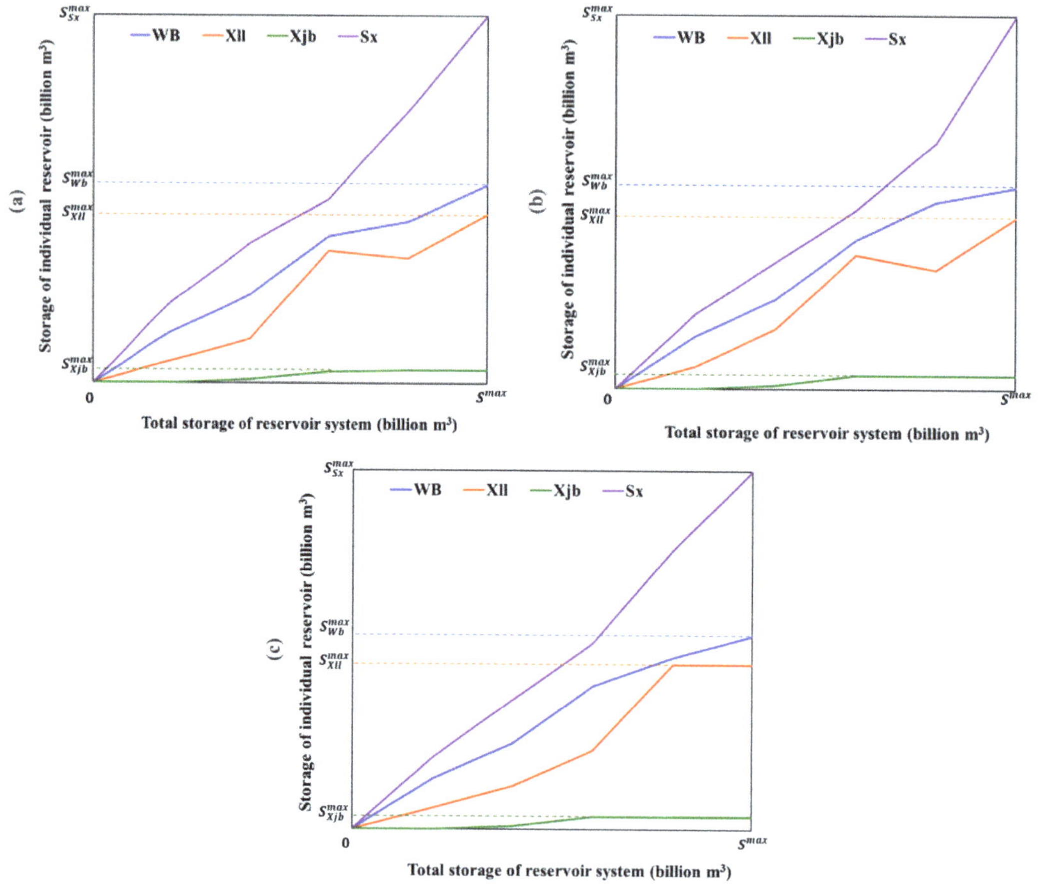

Figure 7. Carryover storage distribution of the cascade reservoir system in September (**a**) SDP; (**b**) RL; (**c**) IRL.

6. Conclusions

Based on the synchronicity and substitutability property of storage distribution in the cascade reservoir system, the first-order derivative relationship between individual reservoir water release decisions and the water availability of the cascade system or a specific individual reservoir is derived. Theoretical analysis results show that: (1) the optimal carryover storage and water release for the tributary use from each individual reservoir is a non-decreasing monotonic function of the cascade reservoir system's overall water availability; (2) the water release for mainstream use from a specific reservoir, or from any downstream reservoir, is a non-decreasing monotonic function of the water availability of that individual reservoir. Conversely, the relationship between upstream water release and water availability is characterized by non-increasing monotonicity.

Considering the derived monotonic properties among carryover storage, water release decisions from each individual reservoir, and the water availability of the cascade reservoir system, an improved reinforcement learning method with a reduced searching region has been developed. To demonstrate the effectiveness and efficiency of the proposed algorithm, the WB-Xll-Xjb-Sx reservoir system in the upper reaches of the Yangtze River in China is employed as a case study. Numerical results from this case study indicate that, compared to standard dynamic programming (SDP) and traditional reinforcement learning (RL), the improved method significantly reduces computation time without substantial loss of solution accuracy, particularly in preserving highly interpretable optimization results.

Author Contributions: Conceptualization, X.L.; Methodology, X.L.; Software, S.C.; Validation, H.M.; Formal analysis, X.L.; Investigation, H.M.; Resources, S.C.; Data curation, Y.X.; Writing—original draft, X.L.; Writing—review & editing, X.Z.; Visualization, Y.X.; Supervision, X.Z.; Project administration, X.Z.; Funding acquisition, X.Z. All authors have read and agreed to the published version of the manuscript.

Funding: This research is supported by the Major Science and Technology Projects of Ministry of Water Resources (SKS-2022119), the Open Research Fund of Hubei Key Laboratory of Intelligent Yangtze and Hydroelectric Science, China Yangtze Power Co., Ltd. (2422020009), National Key Research and Development Program of China (2023YFC3208403) and National Natural Science Foundation of China (52209016; 52479025).

Data Availability Statement: Some or all data, models, or code that support the findings of this study is available from the corresponding author upon reasonable request due to privacy and legislation.

Conflicts of Interest: Authors Xiang Li, Haoyu Ma and Yang Xu were employed by China Yangtze Power Co., Ltd. The research design, analysis, and conclusions presented in this paper were conducted independently and do not represent the official views or interests of the company. The remaining authors declare that the research was conducted in the absence of any commercial or financial relationships that could be construed as a potential conflict of interest.

References

1. Yeh, W.W.G. Reservoir Management and Operations Models: A State-of-the-Art Review. *Water Resour. Res.* **1985**, *21*, 1797–1818. [CrossRef]
2. Oliverira, R.; Loucks, D.P. Operating Rules for Multireservoir Systems. *Water Resour. Res.* **1997**, *33*, 839–852. [CrossRef]
3. Ahmad, A.; El-Shafie, A.; Razali, S.F.M.; Mohamad, Z.S. Reservoir Optimization in Water Resources: A Review. *Water Resour. Manag.* **2014**, *28*, 3391–3405. [CrossRef]
4. Wurbs, R.A. Reservoir System Simulation and Optimization Models. *J. Water Resour. Plan. Manag.* **1993**, *119*, 455–472. [CrossRef]
5. Labadie, J. Optimal Operation of Multi-Reservoir Systems: State-of-the-Art Review. *J. Water Resour. Plan. Manag.* **2004**, *130*, 93–111. [CrossRef]
6. Rani, D.; Moreira, M.M. Simulation-Optimization Modeling: A Survey and Potential Application in Reservoir Systems Operation. *Water Resour. Manag.* **2010**, *24*, 1107–1138. [CrossRef]
7. Dobson, B.; Wagener, T.; Pianosi, F. An Argument-Driven Classification and Comparison of Reservoir Operation Optimization Methods. *Adv. Water Resour.* **2019**, *128*, 74–86. [CrossRef]
8. Giuliani, M.; Lamontagne, J.R.; Reed, P.M.; Castelletti, A. A State-of-the-Art Review of Optimal Reservoir Control for Managing Conflicting Demands in a Changing World. *Water Resour. Res.* **2021**, *57*, e2021WR029927. [CrossRef]
9. Mendoza-Ramírez, R.; Silva, R.; Domínguez-Mora, R.; Juan-Diego, E.; Carrizosa-Elizondo, E. Comparison of Two Convergence Criterion in the Optimization Process Using a Recursive Method in a Multi-Reservoir System. *Water* **2022**, *14*, 2952. [CrossRef]
10. Powell, W. *Approximate Dynamic Programming: Solving the Curses of Dimensionality*, 2nd ed.; Wiley: Hoboken, NJ, USA, 2011; ISBN 9780470604458.
11. Sutton, R.; Barto, A. *Reinforcement Learning: An Introduction*, 2nd ed.; MIT Press: Cambridge, MA, USA, 2018; ISBN 9780262039246.
12. Tilmant, A.; Kelman, R. A Stochastic Approach to Analyze Trade-Offs and Risks Associated with Large-Scale Water Resources Systems. *Water Resour. Res.* **2007**, *43*, W06425. [CrossRef]

13. Howson, H.R.; Sancho, N.G.F. A New Algorithm for the Solution of Multistate Dynamic Programming Problems. *Math. Program.* **1975**, *8*, 104–116. [CrossRef]
14. Turgeon, A. Optimal Operation of Multireservoir Power Systems with Stochastic Inflows. *Water Resour. Res.* **1980**, *16*, 275–283. [CrossRef]
15. Turgeon, A.; Charbonneau, R. An Aggregation-Disaggregation Approach to Long-Term Reservoir Management. *Water Resour. Res.* **1998**, *34*, 3585–3594. [CrossRef]
16. Zeng, X.; Hu, T.S.; Xiong, L.H.; Cao, Z.X.; Xu, C.Y. Derivation of Operation Rules for Reservoirs in Parallel with Joint Water Demand. *Water Resour. Res.* **2015**, *51*, 9539–9563. [CrossRef]
17. Zeng, X.; Hu, T.S.; Cai, X.M.; Zhou, Y.L.; Wang, X. Improved Dynamic Programming for Parallel Reservoir System Operation Optimization. *Adv. Water Resour.* **2019**, *131*, 103373. [CrossRef]
18. Beiranvand, B.; Ashofteh, P.S. A Systematic Review of Optimization of Dams Reservoir Operation Using the Meta-heuristic Algorithms. *Water Resour. Manag.* **2023**, *37*, 3457–3526. [CrossRef]
19. Emami, M.; Nazif, S.; Mousavi, S.F.; Karami, H.; Daccache, A. A hybrid constrained coral reefs optimization algorithm with machine learning for optimizing multi-reservoir systems operation. *J. Environ. Manag.* **2021**, *286*, 112250. [CrossRef] [PubMed]
20. Sutton, R.S. Learning to Predict by the Methods of Temporal Differences. *Mach. Learn.* **1988**, *3*, 9–44. [CrossRef]
21. Watkins, C.J.C.H.; Dayan, P. Technical Note: Q-Learning. *Mach. Learn.* **1992**, *8*, 279–292. [CrossRef]
22. Castro Freibott, R.; García Sánchez, Á.; Espiga-Fernández, F.; González-Santander de la Cruz, G. Intraday Multireservoir Hydropower Optimization with Alternative Deep Reinforcement Learning Configurations. In *Decision Sciences (DSA ISC 2024)*; Lecture Notes in Computer Science; Juan, A.A., Faulin, J., Lopez-Lopez, D., Eds.; Springer: Cham, Switzerland, 2025; Volume 14778.
23. Luo, W.; Wang, C.; Zhang, Y.; Zhao, J.; Huang, Z.; Wang, J.; Zhang, C. A Deep Reinforcement Learning Approach for Joint Scheduling of Cascade Reservoir System. *J. Hydrol.* **2025**, *651*, 132515. [CrossRef]
24. Lee, J.H.; Labadie, J.W. Stochastic Optimization of Multireservoir Systems via Reinforcement Learning. *Water Resour. Res.* **2007**, *43*, W11408. [CrossRef]
25. Castelletti, A. A Reinforcement Learning Approach for the Operational Management of a Water System. In *Proceedings of IFAC Workshop Modelling and Control in Environmental Issues*; Elsevier: Yokohama, Japan, 2001; pp. 22–23.
26. Xu, W.; Zhang, X.; Peng, A.B.; Liang, Y. Deep Reinforcement Learning for Cascaded Hydropower Reservoirs Considering Inflow Forecasts. *Water Resour. Manag.* **2020**, *34*, 3003–3018. [CrossRef]
27. Wu, R.; Wang, R.; Hao, J.; Wu, Q.; Wang, P. Multiobjective Hydropower Reservoir Operation Optimization with Transformer-Based Deep Reinforcement Learning. *arXiv* **2023**, arXiv:230705643.
28. Mitjana, F.; Denault, M.; Demeester, K. Managing Chance-Constrained Hydropower with Reinforcement Learning and Backoffs. *Adv. Water Resour.* **2022**, *169*, 104308. [CrossRef]
29. Zhao, T.T.G.; Cai, X.M.; Lei, X.H.; Wang, H. Improved Dynamic Programming for Reservoir Operation Optimization with a Concave Objective Function. *J. Water Resour. Plan. Manag.* **2012**, *138*, 590–596. [CrossRef]
30. Zhao, T.T.G.; Zhao, J.S.; Yang, D.W. Improved Dynamic Programming for Hydropower Reservoir Operation. *J. Water Resour. Plan. Manag.* **2014**, *140*, 365–374. [CrossRef]
31. Zhao, T.T.G.; Zhao, J.S.; Lei, X.H.; Wang, X.; Wu, B.S. Improved Dynamic Programming for Reservoir Flood Control Operation. *Water Resour. Manag.* **2017**, *31*, 2047–2063. [CrossRef]
32. You, J.Y.; Cai, X.M. Hedging Rule for Reservoir Operations: 1. A Theoretical Analysis. *Water Resour. Res.* **2008**, *44*, W01415. [CrossRef]
33. Shiau, J.T. Analytical Optimal Hedging with Explicit Incorporation of Reservoir Release and Carryover Storage Targets. *Water Resour. Res.* **2011**, *47*, W01515. [CrossRef]
34. Zeng, X.; Lund, J.R.; Cai, X.M. Linear versus Nonlinear (Convex and Concave) Hedging Rules for Reservoir Optimization Operation. *Water Resour. Res.* **2021**, *57*, e2020WR029160. [CrossRef]
35. Fama, E.F. Multiperiod Consumption-Investment Decisions. *Am. Econ. Rev.* **1970**, *60*, 163–174.
36. Draper, A.J.; Lund, J.R. Optimal Hedging and Carryover Storage Value. *J. Water Resour. Plan. Manag.* **2004**, *130*, 83–87. [CrossRef]
37. Carroll, C.D.; Kimball, M.S. On the Concavity of the Consumption Function. *Econometrica* **1996**, *64*, 981–992. [CrossRef]
38. Johnson, S.A.; Stedinger, J.R.; Staschus, K. Heuristic Operating Policies for Reservoir System Simulation. *Water Resour. Res.* **1991**, *27*, 673–685. [CrossRef]
39. Sand, G.M. An Analytical Investigation of Operating Policies for Water-Supply Reservoirs in Parallel. Ph.D. Dissertation, Cornell University, Ithaca, NY, USA, 1984.
40. Lund, J.R.; Ferreira, I. Operating Rule Optimization for Missouri River Reservoir System. *J. Water Resour. Plan. Manag.* **1996**, *122*, 287–295. [CrossRef]

41. Perera, B.J.C.; Codner, P.G. Reservoir Targets for Urban Water Supply Systems. *J. Water Resour. Plan. Manag.* **1996**, *122*, 270–279. [CrossRef]
42. Nalbantis, I.; Koutsoyiannis, D. A Parametric Rule for Planning and Management of Multiple Reservoir Systems. *Water Resour. Res.* **1997**, *33*, 2165–2177. [CrossRef]
43. Zhao, J.S.; Cai, X.M.; Wang, Z.J. Optimality conditions for a two-stage reservoir operation problem. *Water Resour. Res.* **2011**, *47*, W08503. [CrossRef]
44. You, J.Y.; Cai, X.M. Hedging Rule for Reservoir Operations: 2. A Numerical Model. *Water Resour. Res.* **2008**, *44*, W01416. [CrossRef]
45. Wang, C.; Jiang, Z.; Xu, Y.; Wang, S.; Wang, P. Discussion on the monotonicity principle of the two-stage problem in joint optimal operation of cascade hydropower stations. *J. Hydrol.* **2023**, *623*, 129803. [CrossRef]

water

MDPI

Article

Identifying Worst Transient Cases and Optimizing Surge Protection for Existing Water Networks

Hossam Mohamed Ahmed [1,*], Yehya Emad Imam [1,*], Hamdy Ahmed El-Ghandour [2] and Amgad Saad Elansary [1]

[1] Faculty of Engineering, Irrigation and Hydraulics Department, Cairo University, Giza 12613, Egypt; amgad.ansary@gmail.com
[2] Faculty of Engineering, Civil Engineering Department, Damietta University, Damietta 34511, Egypt; hamdy_elghandour@yahoo.com
* Correspondence: hossam.mohamed.j@eng-st.cu.edu.eg (H.M.A.); yimam@eng.cu.edu.eg (Y.E.I.)

Abstract: Previous studies of transients in existing water distribution networks (WDNs) accounted for only single worst cases in optimizing surge protection measures, considered only pressure at pipe end nodes, and did not examine the effect of regulating the duration of demand increase. This study presents a comprehensive model for identifying the minimal set of worst transient cases for which optimized surge protection achieves zero Surge Damage Potential Factor (SDPF) for all transient loading cases. The model introduces $SDPF_P$ to account for pressure at all computational nodes along pipes, as opposed to relying on $SDPF_N$, which considers pressure at pipe end nodes only. The existing New York Tunnel network was used for model validation and for determining the optimal diameters for additional duplicate pipes to achieve higher pressure under steady-state conditions and protect the network from transients due to demand increase. Compared to previous studies, the model achieved $SDPF_N = 0$ with a lower cost for sudden demand increase at a single predefined node. For concurrent sudden demand increase at two nodes, the model identified a total of 8 critical transient cases and corresponding optimum duplicate pipe diameters that achieved $SDPF_N = 0$ and $SDPF_P = 0$ with 46% and 74% higher costs than previous studies, respectively. The higher costs are necessary; previous studies did not achieve zero $SDPF_N$ and $SDPF_P$ in 39% and 91% of transient cases, respectively. To reduce duplicate pipe costs, the model was used to examine the effect of regulating the duration for a gradual demand increase. Using only the pipes optimized for steady-state service, the minimum duration for satisfying the transient pressure constraints was identified as ~260 s for the concurrent demand increase scenario. Slight relaxation of the minimum allowable pressure constraint allows a reduction in the duration to 150 s. For applying a demand increase over a smaller duration, duplicate pipes would be needed and can be optimized using the model. These results indicate the advantage of the proposed model in achieving full protection of existing WDNs while maintaining computational efficiency and cost-effectiveness.

Keywords: water distribution networks; hydraulic transients; worst transient cases; optimization; genetic algorithm

1. Introduction

The optimal design of a water distribution network (WDN) aims to minimize cost while satisfying hydraulic constraints pertaining to pressure, flow velocity, and water demands. Numerous studies have applied various optimization techniques to determine optimum WDN pipe diameters based on steady-state conditions for several benchmark WDNs, including the two-loop, Hanoi, New York Tunnel (NYT), and GoYang networks. Many of these studies—particularly recent ones—have relied on evolutionary algorithms given their efficiency and ability to solve complex problems to obtain near-optimal solutions [1,2]. Examples include the application of Ant Colony optimization (Maier et al. [3], Zecchin et al. [4], Zecchin et al. [5], Mehdi and Massoud [6]), Particle Swarm (PSO) optimization (Babu and Vijayalakshmi [7], Ezzeldin et al. [8], Surco et al. [9]), the Whale Optimization algorithm (Ezzeldin and Djebedjian [10]), and Differential Evolution (Vasan and Simonovic [11]; Mansouri et al. [12]; Yazdi et al. [13]). Among the various evolutionary optimization algorithms, numerous studies applied the genetic algorithm (GA) due to its robust and efficient performance. Recent examples include Jung and Karney [14], EL-Ghandour and Elansary [15], Kim and Kim [16], Huang et al. [17], Jung [18], and Sangroula et al. [19] as well as Ghobadian and Mohammadi [20] and Naidu et al. [21], who applied the non-dominated genetic algorithm with multi-objective optimization to minimize the network cost and the total network pressure deficiency.

In contrast to the large number of studies that applied optimization methods to WDNs under steady-state conditions, relatively few studies focused on the optimization of protection measures against transients in existing WDNs. Four different protection strategies were considered including pipe replacement (Djebedjian et al. [22], and El-Ghandour and Elansary [23]), addition of parallel pipes (Jung and Karney [14,24], Jung [18], El-Ghandour et al. [25]), use of surge protection devices such as surge tanks and pressure relief valves (EL-Ghandour and Elansary [15], Jung and Karney [26,27], Skulovich et al. [28]), and controlling valve closure pattern (Skulovich et al. [29,30], Bohorquez and Saldarriaga [31], Carvajal and Bohorquez [32]). Most of these studies assumed a critical scenario for one or more transient event types. For example, Jung and Karney [33] and Jung et al. [34] arbitrarily selected a transient scenario for a sudden demand increase. Their arbitrarily selected scenario was also used later by El-Ghandour and Elansary [23] and El-Ghandour et al. [25]. Another example is the study by Djebedjian et al. [22], who selected a scenario for three different transient events, including pump power failure, valve sudden closure, and sudden demand change. No justification was given for the selected scenarios. Also, a separate set of optimum pipe diameters was determined for each transient scenario without indication of the overall optimum diameters. In contrast, Jung and Karney [14,24,27] and Jung [18] identified the worst scenario for a suggested transient event by maximizing a Surge Damage Potential Factor (*SDPF*) based on nodal pressures simulated with only the existing pipes. Additional pipes running in parallel to the existing pipes were used for surge protection in the existing New York Tunnel (NYT) network. The optimal diameters for the additional pipes were determined by applying an evolutionary algorithm considering the identified worst case.

The deficiency in this approach is the consideration of only a single worst transient scenario and the assumption that the proposed surge protection measures that are determined based on this scenario are adequate for other transient scenarios that originally had a lower *SDPF*. This assumption is not necessarily valid and requires a more comprehensive analysis. Another deficiency in many—if not all—previous studies (e.g., Jung and Karney [14,24], and Jung [18], El-Ghandour and Elansary [23], El-Ghandour et al. [25]) is that only transient pressure heads at pipe end nodes were considered when examining the deviation from the minimum and maximum allowable heads. However, pressure heads along the pipes may have been higher or lower than at the pipe end nodes. Accordingly, pressure head constraints were not necessarily satisfied along the pipes and identified surge protection strategies may be insufficient to fully protect WDNs from water hammer events.

Another deficiency in previous studies that considered transients due to demand increase in WDNs is the assumption of an almost sudden, abrupt increase. While this assumption gives the most critical scenario for creating extreme transient pressures, it is perhaps overly conservative; control devices for large users can be regulated to extend the opening and closure durations. Even for the operation of fire hydrants, firefighters are instructed and trained on gradually opening and closing the hydrants to reduce transients [35,36]. Accordingly, it is important to include the duration over which the demand increase occurs as a decision variable when identifying optimal surge protection measures.

This study aims to overcome these deficiencies by proposing an efficient and reliable model for identifying the minimum set of worst transient scenarios that should be considered for the protection of existing WDNs against transient loadings. The proposed model, referred to hereafter as optWDN, applies the genetic algorithm to achieve minimum protection cost while satisfying the constraints on allowable steady-state and transient pressures along the pipes and not just at pipe end nodes. The model's performance is demonstrated through application to the existing NYT network. In comparison to previous studies, the focus was on determining the optimal diameters for additional duplicate pipes to achieve higher pressure under steady-state conditions and to protect the network from transients due to sudden demand increase. The optWDN model also determines the minimum duration for the application of gradual demand increase to avoid the need for additional or larger duplicate pipes beyond what is needed to satisfy steady-state requirements. Depending on the viability of this minimum duration, a hybrid transient protection scheme can subsequently be used by employing a shorter feasible duration for applying demand increase and using the model to identify the corresponding optimal duplicate pipe diameters.

This paper begins by presenting the formulation and components of the proposed optWDN model. Afterwards, the essential information about the NYT network is given, and details about the model validation and application to this case study are described. The results section includes comparisons with previous studies, *SDPF* evaluation based on simulated pressure along pipes, identification of worst transient cases, final optimized diameters for additional duplicate pipes, and minimum duration for demand increase without transient protection. Finally, the benefits provided by the model are summarized, and recommendations regarding further model improvements are given.

2. Model Formulation

2.1. Model Scope and Components

The optWDN model applies a genetic algorithm (GA) to determine the optimum diameters for pipes to be added in parallel to existing WDN pipes. The objective is to identify the pipe diameters that provide surge protection for the WDN and give the least cost. The optWDN model was coded in MATLAB version 2018a and consists of two hydraulic simulators and two modules (Figure 1). The hydraulic simulators are the steady-state simulator (SS) and the transient simulator (TS). The modules are the Optimization (OPT) module and the Worst Transient Case (WTC) module. The model workflow is summarized next, and a detailed description of the model components is given in the following sections.

As shown in Figure 1, the optWDN model starts by applying the WTC module, which runs the steady-state simulator for the selected existing network. The steady-state results are then utilized as initial conditions for simulating the transients that arise from all possible scenarios of a specified sudden demand increase at one or more nodes. The worst transient case with the highest *SDPF* is identified and used by the OPT module to determine the optimal diameters for the additional duplicate pipes. The cycle of applying WTC and OPT modules repeats to identify additional worst transient cases and obtain a new set of optimum pipe diameters. The model run cycles are continued until the diameters determined based on the relatively few identified worst cases give zero *SDPF* for all possible transient loading cases, ensuring full protection. This iterative approach is more efficient than optimizing for all possible transient cases simultaneously, which requires substantial runtime and computational resources, particularly for large networks with simultaneous sudden demand changes at multiple nodes.

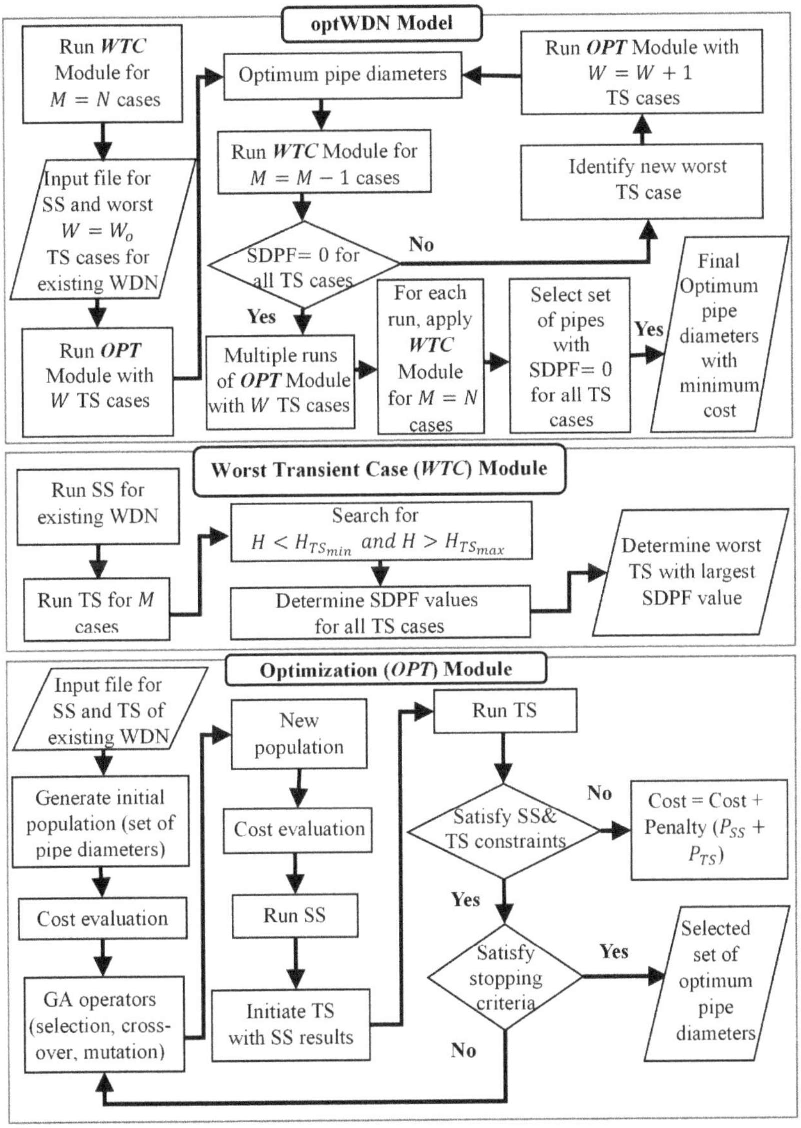

Figure 1. Flowchart for the proposed optWDN model. N denotes the total number of possible transient cases. M denotes a monotonically decreasing number of transient cases in each model run. W denotes a monotonically increasing number of worst transient cases in each model run. W_o denotes the initial number of worst transient cases that are selected for application in the first model run cycle.

2.2. Steady-State Simulator

The steady-state simulator (SS) conducts a hydraulic analysis of the existing WDN under steady-state demands using the EPANET v2.2 hydraulic model [37]. Using the EPANET MATLAB Toolkit v2.1, the SS modifies input data files exported from the EPANET program, runs the model, and transfers results to the transient simulator. The input data file includes design parameters for all network elements, including pipes, nodes, and reservoirs.

In performing steady-state simulations, the friction energy head loss is determined based on the Darcy–Weisbach equation [38],

$$h_f = \frac{fLV^2}{2gD} \tag{1}$$

where h_f is the friction head loss over a WDN pipe, V is flow velocity in the pipe, f is the Darcy–Weisbach friction factor, L is the pipe length, D is the pipe diameter, and g is the gravitational acceleration.

2.3. Transient Simulator

The transient simulator (TS) calculates pressure heads H and flow velocities V in a WDN for a specified transient scenario by solving the unsteady mass balance and momentum equations [39],

$$\frac{\partial H}{\partial t} + \frac{a^2}{g}\frac{\partial V}{\partial x} + Vsin\theta + V\frac{\partial H}{\partial x} = 0 \tag{2}$$

$$\frac{\partial V}{\partial t} + g\frac{\partial H}{\partial x} + \frac{fV|V|}{2D} + V\frac{\partial V}{\partial x} = 0 \tag{3}$$

where x is the along-pipe length, t is time, θ is the pipe angle with the horizontal, and a is the wave speed. Equations (2) and (3) are solved subject to boundary conditions imposed on velocity and/or pressure heads at reservoirs, valves, demand nodes, and junctions connecting multiple pipes. With the imposed boundary conditions and initial conditions based on steady-state simulations, the equations are solved numerically using the method of characteristics (MOC) [39,40]. The results are used within the OPT and WTC modules to identify the worst transient cases and the optimum pipe diameters, respectively.

For performing the transient analysis using MOC, the temporal and spatial domain is simultaneously discretized [41]. Each pipe i in the WDN is divided into a number of segments N_{s_i} such as that of [39],

$$\Delta t \leq \min\left(\frac{L_i}{N_{s_i}(V_i + a_i)}\right) \tag{4}$$

where Δt is the computational time step, L_i is the pipe length, V_i is the flow velocity in the pipe, and a_i is the wave speed in the pipe, and the subscript $i = 1, 2, 3 \ldots N_P$ denotes the pipe number with N_P representing the total number of pipes in the WDN. Equation (4) ensures that the time step Δt is set to or lower than the shortest wave travel time for pipe segments [33].

2.4. Optimization Module

The Optimization (OPT) module applies a single objective function that aims to minimize cost by identifying the optimum diameters of additional pipes. Optimization is performed subject to hydraulic constraints on steady-state and transient pressure heads along pipes. The objective function is,

$$F = C_p + P_{SS} + P_{TS} \quad (5)$$

where $C_p = \Sigma_i U(D_i) \times L_i$ is the cost of the additional pipelines ($i = 1, 2, \ldots$) in which L_i is the length of pipe i, and $U(D_i)$ is the diameter-dependent unit cost of pipes. In Equation (5), P_{SS} and P_{TS} are penalties applied to account for violation of pressure head constraints under steady-state and transient conditions, respectively. The penalties are calculated using,

$$P_{SS} = (C_p) \times \sum_{i=1}^{N_n} \{\max(0, h_i - h_{SS_{max}}) + \max(0,\ h_{SS_{min}} - h_i\ )\} \quad (6)$$

$$P_{TS} = (C_p + C_{po}) \times \frac{SDPF}{\Delta t} \quad (7)$$

where C_{po} is the cost of the existing pipes, N_n is the total number of network nodes, h_i is the steady pressure head at the end node of pipe i; and $h_{SS_{min}}$ and $h_{SS_{max}}$ are the minimum and maximum allowable pressure heads during steady-state, respectively. The definition of *SDPF* is provided below in Section 2.5.

A genetic algorithm (GA) is applied to minimize the objective function. The optimization decision variables are the diameters for duplicate pipes running parallel to all existing WDN pipes. Possible values for the decision variables are the commercially available diameters, which are represented in GA using integers in the range from 1 to the total number of available diameters. In addition, a value of 0 is allowed to include the possibility of not having a duplicate pipe. Starting with an initial population that satisfies the bounds on decision variables, GA applies selection, crossover, and mutation operators to build new generations. The cycle continues until a stopping criterion is satisfied and a near-optimal solution is achieved.

2.5. Worst Transient Case (WTC) Module

The Worst Transient Case (WTC) module determines the Surge Damage Potential Factor (*SDPF*) and identifies the worst transient case with the highest *SDPF* by searching among a predefined set of possible sudden demand increase scenarios for one or more nodes. The objective for determining the worst transient cases is to maximize the *SDPF* calculated using,

$$SDPF = \sum_{k=1}^{N_t} \sum_{j=1}^{N_n} \sum_{i=1}^{N_p} \sum_{m=1}^{N_{s_i}} \Delta t\ \{\max(0, H_{m,i,k} - H_{TS_{max}}) + \max(0,\ H_{TS_{min}} - H_{m,i,k}\ )\} \quad (8)$$

where N_n and N_P are the total number of network nodes and pipes, respectively; N_t is the total number of time steps in the transient simulation; N_{s_i} is the total number of segments in pipe i; $H_{m,i,k}$ is the transient pressure head at the end node of segment m of pipe i at time step k; and $H_{TS_{min}}$ and $H_{TS_{max}}$ are the minimum and maximum allowable pressure heads during transients, respectively.

As defined in Equation (8), the *SDPF* represents the sum of magnitudes of deviations of simulated pressure heads $H_{m,i,k}$ from the minimum and maximum allowable transient pressure heads $H_{TS_{min}}$ and $H_{TS_{max}}$ when these limits are exceeded. Previous researchers, including Jung and Karney [14,24], Jung [18], El-Ghandour and Elansary [23], and Jung et al. [34], calculated the *SDPF* based only on transient pressure heads at pipe end nodes, referred to hereafter as $SDPF_N$. This study introduces an additional *SDPF* version, $SDPF_P$, calculated from transient pressure heads at all computational nodes along pipes, including pipe end nodes. An $SDPF_N = 0$ indicates no exceedance at pipe end nodes, while an $SDPF_P = 0$ indicates no exceedance along the full length of WDN pipes. In Equation (8), N_n equals 2 for $SDPF_N$ and the total number of all computational nodes for $SDPF_P$.

3. Model Validation and Application

3.1. Case Study Description

The optWDN model was applied to the New York Tunnel (NYT) network, examined in several previous studies [14,18,23–25,27,34]. The network includes a single reservoir, 19 nodes, and 21 pipes (Figure 2). The reservoir water level is 91.44 m above the datum, node elevations are 0.00 m, pipe diameters range between 1500 and 5100 mm, pipe lengths range between 2225 and 11,707 m, and the Hazen–Williams roughness coefficient for all pipes is 100 [2,25,42]. Under peak steady-state conditions, node demand ranges between 0.0283 and 4.818 m^3/s with a total of 57.13 m^3/s.

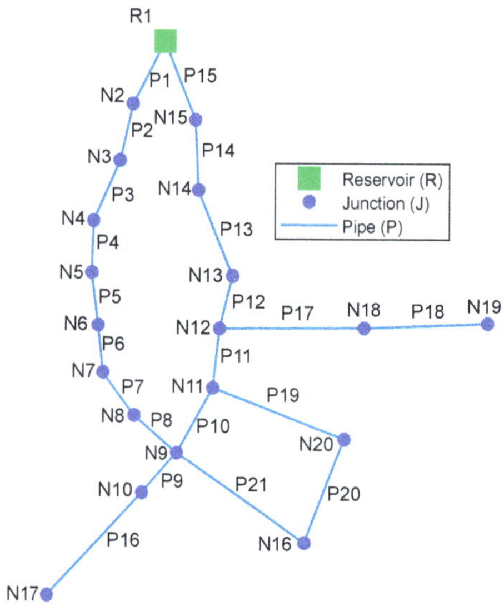

Figure 2. Layout of the existing NYT network (after Jung and Karney [14]).

According to Maier et al. [3], population growth in the region served by the aging NYT network has led to insufficient steady-state pressure heads at several locations. To restore adequate pressure levels, numerous studies proposed adding duplicate pipes and applied optimization algorithms to identify optimum diameters [2,3,43]. The objective was to minimize cost while satisfying pressure head constraints. Dany et al. [40], El-Ghandour and Elbeltagi [2], Maier et al. [3], and Eusuff and Lansay [43] identified 6 duplicate pipes with a total cost ranging between 38.13 and 38.79 million USD (Table 1).

Table 1. Optimized diameters (mm) given by previous studies for duplicate pipes under steady-state conditions. Also shown is the total cost for the duplicate pipes.

Pipe Number	Dany et al. [38]	Ghandour and Elbeltagi [2]	Maier et al. [3]	Eusuff and Lansay [43]
7	-	-	3600	3300
15	3000	3000	-	-
16	2100	2100	2400	2400
17	2400	2400	2400	2400
18	2100	2100	2100	2100
19	1800	1800	3000	1800
21	1800	1800	1500	1800
Total Cost (million USD)	38.79	38.79	38.64	38.13

Few studies focused on optimizing protection against transient events, considering only sudden demand increase. As the NYT network is gravity-driven with no pumps, pump power failure events are not relevant. For sudden valve closure, transients are most effectively mitigated by controlling the valve closure duration and applying an appropriate closure pattern [23,29–32,44].

3.2. Model Parameters

The optWDN model was applied to the NYT network with a computational time step of 0.55 s, and a simulation duration of 300 s. The node-dependent minimum allowable steady-state pressure head ranges between 77.7 m and 91.5 m [2], with minimum and maximum allowable transient pressure heads of 54.9 m and 304.9 m, respectively [23]. Similar to El-Ghandour et al. [25] and Jung [18], new duplicate pipe sizes were selected from a set of 15 diameters ranging from 900 mm to 5100 mm with a step of 300 mm. The unit cost (USD/m) for different pipe diameters (mm) was calculated using

$$U(D_i) = 0.06665 \times D_i^{1.2401} \tag{9}$$

where the same material for existing pipes was used for duplicate pipes to avoid multiple pressure wave reflections due to material mismatch at junctions, which may lead to cyclic stress loading and fatigue failure for pipes [45–47]. In line with previous NYT studies, a constant transient wave speed of 1000 m/s was applied for both existing and new duplicate pipes [23]. For application of the optWDN application, the Hazen–Williams roughness coefficient for the pipes was converted to the Darcy–Weisbach friction factor using

$$f = \frac{133.32}{C_{HW}^{1.852} V^{0.15} D^{0.17}} \tag{10}$$

where the GA was applied within the optWDN model, population sizes of 20, 30, 50, and 100 were examined. For each initial population, 30 trial runs were conducted with a maximum number of generations of 1000. Results given by the different population sizes were evaluated using multiple metrics, including a Reliability indicator R, Efficiency-Rate metric E, and Performance indicator η_{alg} [2,10,48]. The metrics were calculated using

$$R = \frac{N_{succ}}{N_{sim}} (\%) \tag{11}$$

$$E = R \times \frac{TSP}{FENOT} \times f_c \cdot (\%) \tag{12}$$

$$\eta_{alg} = 100 - \begin{bmatrix} 0.99\log_{10} N_{gen} \\ +0.01 \log_{10}\left(N_{gen} - N_{obj-eval} + 1\right) \end{bmatrix} (\%) \tag{13}$$

where N_{succ} is the number of trials that obtained the minimum cost, N_{sim} is the total number of trials, TSP is the total solution space given by the number of commercial pipes to the power of the number of WDN pipes, $FENOT$ is the number of objective functions to reach the optimum results during the first simulation trial, f_c is an arbitrary constant set to 10^{-21} [2], N_{gen} is the product of the maximum number of generations and the population size, and $N_{obj-eval}$ is the number of minimum function evaluations. For the NYT network, TSP was set to $16^{21} = 1.93 \times 10^{25}$ [2].

As given by Equations (11)–(13), the values for the three metrics R, E, and η_{alg} range between 0 and 100%. Higher values indicate better performance, with 100% indicating an ideal optimization model.

3.3. Model Validation

For validating the optWDN model, the model was applied to the NYT network to determine optimum diameters for new parallel pipes needed for protection against two sudden demand increase scenarios. Scenario S1 included a demand increase from 0.0283 m^3/s to 4.818 m^3/s over 1 s at node 10 and results were compared with those of El-Ghandour and Elansary [23], El-Ghandour et al. [25], and Jung et al. [34]. Scenario S2 involved a simultaneous demand increase by 3 m^3/s over 1 s at two nodes. All possible combinations of the two nodes were tested to identify the worst case with the highest *SDPF*. Results were compared with those of Jung and Karney [14,24] and Jung [18]. Comparison included calculating the Nash–Sutcliffe model efficiency coefficient (NSE) using [49],

$$NSE = 1 - \frac{\sum_{l=1}^{n}\left(y_{p_l} - y_{c_l}\right)^2}{\sum_{l=1}^{n}\left(y_{p_l} - \overline{y}_p\right)^2} \tag{14}$$

where n is the number of points used for comparing the two-time series (n is about 540 in this study), $\overline{y}_p$ is the average simulated pressure head in the previous study, y_{p_l} is the head simulated in the previous study at a given time step and node, and y_{c_l} is the corresponding head from the current study. As given by Equation (14), NSE values range from negative infinity to 1, with 1.0 representing perfect agreement with previous results.

3.4. Testing Model Performance

The effect of including pressure at computational nodes along pipes on identifying pressure extremes and calculating *SDPF* was examined using scenario S1. Minimum and maximum pressure heads along pipes were compared to those pressure extremes at pipe end nodes without duplicate pipes. The optWDN model was then applied to identify duplicate pipe diameters that minimized cost and achieved zero *SDPF* calculated based on pressure along all computational nodes along the network pipes. The identified optimal pipe diameters were compared to those identified considering only pressure heads at pipe ends to examine if the latter approach was sufficient for transient protection.

Finally, the performance of the optWDN model in identifying the worst transient cases and the optimal duplicate pipe diameters for full protection was examined using scenario S2 with simultaneous sudden demand increase at two nodes. The model was run in successive cycles (Section 2.1), with two alternatives for the initial cycle: one using a single worst transient case, and another using 9 worst cases, amounting to about 5% of the total number of possible transient cases for scenario S2. Results were analyzed by the total number of cycles to achieve zero *SDPF* at pipe end nodes and along pipes for all possible node combinations. Final diameters and costs were compared to those of Jung [18].

3.5. Examining the Effect of Gradual Demand Increase

The optWDN transient simulator was used to determine the duration for demand increase that does not require additional surge protection and can be accommodated by the duplicate pipes optimized for only steady-state conditions (as given by Dany et al. [40], El-Ghandour and Elbeltagi [2], and shown in Table 1). This duration was determined by conducting transient simulations that started with a sudden demand increase, extending the duration by 1 s increments until the simulated pressure head reached the minimum allowable transient head. This scheme was applied to the single worst transient case of scenarios S1 and S2. For S2, the final demand increase duration was applied to all 171 transient cases to ensure that the minimum allowable transient head was satisfied in all cases.

4. Results and Discussion

4.1. Validation of Transient Simulator

Compared to Jung [18] and Jung et al. [34], the transient pressure heads simulated in this study show similar trends for scenarios S1 and S2 (Figures 3 and 4). NSE values were 0.633 for node 17 and 0.843 for node 19 compared to Jung et al. [34]. Furthermore, NSE values were 0.889 for node 17 and 0.916 for node 18 compared to Jung [18]. These relatively high NSE values indicate that the transient simulator in the optWDN model provides reasonable transient results. The lack of perfect agreement with previous results may be due to slight differences in initial steady-state conditions, simulation time steps, and estimated transient wave speed, which were not specified explicitly by these authors.

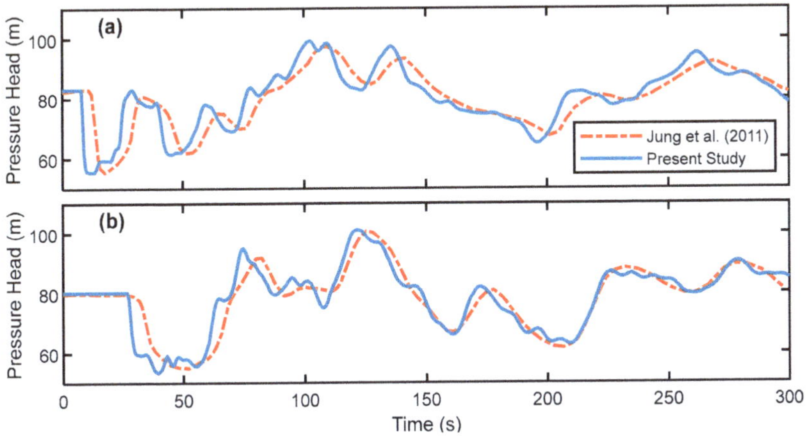

Figure 3. Variation in simulated pressure head with time at (**a**) node 17 and (**b**) node 19 due to sudden demand change at node 10 (scenario S1). Shown are results from the optWDN model and from Jung et al. [34].

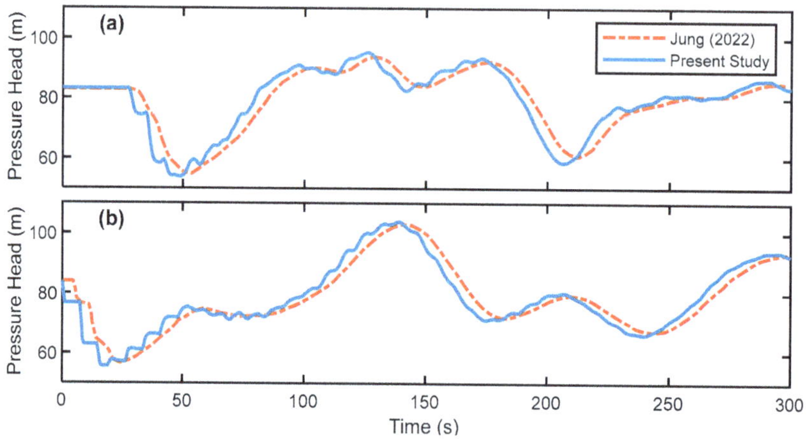

Figure 4. Variation in simulated pressure head with time at (**a**) node 17 and (**b**) node 18 due to simultaneous, identical sudden demand changes at nodes 18 and 19 (scenario S2). Shown are results from the optWDN model and from Jung [18].

4.2. Validation of Optimization Module

For scenario S1 with sudden demand change at node 10 [34], applying different GA population sizes to optimize duplicate pipe diameters gave identical reliability metric R values and similar performance indicator η_{alg} values. In contrast, the efficiency-rate metric E ranged between 2.7% and 15.4% (Table 2). The population size of 30 gave the highest value of 15.4% for the metric E and was selected to be applied in the rest of this study.

Table 2. Performance metrics for different population sizes used in applying GA to the NYT Network case study.

Population Size	20	30	50	100
Reliability, R (%)	3.34	3.34	3.34	3.34
Efficiency-Rate metric, E (%)	13.3	15.4	5.5	2.7
Performance indicator, η_{alg} (%)	95.70	95.52	95.30	95.00

From 30 independent GA runs, the optWDN model identified 7 duplicate pipes out of the 21 existing pipes, similar to El-Ghandour and Elansary [23] and Jung et al. [34], and less than the 8 duplicate pipes given by El-Ghandour et al. [25]. The identified pipes were identical to those by El-Ghandour and Elansary [23], with similar or smaller diameters except for pipe 7 (Table A1).

The total costs of the duplicate pipes identified by El-Ghandour and Elansary [23] and Jung et al. [34] were 47.2 and 49.1 million USD, respectively. The total cost identified by the optWDN model was 46.37 million USD, less by 5.7% than the cost in Jung et al. [34], less by 1.9% than the cost in El-Ghandour and Elansary [23], and ~0.3% higher than the 46.25 million USD in El-Ghandour et al. [25]. The solution with the slightly lower cost was not reproduced, as the corresponding duplicate pipe diameters violated the minimum allowable head constraint of 54.90 m at node 19 with the minimum head of 54.81 m.

The optimal duplicate pipe diameters identified by the optWDN model gave an $SDPF_N$ of zero, indicating compliance with allowable pressure head constraints at all the network nodes. As an example, the pressure heads simulated at nodes 17 and 19 varied between 55.0 and 101.0 m similar to the range in previous studies (Figure 5).

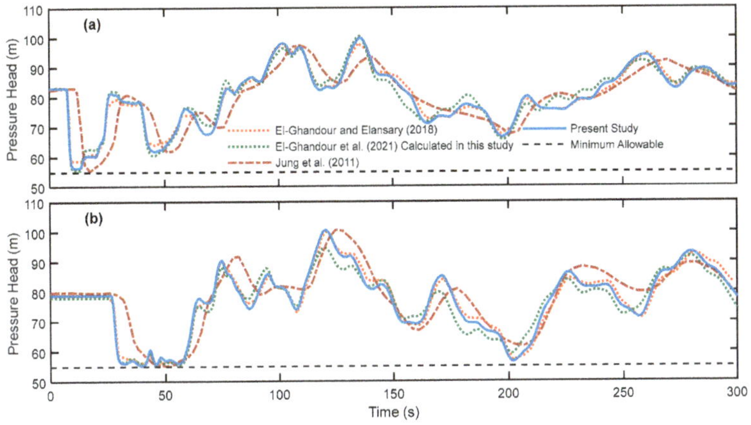

Figure 5. Variation in simulated pressure head with time at (**a**) node 17 and (**b**) node 19 due to sudden demand change at node 10 (scenario S1). Simulated pressure is for the existing network with duplicate pipes identified in the present study, as well as by El-Ghandour and Elansary [23], El-Ghandour et al. [25], and Jung et al. [34].

4.3. Validation of Worst Transient Case Module

The WTC module in the optWDN model was applied to scenario S2 to identify the worst transient case with the highest $SDPF_N$. The worst transient case consisted of a simultaneous sudden demand increase at nodes 18 and 19, consistent with Jung [18] since both nodes are located in dead-end pipes outside loops. The optWDN model gave an $SDPF_N$ of 8.04×10^4 m-s for the worst transient case, slightly higher than the value of 7.90×10^4 m-s reported by Jung [18]. Because node 19 is at a dead-end, the optWDN model results showed that, besides the worst transient case, all other cases with sudden demand change at node 19 also gave relatively high $SDPF_N$ values ranging between 4.86×10^4 and 5.61×10^4 m-s (Figure 6). This finding highlights the importance of considering multiple transient cases as opposed to just a single worst case when optimizing duplicate pipe diameters to achieve comprehensive transient protection.

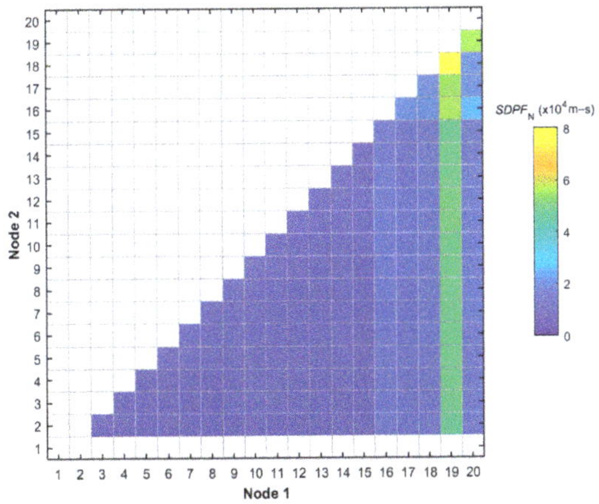

Figure 6. $SDPF_N$ for all possible combinations of simultaneous sudden demand increase at two nodes (scenario S2).

4.4. Effect of SDPF Calculation Approach

For scenario S1 with sudden demand change at node 10, transient simulations in the optWDN model showed that the pressure heads simulated at internal nodes along a number of pipes were more extreme than at the pipe end nodes for 9 out of 21 pipes (pipes 2, 4, 6, 7, 8, 12, 19, 20, and 21). The minimum pressure head at the internal nodes of the three pipes 16, 18, and 20 was also less than the minimum head at the end nodes of these pipes (Figure 7a) with deviations limited to about 4.45 m except for pipe 18, which connects nodes 18 and 19, where the internal minimum pressure head was 10.4 m lower. Given that pressure heads along pipes may be more extreme than at pipe end nodes, it is important to consider the pressure heads along pipes for transient protection.

As shown in Figure 7a for the existing pipes of the NYT network, simulated transient pressure heads within pipes 17 to 21 were less than the minimum allowable head. For the set of optimal duplicate pipe diameters identified in Section 4.2 with $SDPF_N = 0$, the pressure heads at pipe ends satisfied the minimum allowable as expected (Figure 7(b1,b2)). However, the pressure head along pipe 20 dropped below the end-node pressure for this pipe and up to 7.4 m below the minimum allowable head. These results indicate that satisfying the allowable pressure heads at pipe-end nodes does not guarantee that the allowable heads at pipe-internal nodes are also achieved. Indeed, optimal duplicate pipe diameters from Jung et al. [34], El-Ghandour and Elansary [23], and El-Ghandour et al. [25] satisfied the minimum allowable pressure head constraint at the end nodes of pipe 20 but allowed most of the pipe to experience minimum pressures that were lower than the minimum allowable value (Figure 8). Therefore, setting $SDPF_N = 0$ alone is likely insufficient to fully protect WDNs from water hammer events.

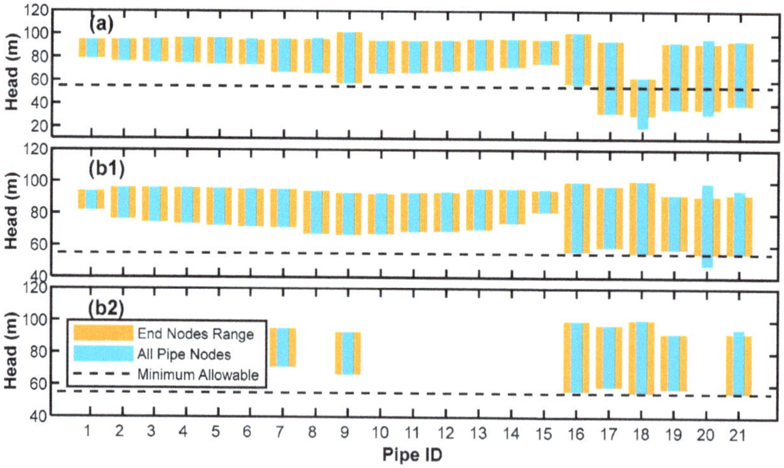

Figure 7. Range of simulated transient pressure heads for the (**a**) existing pipes of the NYT network without duplicate pipes and for the (**b1**) existing pipes of the NYT network with (**b2**) optimal duplicate pipes identified for scenario S1 using $SDPF_N = 0$. Orange bars show the range of transient pressure heads at the two end nodes of each pipe, while cyan bars show the range along each pipe. The dashed line indicates the minimum allowable pressure head.

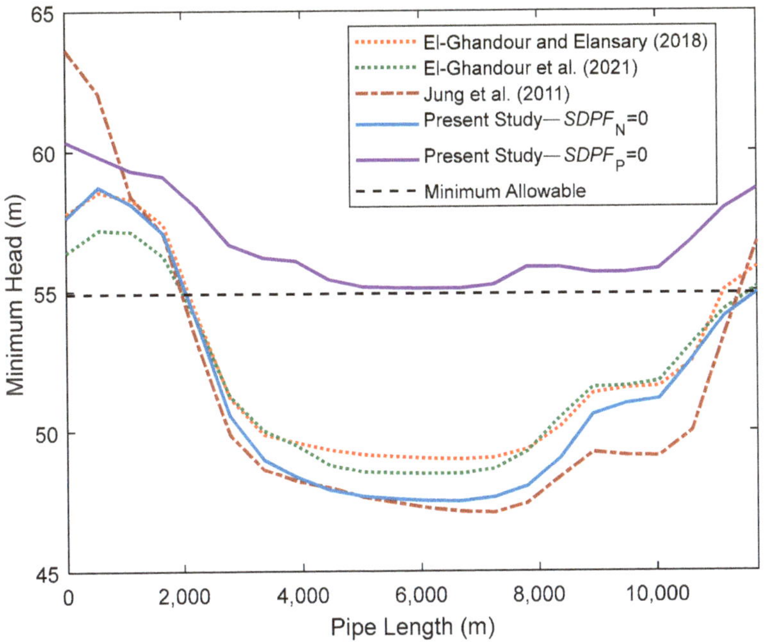

Figure 8. Lower envelope of simulated transient pressure head along pipe 20 under scenario S1. Simulations are for the existing network with different sets of optimal duplicate pipes identified in this study and in previous studies [23,25,34]. The dashed line indicates the minimum allowable transient pressure head.

Accordingly, the optWDN model was reapplied to scenario S1 by setting $SDPF_P = 0$ to satisfy the allowable pressure head constraints at all computational nodes along pipes. The optimization module selected duplicate pipes for 8 of 21 existing pipes, similar to the number given by El-Ghandour et al. [25] and higher than those in El-Ghandour and Elansary [23], Jung et al. [34], and the application of $SDPF_N = 0$ in this study, all of which required 7 duplicate pipes. The identified pipes were identical to those by El-Ghandour et al. [25], with mostly similar or higher diameters except for pipes 16 and 17 (Figure 9 and Table A1). The resulting total cost was 52.83 million USD, higher than all previous estimates and 14% greater than $SDPF_N = 0$ solution. The higher cost for $SDFP_P = 0$ was essential to satisfy the transient pressure constraints at all computational nodes along pipes. The optimized duplicate pipe diameters eliminated violations of the minimum allowable pressure along the entire length of pipe 20 (Figure 8). These results indicate that imposing $SDPF_P = 0$ provides optimum duplicate pipe diameters that can fully protect existing WDNs from water hammer effects. The additional computational effort associated with imposing $SDPF_P = 0$ was relatively minor, requiring 15% more function evaluations across 30 runs, with a median increase of 11.5% (Figure 10).

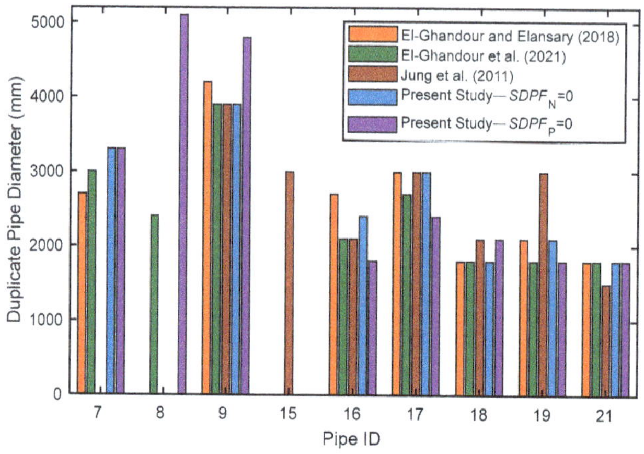

Figure 9. Optimized diameters of additional duplicate pipes as identified in the present study, as well as by El-Ghandour and Elansary [23], El-Ghandour et al. [25], and Jung et al. [34].

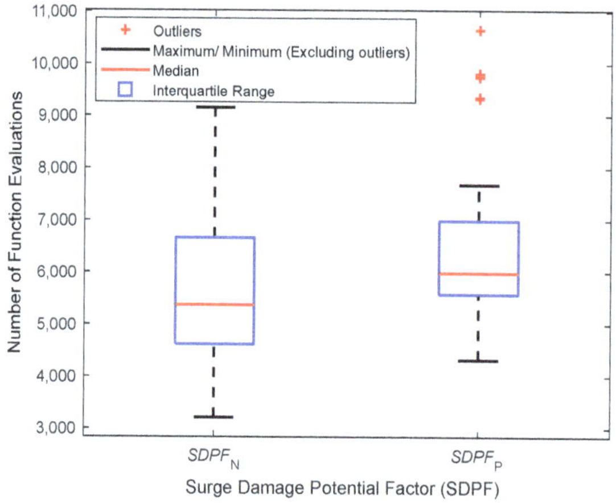

Figure 10. Box plot of the number of function evaluations required to achieve zero $SDPF$ values ($SDPF_N$ and $SDPF_P$) for 30 independent runs under scenario S1.

4.5. Optimum Protection for Worst Transient Cases

For scenario S2 with simultaneous sudden demand increase at two nodes, the optWDN model took 8 run cycles to achieve $SDPF_N = 0$ for all 171 possible cases when initialized with the worst transient case (nodes 18 and 19). Within the next two run cycles, the total number of cases with non-zero $SDPF_N$ diminished rapidly to 3, but increased to 15 in run cycle 4, before decreasing again to 3 in the next cycle, and then steadily decreased, getting to zero by the end of cycle 8 (Table 2 and Figure 11). A similar pattern was observed for the maximum $SDPF_N$ for the selected worst cases in each cycle (Table 3). By the end of run cycle 8, the worst transient cases identified included three cases in which one of the locations for sudden demand change was nodes 16, 17, 19, and 20. Sudden demand changes at nodes 7, 8, 10, and 18 occurred in single cases. Compared to a cost of 103.36 million USD for the duplicate pipe diameters in cycle 8, applying the model over 30 trials with the set of cases from cycle 8 produced a more optimal selection of duplicate pipe diameters with a slightly lower cost of 102.11 million USD.

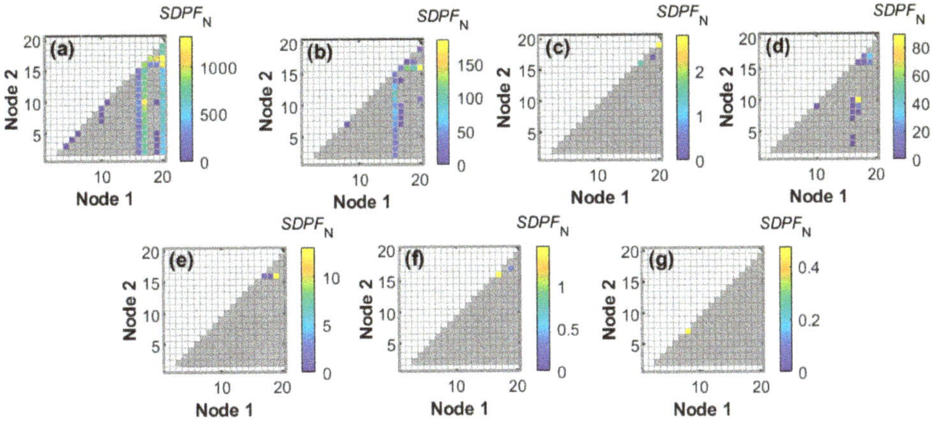

Figure 11. $SDPF_N$ for all possible combinations of simultaneous sudden demand increase at two nodes (scenario S2) corresponding to optimized duplicate pipe diameters in each run cycle of the optWDN model. Panels (**a**) Run 1; (**b**) Run 2; (**c**) Run 3; (**d**) Run 4; (**e**) Run 5; (**f**) Run 6; (**g**) Run 7 correspond to run cycles 1 and 7 with 1 and 7 worst cases, respectively. Grey shading denotes node combinations giving $SDPF_N = 0$.

Table 3. Outcome of run cycles of the optWDN model with $SDPF_N = 0$ for scenario S2. The number of considered worst transient cases was initially set to 1 and was subsequently increased from one cycle to the next.

Run Cycle	Number of Worst Transient Cases	Set of Nodes for Sudden Demand Change Considered in the *OPT* Module	Number of Transient Cases with Non-Zero $SDPF_N$	Maximum $SDPF_N$ over All 171 Cases	Total Duplicate Pipe Cost (Million USD)
0	0	-	171	80.5×10^3	-
1	1	(18, 19)	63	1.32×10^3	73.68
2	2	(18, 19); (17, 20)	28	0.19×10^3	97.30
3	3	(18, 19); (17, 20); (16, 20)	3	2.81	103.65
4	4	(18, 19); (17, 20); (16, 20); (19, 20)	15	90.0	95.29
5	5	(18, 19); (17, 20); (16, 20); (19, 20); (10, 17)	3	13.1	142.46
6	6	(18, 19); (17, 20); (16, 20); (19, 20); (10, 17); (16, 19)	2	1.47	104.47
7	7	(18, 19); (17, 20); (16, 20); (19, 20); (10, 17); (16, 19); (16, 17)	1	0.47	108.75
8	8	(18, 19); (17, 20); (16, 20); (19, 20); (10, 17); (16, 19); (16, 17); (7, 8)	0	0	103.36
30 GA independent trial runs		(18, 19); (17, 20); (16, 20); (19, 20); (10, 17); (16, 19); (16, 17); (7, 8)	0	0	102.11

As a slight modification to the above model application, the first run cycle of the optWDN model accounted for the top 5% (9 out of 171 cases) with the highest $SDPF_N$ values as obtained in Section 4.3. An additional worst case was considered in each subsequent run cycle, $SDPF_N = 0$ was achieved in 3 cycles (Table 4 and Figure 12). The last cycle considered 11 worst transient cases and yielded a cost of 114.68 million USD, which was reduced to 102.26 million USD after 30 GA trials, only 0.15% higher than the cost from the single initial worst-case alternative. Although the alternative initialized with 9 cases accounted for a total of 30 cases over all run cycles compared to 36 cases for the earlier alternative with a single initial worst case, the single initial case alternative resulted in a shorter overall run time as well as fewer final worst cases, enabling a slightly lower cost solution.

Table 4. Outcome of run cycles of the optWDN model with $SDPF_N = 0$ for scenario S2. The number of considered worst transient cases was initially set to 5% of the total number of possible cases and was subsequently increased by 1 from one cycle to the next.

Run Cycle	Number of Worst Transient Cases	Set of Nodes for Sudden Demand Change Considered in the *OPT* Module	Transient Cases Non-Zero $SDPF_N$	Maximum $SDPF_N$ over All 171 Cases	Total Duplicate Pipe Cost (Million USD)
0	0	-	171	80.5×10^3	-
1	9	(18, 19) & (19, 20) & (16, 19) & (17, 19) & (12, 19) & (11, 19) & (9, 19) & (10, 19) & (13, 19)	1	1.29	113.04
2	10	(18, 19) & (19, 20) & (16, 19) & (17, 19) & (12, 19) & (11, 19) & (9, 19) & (10, 19) & (13, 19) & (16, 17)	1	2.77	107.58
3	11	(18, 19) & (19, 20) & (16, 19) & (17, 19) & (12, 19) & (11, 19) & (9, 19) & (10, 19) & (13, 19) & (16, 17) & (10, 17)	0	0	114.68
30 GA independent trial runs		(18, 19) & (19, 20) & (16, 19) & (17, 19) & (12, 19) & (11, 19) & (9, 19) & (10, 19) & (13, 19) & (16, 17) & (10, 17)	0	0	102.26

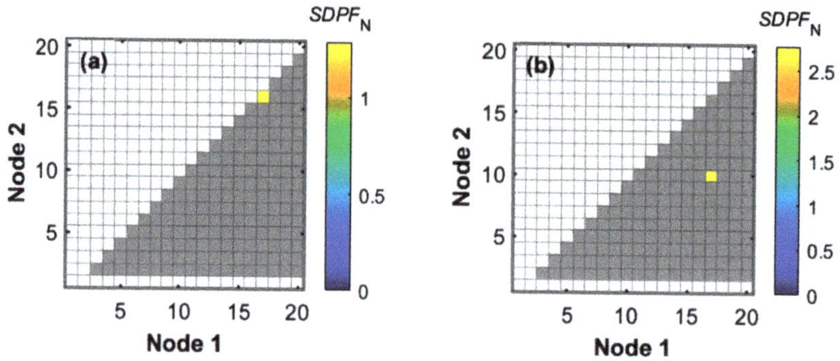

Figure 12. $SDPF_N$ for all possible combinations of simultaneous sudden demand increase at two nodes (scenario S2) corresponding to optimized duplicate pipe diameters in each run cycle of the optWDN model. Panels (**a**,**b**) correspond to run cycles 1 and 2 with 9 and 10 worst cases, respectively. Grey shading denotes node combinations giving $SDPF_N = 0$.

The final optimal diameters with a cost of 102.11 million USD gave $SDPF_N = 0$ for all possible transient cases under scenario S2, indicating that the model results satisfied the allowable pressure head constraints at all pipe end nodes. As an example, the pressure heads simulated at nodes 17 and 18 due to the worst transient case at nodes 18 and 19 varied between 55.0 and 105.0 m, similar to previous studies (Figure 13). Simulations for another worst transient case at nodes 17 and 20 ensured that the identified final duplicate pipes satisfied the allowable pressure heads at all the network nodes for all 171 cases (total $SDPF_N$ equal zero). The pressure heads simulated based on the final optimum diameters varied between 55.0 and 105.0 m, similar to previous studies (Figure 13).

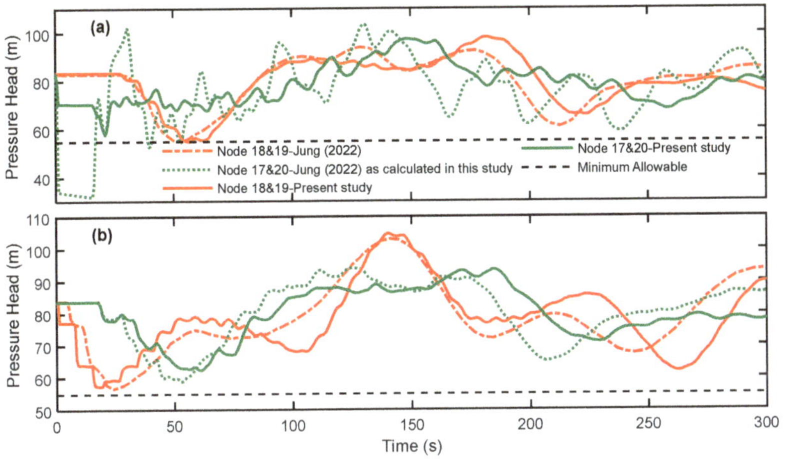

Figure 13. Variation in pressure head with time at (**a**) node 17 and (**b**) node 18 due to simultaneous, identical sudden demand changes at nodes 18 and 19 and at nodes 17 and 20, with duplicate pipes identified in the present study as well as by Jung [18].

For comparison, the OptWDN model was applied to the optimum pipe diameters identified by Jung [18] under the same transient case. While the resulting transient head at node 18 satisfied the allowable transient head constraints, the allowable minimum head constraint was violated at node 17 (Figure 13). Overall, 66 transient cases had non-zero $SDPF_N$, indicating that the optimum pipe diameters based only on a single worst transient case, as given by Jung [18], were insufficient to satisfy all other transient cases.

The above results correspond to $SDPF_N = 0$. However, as concluded in the previous section, this condition does not guarantee that pressure head constraints are satisfied along the full pipe length. Accordingly, the optWDN model was applied for scenario S2 to achieve zero $SDPF_P$ for all 171 possible cases. First, the WTC module identified the initial worst transient case with the highest $SDPF_P$ value. Similar to the result of $SDPF_N = 0$, nodes 18 and 19 were identified as the initial worst transient cases for simultaneous sudden demand increase. The calculated $SDPF_P$ was 80.22×10^4 m-s for the worst transient case, which is higher compared to the $SDPF_N$ of 8.04×10^4 m-s calculated earlier.

Achieving $SDPF_P = 0$ required 8 run cycles, similar to the results for $SDPF_N = 0$. However, only 5 out of the total 8 worst transient cases were similar to the results for $SDPF_N = 0$ (Table 5 and Figure 14). By the end of run cycle 8, the worst transient cases included three cases in which one of the locations for sudden demand change was nodes 16, 19, and 20. Sudden demand changes at nodes 10 and 17 occurred in two cases, while each of nodes 2, 15, and 18 occurred in a single case. Compared to a cost of 138.54 million USD for the duplicate pipe diameters in cycle 8, applying GA over 30 trials with the set of cases from cycle 8 produced a lower cost of 124.38 million USD. While this total cost was 22% higher than for $SDPF_N = 0$, the increase was necessary to achieve $SDPF_P = 0$ and satisfy the allowable minimum and maximum pressure constraints along all pipes.

Table 5. Outcome of run cycles of the optWDN model with $SDPF_P = 0$ for scenario S2.

Run Cycle	Number of Worst Transient Cases	Set of Nodes for Sudden Demand Change Considered in the *OPT* Module	Transient Cases with Non-Zero $SDPF_P$	Maximum $SDPF_P$ for All 171 Cases	Total Duplicate Pipe Cost (Million USD)
0	0		171	802,284.7	-
1	1	(18, 19)	124	14,507	87.67
2	2	(18, 19); (17, 20)	115	8943.62	97.90
3	3	(18, 19); (17, 20); (16, 20)	36	7.95	134.44
4	4	(18, 19); (17, 20); (16, 20); (10, 16)	93	73.24	112.17
5	5	(18, 19); (17, 20); (16, 20); (10, 16); (10, 17)	90	4.58	115.31
6	6	(18, 19); (17, 20); (16, 20); (10, 16); (10, 17); (19, 20)	86	4.11	118.00
7	7	(18, 19); (17, 20); (16, 20); (10, 16); (10, 17); (19, 20) (2, 15)	23	0.37	130.59
8	8	(18, 19); (17, 20); (16, 20); (10, 16); (10, 17); (19, 20) (2, 15); (9, 16)	0	0	138.54
30 GA independent trial runs		(18, 19); (17, 20); (16, 20); (10, 16); (10, 17); (19, 20) (2, 15); (9, 16)	0	0	122.24

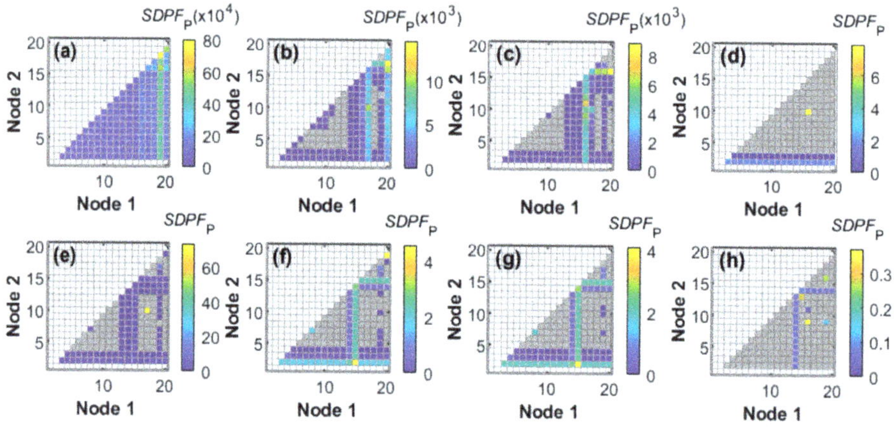

Figure 14. $SDPF_P$ for all possible combinations of simultaneous sudden demand increase at two nodes (scenario S2) corresponding to optimized duplicate pipe diameters in each run cycle of the optWDN model. Panels (**a**) Run 0; (**b**) Run 1; (**c**) Run 2; (**d**) Run 3; (**e**) Run 4; (**f**) Run 5; (**g**) Run 6; (**h**) Run 7 correspond to run cycles 0 to 7 with 0 and 7 worst cases, respectively. Grey shading denotes node combinations giving $SDPF_P = 0$.

The final optimum duplicate pipe diameters identified by the optWDN model to achieve an $SDPF_N = 0$ included 7 duplicate pipes out of 21 existing pipes, similar to the number in Jung [18], with similar or higher diameters except for pipe 7 (Figure 15 and Table A2). The total cost was 102.11 million USD, higher by 46% than the 70 million USD cost reported by Jung [18] to satisfy all 171 cases.

For $SDPF_P = 0$, the final optimum results included 9 duplicate pipes higher than the number in Jung [18], with lower or similar diameters except for pipes 1, 15, 19, and 20 (Figure 15 and Table A2). The total cost was 122.24 million USD, 74% higher than Jung [18] and 22% higher than the cost with $SDPF_N = 0$ in this present study. This higher cost satisfied the allowable minimum and maximum transient conditions at all computational nodes for all 171 cases. In contrast, the set of duplicate pipe diameters given by Jung [18] gave non-zero $SDPF_P$ in about 91% of the possible transient cases, confirming again that relying solely on a single worst transient case fails to provide adequate protection.

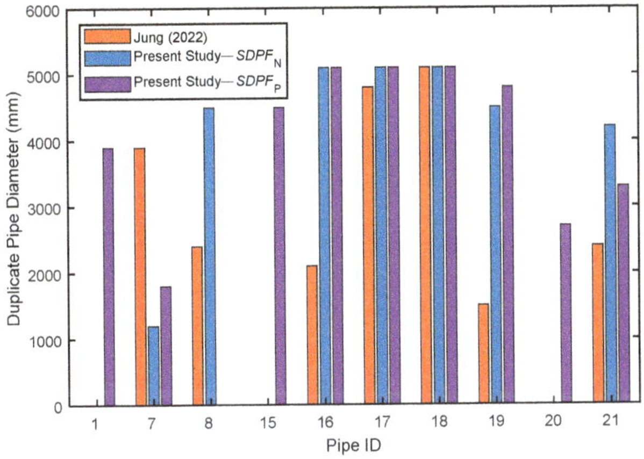

Figure 15. Optimized diameters for duplicate pipes identified by the present studies, as well as by Jung [18].

4.6. Cost Reduction for Surge Protection

Based on unit costs from Equation (9), the total cost for the original pipes in the NYT network amounts to 179.80 million USD. The lowest cost of duplicate pipes for improving network performance under steady-state conditions without surge protection was 38.13 million USD [43], giving a total cost of 217.93 million USD (Table 1). In comparison, the cost for the optimal solution given by the optWDN model, considering both steady-state and transients (with $SDPF_P = 0$) was 52.83 million USD for scenario S1 and 122.24 million USD for S2. By subtracting the cost of 38.13 million USD for steady-state service improvement, the incremental cost for protection against transients amounts to 14.70 and 84.11 million USD for scenarios S1 and S2, respectively. These costs represent about 6.7% for S1 and 38.6% for S2 compared to the total cost of 217.93 million USD for steady-state service.

To eliminate these additional costs, the optWDN was used to examine the effect of the duration of demand increase on transient pressures using only the duplicate pipes from Dany et al. [40] and El-Ghandour and Elbeltagi [2] for steady-state service improvement (Figure 16). A minimum duration of 35 s was required to satisfy the transient pressure constraints for scenario S1 with demand change at node 10, and a much longer minimum duration of ~260 s was identified for scenario S2 with simultaneous demand increase at nodes 18 and 19. These durations achieved zero $SDPF_P$ values, indicating that the minimum and maximum allowable pressure heads were satisfied over the entire network.

The long duration of 260 s for the demand increase is likely impractical, particularly with the opening of fire hydrants. By slightly relaxing the minimum pressure constraint from 54.9 m to 53.72 m, a shorter duration of 150 s was achieved, though this remains too long for practical application. A more viable solution could be to use a combination of duplicate pipes along with the regulation of the demand increase duration. There are two possible approaches to applying this hybrid solution. In the first approach, the designer selects a feasible duration to be used for regulating the demand increase (e.g., 15 s for operating fire hydrants [36]) and applies the optWDN model to optimize the duplicate pipe diameters needed to satisfy pressure constraints. Alternatively, the pressure envelope versus duration, as shown in Figure 16, can be utilized to select a suitable value for the demand increase duration that minimizes pressure violations or is within relatively flat regions of the envelope curve where slight variations in duration would not cause excessive changes in the corresponding minimum pressure. For example, as indicated by the envelope curve for S2, a duration of about 80 s would cause a minimum simulated pressure of 35.65 m, less by 19.25 m below the 54.9 m limit. Even if the duration control was not completely successful and the duration decreased by ~12% to 70 s, the resulting minimum simulated pressure would not decrease below the allowable threshold.

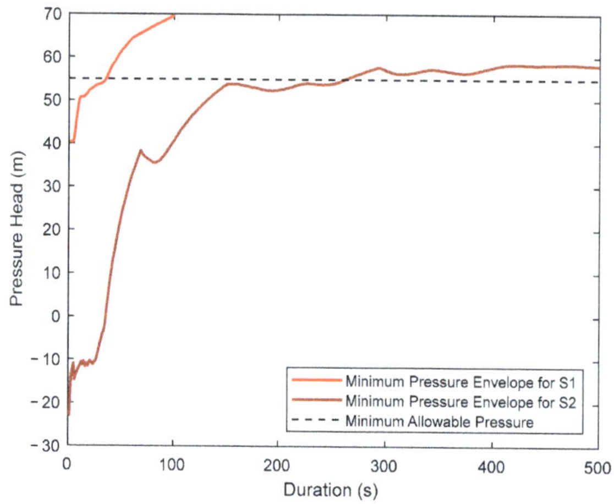

Figure 16. Envelope for minimum pressure head versus duration for demand increase at node 10 (scenario S1) and, simultaneously, identical demand change at nodes 18 and 19 (scenario S2).

Aside from the cost savings, controlling demand increases duration, minimizes cyclic pressure fluctuations, and reduces fatigue. As shown in Figure 17, the standard deviation for the pressure head at node 19 over 1000 s amounted to 8.7 m and 3.1 m for the durations of 1 and 261 s, respectively. The nearly threefold decrease in standard deviation for the longer duration clearly demonstrates the reduction in pressure fluctuations through gradual sudden demand.

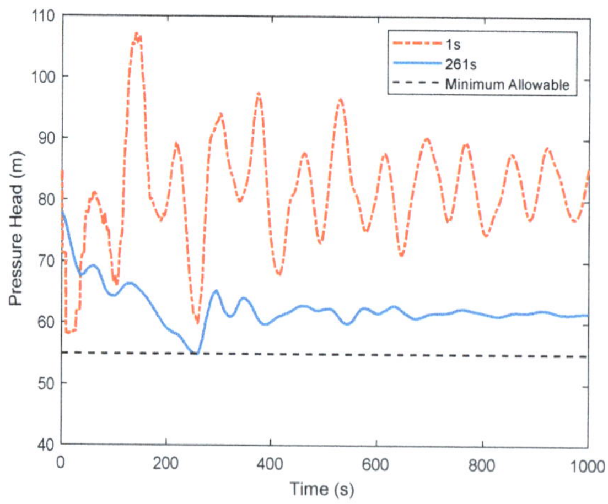

Figure 17. Variation in simulated pressure head at node 19 with time due to simultaneous, identical sudden demand change at nodes 18 and 19 (scenario S2) for demand increase durations of 1 s and 261 s.

5. Conclusions

This study developed a comprehensive model, optWDN, to identify the set of worst transient cases instead of relying on a single arbitrarily chosen case as in previous studies. The model achieves efficiency by identifying the set of cases that are most critical instead of accounting for all possible cases. For quantifying violations of pressure constraints, the model relied on the calculation of the Surge Damage Potential Factor (*SDPF*). Two versions of *SDPF* were considered: $SDPF_N$, based on the pressure at pipe end nodes (similar to previous studies), and $SDPF_P$, based on pressure at all computational nodes along pipes.

The existing New York Tunnel network was selected as a case study for model validation and for demonstration of model performance. The focus was on identifying the optimal diameters for additional duplicate pipes to achieve higher pressure under steady-state conditions and to protect the network from transients caused by sudden demand increases while minimizing cost. For a sudden demand increase event at a single predefined node, optWDN achieved $SDPF_N = 0$ at a lower cost for duplicate pipes compared to previous studies. However, the corresponding set of optimum duplicate pipe diameters, as well as the optimum diameters from previous studies, were found to be insufficient to fully protect the WDN from transient events. To achieve full protection ($SDPF_P = 0$), the optWDN model identified a different set of duplicate pipe diameters, resulting in a 14% higher cost and an 11.5% longer computational time.

The optWDN model was also applied to determine the worst transient cases and the corresponding optimum duplicate pipe diameters for a scenario with concurrent sudden demand increase at two nodes. The model reproduced the worst transient case identified by previous studies, considering the highest $SDPF_N$, but also revealed numerous transient cases with very similar high *SDPF* values, highlighting the importance of considering multiple transient cases when optimizing duplicate pipe diameters. Through an iterative scheme starting with a single worst case and adding a single case per run cycle, the optWDN identified a total of 8 cases that were critical for optimization. This approach was more efficient than starting with a fixed percentage of about 5% (9 cases) of all possible cases. The final set of optimum duplicate pipe diameters was 46% and 74% more costly than previous studies to achieve $SDPF_N = 0$ and $SDPF_P = 0$, respectively. However, the higher cost given by the optWDN model was crucial to provide full protection; optimum diameters identified by previous studies were inadequate.

The above results were based on a nearly sudden demand increase. Cost savings can be achieved by regulating the duration of the demand increase. Using only steady-state optimized pipes, minimum durations for satisfying the transient pressure constraints were identified as 35 and 261 s for the single and concurrent demand increase scenarios, respectively. For the latter scenario, a slight relaxation of the minimum allowable pressure constraint allows the reduction in the duration to 150 s. Further reduction would require the use of duplicate pipes along with control of the demand increase duration.

Besides demand increase, the optWDN framework can be extended to account for other critical transient events, such as pump failure and valve closure, by identifying the minimum set of worst transient cases necessary to achieve zero *SDPF* for all possible transient loading cases under different transient events. Additional future improvements could include allowing different surge protection devices, incorporating extended period simulations, employing network skeletonization to effectively manage the complexity of real WDNs, and extending the optimum design of new proposed WDNs.

Author Contributions: Conceptualization, H.M.A.; methodology, software, validation, formal analysis, investigation, data curation, writing—original draft preparation, H.M.A. and Y.E.I.;

writing—review and editing, supervision, Y.E.I., H.A.E.-G. and A.S.E. All authors have read and agreed to the published version of the manuscript.

Funding: This research received no external funding.

Data Availability Statement: All the data used in this study are available upon request.

Conflicts of Interest: The authors declare that they have no conflicts of interest.

Appendix A

Table A1. Optimized diameters (mm) for duplicate pipes with total cost identified by the present studies as well as by Jung et al. [34] for scenario S1.

Pipe ID	Jung et al. [34]	Present Study with $SDPF_N$=0	Present Study with $SDPF_P$ = 0
7	-	3300	3300
8	-	-	5100
9	3900	3900	4800
15	3000	-	-
16	2100	2400	1800
17	3000	3000	2400
18	2100	1800	2100
19	3000	2100	1800
21	1500	1800	1800
Total optimized pipe Cost (million USD)	49.1	46.37	52.83

Table A2. Optimized diameters (mm) for duplicate pipes with total cost identified by the present study as well as by Jung [18] for scenario S2.

Pipe ID	Jung [18]	Present Study with $SDPF_N$=0	Present Study with $SDPF_P$ = 0
1	-	-	3900
7	3900	1200	1800
8	2400	4500	-
15	-	-	4500
16	2100	5100	5100
17	4800	5100	5100
18	5100	5100	5100
19	1500	4500	4800
20	-	-	2700
21	2400	4200	3300
Total optimized pipe Cost (million USD)	70	102.11	122.24

References

1. Zhou, X.; Gao, D.Y.; Simpson, A.R. Optimal design of water distribution networks by a discrete state transition algorithm. *Eng. Optim.* **2016**, *48*, 603–628. [CrossRef]
2. El-Ghandour, H.A.; Elbeltagi, E. Comparison of five evolutionary algorithms for optimization of water distribution networks. *J. Comput. Civ. Eng.* **2018**, *32*, 04017066. [CrossRef]
3. Maier, H.R.; Simpson, A.R.; Zecchin, A.C.; Foong, W.K.; Phang, K.Y.; Seah, H.Y.; Tan, C.L. Ant colony optimization for design of water distribution systems. *J. Water Resour. Plan. Manag.* **2003**, *129*, 200–209. [CrossRef]
4. Zecchin, A.C.; Simpson, A.R.; Maier, H.R.; Nixon, J.B. Parametric study for an ant algorithm applied to water distribution system optimization. *IEEE Trans. Evol. Comput.* **2005**, *9*, 175–191. [CrossRef]

5. Zecchin, A.C.; Maier, H.R.; Simpson, A.R.; Leonard, M.; Nixon, J.B. Ant colony optimization applied to water distribution system design: Comparative study of five algorithms. *J. Water Resour. Plan. Manag.* **2007**, *133*, 87–92. [CrossRef]
6. Mehdi, D.; Massoud, T. Optimal renovation planning of water distribution networks considering hydraulic and quality reliability indices. *Urban Water J.* **2019**, *16*, 249–258.
7. Babu, K.J.; Vijayalakshmi, D.P. Self-adaptive PSO-GA hybrid model for combinatrial water distribution network design. *J. Pipeline Syst. Eng. Pract.* **2013**, *4*, 57–67. [CrossRef]
8. Ezzeldin, R.M.; Djebedjian, B.; Saafan, T. Integer discrete particle swarm optimization of water distribution networks. *J. Pipeline Syst. Eng. Pract.* **2014**, *5*, 04013013. [CrossRef]
9. Surco, D.F.; Vecchi, T.P.; Ravagnani, M.A. Optimization of water distribution networks using a modified particle swarm optimization algorithm. *Water Sci. Technol. Water Supply* **2018**, *18*, 660–678. [CrossRef]
10. Ezzeldin, R.M.; Djebedjian, B. Optimal design of water distribution networks using whale optimization algorithm. *Urban Water J.* **2020**, *17*, 14–22. [CrossRef]
11. Vasan, A.; Simonovic, S.P. Optimization of water distribution network design using differential evolution. *J. Water Resour. Plan. Manag.* **2010**, *136*, 279–287. [CrossRef]
12. Mansouri, R.; Torabi, H.; Hoseini, M.; Morshedzadeh, H. Optimization of the water distribution networks with differential evolution (DE) and mixed integer linear programming (MILP). *J. Water Resour. Prot.* **2015**, *7*, 715. [CrossRef]
13. Yazdi, J.; Choi, Y.H.; Kim, J.H. Non-dominated sorting harmony search differential evolution (NS-HS-DE): A hybrid algorithm for multi-objective design of water distribution networks. *Water* **2017**, *9*, 587. [CrossRef]
14. Jung, B.S.; Karney, B.W. Revisiting Transient Optimization of Water Supply Systems. In Proceedings of the 38th IAHR World Congress Panama, Panama City, Panama, 1–6 September 2019.
15. El-Ghandour, H.A.; Elansary, A.S. Optimal selection of pressure relief valve parameters for surge pressure control in water distribution systems. *Urban Water J.* **2019**, *16*, 269–276. [CrossRef]
16. Kim, H.; Kim, S. Optimization of pressure relief valve for pipeline system under transient induced cavitation condition. *Urban Water J.* **2019**, *16*, 718–726. [CrossRef]
17. Huang, Y.; Zheng, F.; Duan, H.F.; Zhang, Q. Multi-objective optimal design of water distribution networks accounting for transient impacts. *Water Resour. Manag.* **2020**, *34*, 1517–1534. [CrossRef]
18. Jung, B.S. Water Distribution System Optimization Accounting for Worst-Case Transient Loadings. *J. Environ. Inform. Lett.* **2022**, *7*, 20–29. [CrossRef]
19. Sangroula, U.; Han, K.H.; Koo, K.M.; Gnawali, K.; Yum, K.T. Optimization of water distribution networks using genetic algorithm based SOP–WDN program. *Water* **2022**, *14*, 851. [CrossRef]
20. Ghobadian, R.; Mohammadi, K. Optimal design and cost analysis of water distribution networks based on pressure-dependent leakage using NSGA-II. *Appl. Water Sci.* **2023**, *13*, 92. [CrossRef]
21. Naidu, M.N.; Vasan, A.; Varma, M.R.; Patil, M.B. Multiobjective design of water distribution networks using modified NSGA-II algorithm. *Water Supply* **2023**, *23*, 1220–1233. [CrossRef]
22. Djebedjian, B.; Mohamed, M.S.; Mondy, A.G.; Rayan, M.A. Cost optimization of water distribution systems subjected to water hammer. In Proceedings of the Thirteenth International Water Technology Conference, Hurghada, Egypt, 12–15 March 2009; pp. 491–513.
23. El-Ghandour, H.A.; Elansary, A.S. Optimal transient network rehabilitation using multi-objective ant colony optimization algorithm. *Urban Water J.* **2018**, *15*, 645–653. [CrossRef]
24. Jung, B.S.; Karney, B.W. Pipeline Optimization Accounting For Transient Conditions: Exploring The Connections Between System Configuration, Operation, And Surge Protection. In Proceedings of the World Environmental and Water Resources Congress 2013: Showcasing the Future, Cincinnati, OH, USA, 19–23 May 2013; pp. 903–912.
25. El-Ghandour, H.A.; Elabd, S.M.; Elbeltagi, E. Assessment of optimal water distribution systems design under steady-state and transient conditions due to pipe roughness uncertainty. *Ain Shams Eng. J.* **2021**, *12*, 465–473. [CrossRef]
26. Jung, B.S.; Karney, B.W. Hydraulic optimization of transient protection devices using GA and PSO approaches. *Water Resour. Plan. Manag.* **2006**, *132*, 44–52. [CrossRef]
27. Jung, B.S.; Karney, B.W. Systematic surge protection for worst-case transient loadings in water distribution systems. *J. Hydraul. Eng.* **2009**, *135*, 218–223. [CrossRef]
28. Skulovich, O.; Bent, R.; Judi, D.; Perelman, L.S.; Ostfeld, A. Piece-wise mixed integer programming for optimal sizing of surge control devices in water distribution systems. *Water Resour. Res.* **2015**, *51*, 4391–4408. [CrossRef]
29. Skulovich, O.; Perelman, L.; Ostfeld, A. Modeling and optimizing hydraulic transients in water distribution systems. *Procedia Eng.* **2014**, *70*, 1558–1565. [CrossRef]

30. Skulovich, O.; Perelman, L.; Ostfeld, A. Optimal closure of system actuators for transient control: An analytical approach. *J. Hydroinform.* **2016**, *18*, 393–408. [CrossRef]
31. Bohorquez, J.; Saldarriaga, J. Valve operation optimization for minimizing transient flow effects in water distribution systems (WDS): Application to main pipes in Bogota, Colombia. In Proceedings of the World Environmental and Water Resources Congress, Austin, TX, USA, 17–21 May 2015; pp. 890–901.
32. Carvajal, J.D.; Bohorquez, J. Estimation of Marginal Time in Water Distribution Systems for Valve Closure Considering Transient Flow Effects: (149). In Proceedings of the WDSA/CCWI Joint Conference Proceedings, Kingston, ON, Canada, 23–25 July 2018; Volume 1.
33. Jung, B.S.; Karney, B.W. Fluid transients and pipeline optimization using GA and PSO: The diameter connection. *Urban Water J.* **2004**, *1*, 167–176. [CrossRef]
34. Jung, B.S.; Boulos, P.F.; Altman, T. Optimal transient network design: A multi-objective approach. *J. Am. Water Work. Ass.* **2011**, *103*, 118–127. [CrossRef]
35. Walski, T.M. Not So Fast!: Close Hydrants Slowly. *Opflow* **2009**, *35*, 14–16. [CrossRef]
36. Lingireddy, S.; Ormsbee, L.E.; Kamojjala, S.; Hall, A.M. How Slow Is Slow? Managing Fire Hydrant Operation for Protecting Water Infrastructure. *AWWA Water Sci.* **2022**, *4*, e1290. [CrossRef]
37. Rossman, L.A. *EPANET 2: Users Manual*; United States Environmental Protection Agency: Washington, DC, USA, 2000.
38. Larock, B.E.; Jeppson, R.W.; Watters, G.Z. *Hydraulics of Pipeline Systems*; CRC Press: Boca Raton, FL, USA, 1999.
39. Wylie, E.B.; Streeter, V.L. *Fluid Transients in Systems*; Prentice Hall: Englewood Cliffs, NJ, USA, 1993.
40. Chaudhry, M.H. *Applied Hydraulic Transients New York*; Springer: Berlin/Heidelberg, Germany, 2014; Volume 415, pp. 31–42.
41. Xing, L.; Sela, L. Transient simulations in water distribution networks: TSNet python package. *Adv. Eng. Softw.* **2020**, *149*, 102884. [CrossRef]
42. Dandy, G.C.; Simpson, A.R.; Murphy, L.J. An improved genetic algorithm for pipe network optimization. *Water Resour. Res.* **1996**, *32*, 449–458. [CrossRef]
43. Eusuff, M.M.; Lansey, K.E. Optimization of water distribution network design using the shuffled frog leaping algorithm. *J. Water Resour. Plan. Manag.* **2003**, *129*, 210–225. [CrossRef]
44. Yazdi, J.; Hokmabadi, A.; JaliliGhazizadeh, M.R. Optimal Size and Placement of Water Hammer Protective Devices in Water Conveyance Pipelines. *Water Resour. Manag.* **2019**, *33*, 569–590. [CrossRef]
45. Lee, J.S.; Zeng, W.; Lambert, M.; Hilditch, T.; Gong, J. Fatigue analysis of metallic-plastic-metallic pipeline systems: A numerical study. *Results Eng.* **2023**, *17*, 100986. [CrossRef]
46. Jara-Arriagada, C.; Stoianov, I. Pressure-induced fatigue failures in cast iron water supply pipes. *Eng. Fail. Anal.* **2024**, *155*, 107731. [CrossRef]
47. Konstantinou, C.; Jara-Arriagada, C.; Stoianov, I. Investigating the impact of cumulative pressure-induced stress on machine learning models for pipe breaks. *Water Resour. Manag.* **2024**, *38*, 603–619. [CrossRef]
48. Rao, S.S. *Reliability-Based Design*; McGraw-Hill: New York, NY, USA, 1992.
49. McCuen, R.H.; Knight, Z.; Cutter, A.G. Evaluation of the Nash–Sutcliffe efficiency index. *J. Hydrol. Eng.* **2006**, *11*, 597–602. [CrossRef]

water

MDPI

Article

Advancing Ion Constituent Simulations in California's Sacramento–San Joaquin Delta Using Machine Learning Tools

Peyman Namadi, Minxue He * and Prabhjot Sandhu

Modeling Support Office, California Department of Water Resources, Sacramento, CA 95814, USA
* Correspondence: kevin.he@water.ca.gov

Abstract: This study extends previous machine learning work on ion constituent simulation in California's Sacramento–San Joaquin Delta (Delta) to include three critical water intake locations. The developed Artificial Neural Network models demonstrate exceptional accuracy ($R^2 > 0.96$) in predicting chloride, bromide, and sulfate concentrations at these strategically important facilities. Water intake location models show substantial improvements in prediction accuracy, with MAE reductions of 60.7–74.0% for chloride, 63.3–72.5% for bromide, and 70.4–87.9% for sulfate, compared to existing methods for the Interior Delta. Performance evaluation through comprehensive cross-validation confirms robust model stability across varied conditions, with remarkably consistent metrics (standard deviation in $R^2 \leq 0.006$). Four complementary interactive dashboards were developed, enabling users, regardless of programming expertise, to simulate ion constituents throughout the Delta system. A Model Interpretability Dashboard specifically addresses the complexity of machine learning models by visualizing parameter sensitivity and prediction behavior, thereby enhancing transparency and building stakeholder trust in the modeling approach. For the first time, spatial coverage limitations are addressed through hybrid modeling that combines DSM2 hydrodynamic simulation with machine learning to enable continuous prediction of ion distributions across several points in the Interior Delta. These advancements provide water managers with accessible, accurate tools for informed decision-making regarding agricultural operations, drinking water treatment, and ecosystem management in this vital water resource.

Keywords: ion constituents; machine learning; hybrid modeling; Sacramento–San Joaquin Delta; dashboard tools

1. Introduction

The Sacramento–San Joaquin Delta (Delta) serves as California's most crucial water hub, where the state's two largest rivers, Sacramento River and San Joaquin River, converge to form a 2978-square-kilometer (1150-square-mile) inland Delta and estuary system [1,2]. This vital water resource supports approximately 27 million people, irrigates 1.21 million hectares of farmland, and maintains a diverse ecosystem of over 750 species [3,4]. Beyond water conveyance, the Delta generates significant economic value through recreation, recreational fishing, and transportation activities.

Water quality in the Delta, particularly the concentration levels of ion constituents including dissolved chloride, bromide, and sulfate, directly impacts multiple stakeholders and uses. For agricultural operations, elevated chloride levels can increase soil salinity and reduce crop yields. In urban water treatment, high bromide concentrations lead to the formation of harmful disinfection by-products (DBPs), requiring additional treatment steps to ensure public safety [5,6]. Agricultural runoff, urban development, and tidal influences continuously introduce various pollutants and ions into the system, making water quality monitoring and prediction essential for effective management [7–9].

Managing these water quality challenges requires balancing diverse needs: agricultural productivity, drinking water safety, and ecosystem health [10]. The complex interplay between these demands, combined with Delta's dynamic hydrology, creates a pressing need for accurate and timely water quality monitoring and prediction tools [11].

Physical sampling and laboratory analysis remain the gold standard for monitoring ion constituents in the Delta, providing highly accurate and reliable measurements of water quality parameters [12]. This direct measurement approach ensures precise quantification of individual ion constituents and serves as the foundation for calibrating and validating other estimation methods.

However, traditional sampling methods present several limitations:

1. Time and cost constraints: laboratory analysis often requires days or weeks to complete, involving substantial personnel and equipment resources.
2. Limited coverage: physical sampling provides only point-in-time measurements at specific locations, lacking continuous temporal and spatial coverage.
3. Operational delays: the time lag between collecting samples in the field and receiving laboratory results can impede timely operational decisions for water management.

These limitations have driven the development of alternative approaches for estimating ion constituent levels, particularly through the use of Electrical Conductivity (EC) as a proxy measurement.

Due to the limitations of direct sampling, scientists have explored faster and more cost-effective methods for estimating ion constituent levels in the Delta. Electrical Conductivity (EC) has emerged as a valuable proxy measurement, due to its unique advantages in water quality monitoring. EC can be measured continuously through in situ sensors and simulated effectively through hydrodynamic models, providing real-time monitoring capabilities that are crucial for operational decision-making [13]. Moreover, EC demonstrates strong correlations with various ion concentrations, making it a reliable indicator of the overall ionic content in water [14]. These characteristics have made EC the foundation for most ion constituent estimation methods in the Delta, enabling more frequent and widespread simulation than would be possible through direct sampling alone.

The development of methods for estimating ion constituents from EC measurements spans several decades. Classical approaches primarily rely on parametric regression equations, assuming that the relationship between EC and ion concentrations follows specific mathematical patterns (linear, quadratic, etc.) under different conditions [15–17]. These methods aim to simplify the conversion process from EC measurements to ion constituent levels.

The evolution of these approaches shows increasing sophistication over time. The pioneering work by [18] established the foundation by developing linear regression equations between EC, chloride, and total dissolved solids (TDS). Throughout the 1990s and early 2000s, researchers significantly expanded upon this foundation. Jung [19] investigated correlations between EC and various ions in Delta island return flows, while Suits [20] developed linear relationships between EC, chloride, and bromide at Delta export locations. A significant advancement came with Denton's comprehensive study, which provided multiple methodological approaches for estimating a broader suite of ions [5].

The current state-of-the-art classical approach, developed by Hutton et al. [21], represents the most comprehensive parametric regression method to date. This study introduced a novel decision tree framework that determines appropriate regression equations based on multiple factors, including the sub-region of the Delta, Water Year Type (WYT), Month, and Sacramento X2 location [22], which is a key marker for salinity intrusion. Using these parameters, the method classifies conditions as seawater-dominated, San Joaquin River-dominated, or mixed, applying specific equations accordingly.

While the approach of Hutton et al. [21] significantly improved upon previous methods, it faces several limitations. A fundamental challenge lies in the simplified source assumptions, where classical methods generally consider only two major sources of salinity—seawater and the San Joaquin River. Agricultural drainage, despite being a significant salinity source in the Delta, is often oversimplified or considered aligned with the San Joaquin River source. This simplification can lead to inaccuracies, since agricultural drainage has distinct ionic compositions and behavior patterns compared to river inflows.

The handling of mixed conditions presents another significant challenge. Classical models tend to oversimplify these conditions by assuming a dominant source or using a single equation for mixed conditions. However, the reality is more complex, since salinity in the Delta typically results from multiple simultaneous sources, with their relative influences varying over time. The same EC level can correspond to markedly different ionic compositions depending on the source: seawater-dominated water often contains higher proportions of chloride and bromide, whereas water influenced by agricultural drainage or river flows is generally richer in sulfate.

Perhaps the most challenging aspect is the dynamic nature of these mixed conditions. The Delta exists almost perpetually in a mixed state, with salinity levels influenced by multiple sources simultaneously. The relationship between EC and ion concentrations shifts dynamically based on the varying contributions from different sources, and current parametric approaches struggle to capture these relationships, particularly during periods of rapid hydrological changes.

These limitations highlight the need for more sophisticated approaches that can better capture the complex and dynamic relationships between EC and ion constituents in the Delta system. Machine learning methods, with their ability to identify and adapt to complex patterns in data, offer promising solutions to these challenges.

Recent developments in machine learning have opened new possibilities for addressing these challenges. Following established protocols for water and environmental modeling using machine learning in California [23], our first study [24] served as a proof of concept to evaluate whether machine learning models could outperform traditional regression equations in simulating ion constituents. This pioneering work focused on the South Delta region, using a relatively small dataset of approximately 200 samples collected from seven stations between 2018 and 2020. Using R (https://www.r-project.org/, accessed on 1 June 2022) as the programming platform, we compared four machine learning approaches—Generalized Additive Model, Regression Trees, Random Forest, and Artificial Neural Networks—against conventional regression equations. The results demonstrated that machine learning models, particularly Random Forest, provided superior simulations of ion constituents compared to traditional regression equations. These improvements were especially notable for ions that exhibit pronounced non-linear relationships with EC. This study established the viability of machine learning approaches for simulating ion constituents in the Delta.

Building on these promising initial results, our subsequent research [25] significantly expanded both the geographic and temporal scope of the analysis. Implemented in Python (https://www.python.org/, accessed on 1 June 2022) this comprehensive study covered the entire Interior Delta and incorporated an extensive dataset spanning over 60 years, with sample sizes ranging from approximately 1000 to 2000 measurements for different ion constituents. This study developed and validated Artificial Neural Network (ANN) architectures specifically designed for ion constituent prediction, demonstrating remarkable improvements over the benchmark classical approach [21]. The ANN model showed significant improvements in both accuracy metrics, with R^2 improvements ranging from 0% for well-modeled constituents like TDS, to 85% for more complex constituents like sulfate. More importantly, the ANN model reduced the Mean Absolute Error (MAE) across all constituents, with improvements of 24% for TDS, 33% for magnesium, 32% for sodium, 40% for calcium, 34% for chloride, 59% for sulfate, 25% for bromide, 26% for alkalinity, and 20% for potassium. Additionally, the study introduced an interactive web browser-based dashboard hosted on Microsoft Azure, making these sophisticated prediction tools accessible to users regardless of their programming expertise. This user-friendly interface allows stakeholders to simulate ion levels under various hydrological conditions, compare results across different machine learning models, and visualize outcomes for informed decision-making.

The current study builds on the two studies mentioned above [24,25] and extends them in several important directions. First, we focus on applying ML models to three strategically important drinking water intake locations, where water quality monitoring is crucial for operational decision-making. Second, we enhance the accessibility and utility of our modeling tools through three complementary dashboards: (a) a new publicly available dashboard for intake locations; (b) an improved version of our previous dashboard [25] that incorporates Hutton et al.'s [22] parametric regression equations for direct comparison with ML approaches for the Interior Delta; and (c) a novel Model Interpretability Dashboard that helps demystify the "black box" nature of ML models by visualizing how they respond to different predictors and operating conditions for the Interior Delta. This third dashboard is particularly innovative as it allows users to explore model behavior, assess sensitivity to input variables, and build trust in ML predictions by comparing them with traditional parametric approaches. Finally, to address key limitations of current ML approaches—specifically, their temporal and spatial distribution constraints and inability to simulate potential future scenarios—we develop a novel hybrid water quality model that combines hydrodynamic modeling with machine learning, supported by its own interactive dashboard for scenario exploration and analysis. Together, these advancements not only provide more accurate predictions, but also make complex ML models more transparent and interpretable for water resource managers, helping them to make more informed decisions about monitoring and managing ion constituents in the Delta.

This paper first presents our development of machine learning models for predicting ion constituents (chloride, bromide, and sulfate) at three critical water intake locations. We describe our methodology, including data characteristics and model optimization procedures, followed by performance analyses demonstrating significant improvements over existing approaches.

We then present three complementary web-based tools. The first is a new Intake Locations Dashboard specifically designed for water intake locations. The other two dashboards build upon the previous Interior Delta study of Namadi et al. [25]: an Enhanced Comparison Dashboard that integrates classical methods with machine learning approaches, and a Model Interpretability Dashboard that helps users to understand model behavior across the Interior Delta.

To address spatial coverage limitations in the Interior Delta, we introduce a novel hybrid approach combining DSM2 hydrodynamic modeling with machine learning [26]. This model enables continuous spatial prediction of ion constituents across 162 points throughout the Interior Delta.

Finally, we discuss implications of the current study for Delta water management, acknowledge current limitations, and suggest directions for future research.

2. Materials and Methods

2.1. Study Area and Dataset Characteristics

Our research synthesizes and builds upon two complementary datasets from the Delta: the Interior Delta dataset from our previous work [25] and a new focused dataset from strategic water intake locations. As shown in Figure 1, our study area encompasses both interior monitoring stations and critical water intake facilities. The Interior Delta monitoring network consists of three sub-regions: the Old and Middle River corridor (OMR, shown as red squares), the San Joaquin River corridor (SJRcorridor, marked by black triangles), and the South Delta region (represented by blue circles). The water intake locations, depicted as green stars, include three strategically important pumping plants: the Harvey O. Banks Pumping Plant (HRO; 37°47′53″ N, 121°37′23″ W), the Tracy Pumping Plant (TRP; 37°48′00″ N, 121°35′06″ W), and the O'Neill Forebay at the Gianelli Pumping Plant (ONG; 37°03′60″ N, 121°04′19″ W).

This study encompasses two complementary analyses. First, we expand upon our previous work that developed machine learning models for nine ion constituents (TDS, magnesium, sodium, calcium, chloride, sulfate, bromide, alkalinity, and potassium) across the Interior Delta monitoring stations, providing enhanced analysis and interpretation through our new dashboard tools. Second, we extend our modeling approach to water intake locations, where available monitoring data focus on three key ion constituents (chloride, bromide, and sulfate). These three constituents are particularly crucial for water quality management at pumping facilities, as they directly impact treatment processes and water supply operations.

The temporal and spatial coverage of these datasets reveals distinct characteristics that reflect their different roles in Delta water management. The Interior Delta dataset encompasses data from 1959 to 2022, though the temporal coverage varies by constituent and location. For detailed information about the Interior Delta monitoring timeline and data characteristics, readers are referred to our previous study [25]. In contrast, our water intake locations feature more recent but substantially more intensive monitoring programs. The HRO and TRP data collection spans from 2019 to 2023, while ONG has a longer record, from 2012 to 2023. These locations are characterized by high-frequency sampling, resulting in robust sample sizes: ONG collected 2620–2783 samples, while HRO and TRP gathered 1214–1328 and 1181–1233 samples, respectively. This intensive monitoring reflects the operational importance of these facilities in California's water supply infrastructure.

Analysis of constituent distributions, as illustrated in Figure 2, reveals distinct patterns between the Interior Delta and the water intake locations. Interior Delta locations consistently show higher variability and maximum values across all the measured constituents. The South Delta region exhibits the highest median EC values and widest range, varying from approximately 200 to 1400 µS/cm, while intake locations maintain more consistent EC patterns between 300 and 800 µS/cm. Similar patterns emerge for chloride concentrations, where Interior Delta locations, particularly the South Delta and OMR, show notably higher maximum values and greater variability compared to the more stable concentrations at intake locations.

Bromide distributions follow comparable trends, with OMR exhibiting particularly high outliers and greater variability compared to the relatively consistent levels maintained at intake locations. Sulfate concentrations in the Interior Delta, especially in the South Delta and SJRcorridor, show higher median values and greater variability compared to intake locations, though ONG typically shows slightly higher median values among the intake locations, while still maintaining more stable concentrations overall.

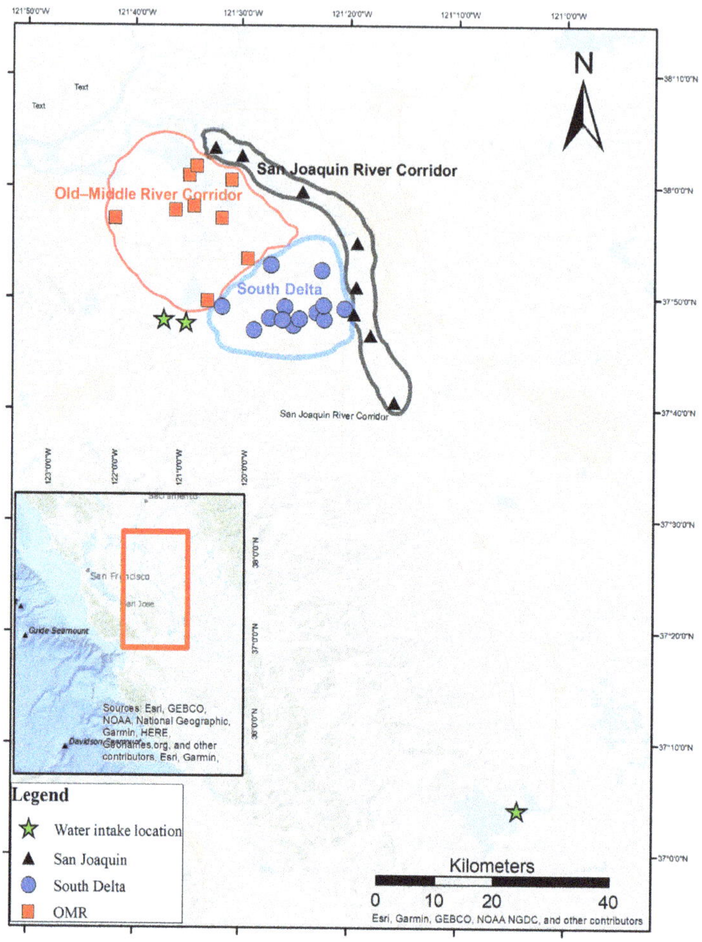

Figure 1. A map of monitoring locations in the Sacramento–San Joaquin Delta, showing Interior Delta stations (OMR—red squares, San Joaquin River corridor—black triangles, South Delta—blue circles) and water intake locations (green stars).

These distinct patterns between the Interior Delta and the intake locations reflect their different roles in the Delta system. The Interior Delta locations capture the system's natural variability and multiple water source influences, resulting in wider ranges of constituent concentrations. The more consistent patterns at intake locations reflect managed operations and water quality control measures, which are crucial for maintaining reliable water supply quality.

This comparison provides crucial context for our modeling approach and dashboard development. The stark differences between the Interior Delta and intake locations—particularly in sampling frequency (moderate vs. intensive) and constituent variability (wide-ranging vs. more consistent)—reveal fundamentally different water quality patterns at these locations. Due to these distinct characteristics, we developed separate simulation models specifically for water intake locations, rather than applying our Interior Delta models. This location-specific modeling approach ensures that predictions accurately reflect the unique water quality dynamics and operational patterns at each type of monitoring site.

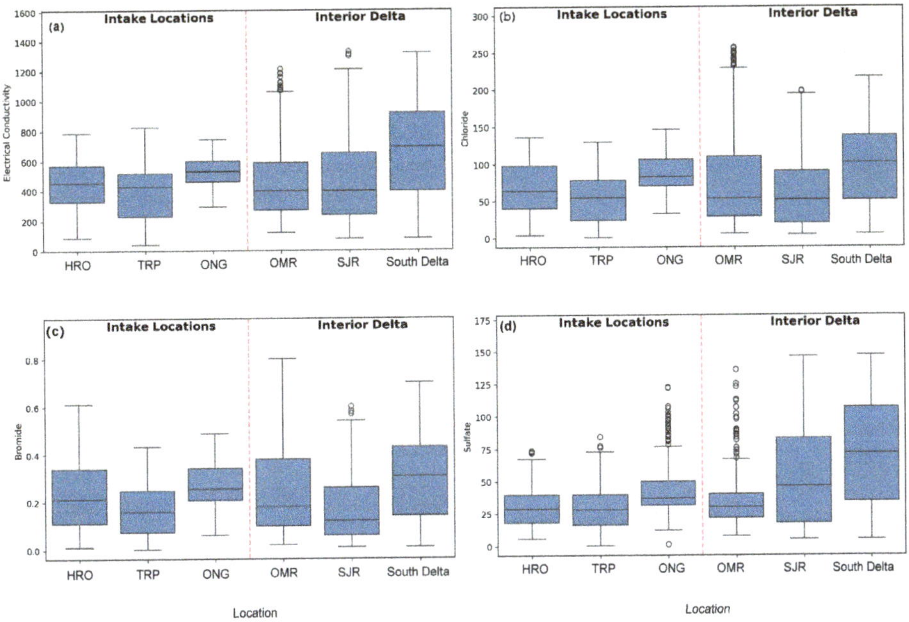

Figure 2. Distribution of (**a**) Electrical Conductivity, (**b**) chloride, (**c**) bromide, and (**d**) sulfate concentrations at water intake locations and Interior Delta monitoring stations. Red dashed line separates intake locations (**left**) from Interior Delta locations (**right**).

2.2. Model Development

2.2.1. Input Variable Selection

Our model development utilized a consistent set of predictor variables across both Interior Delta and water intake locations, maintaining methodological continuity with our previous work [25]. The selected predictors capture key physical and operational factors influencing ion constituent concentrations:

1. Electrical Conductivity (EC): A real-time measurement obtained through in situ sensors that indicates the water's ability to conduct an electrical current, serving as a proxy for total dissolved solids and overall salinity levels [27]. EC measurements provide continuous monitoring capabilities, which are crucial for operational decision-making.

2. Sacramento X2: A key hydrodynamic indicator, measured as the distance (in kilometers) from the Golden Gate Bridge to the location where the tidally averaged near-bottom salinity is 2 practical salinity units (psu) [28]. This metric effectively captures Delta outflow conditions and the extent of seawater intrusion from the Pacific Ocean into the Delta system [29].

3. Water Year Type (WYT): California's water year runs from October 1 to September 30, with types classified as Wet, Above-Normal, Below-Normal, Dry, or Critical, based on measured unimpaired runoff [30]. This classification integrates various hydrological factors, including precipitation, snowmelt, and river flows, providing a comprehensive indicator of water availability and system conditions [31].

4. Month: This is included to capture seasonal patterns in water quality, reflecting cyclical changes in precipitation, temperature, agricultural practices, and water management operations.

5. Location: A categorical variable distinguishing between monitoring sites. For water intake locations, this includes HRO, TRP, and ONG, while the Interior Delta locations are categorized into the OMR, SJRcorridor, and South Delta sub-regions.

These predictors were selected for their demonstrated relationships with ion constituent concentrations and their operational relevance in water management. The combination of continuous (EC, X2) and categorical (WYT, Location, and Month) variables enables our models to capture both direct relationships and complex interaction effects influencing water quality patterns.

While maintaining methodological consistency, our benchmark comparison uses performance metrics from existing approaches. Since no prior models exist for water intake locations, we compared our new models' performance metrics against the documented performance of both ML and classical methods in the Interior Delta [24,25]. For example, if the Interior Delta ML model demonstrated a specific MAE for chloride predictions in that region, we used this as a benchmark to evaluate our new model's MAE for chloride predictions at water intake locations. This comparison, while across different locations, provides context for assessing our models' relative performance in predicting ion constituents.

Data preprocessing and quality control were crucial steps to ensure model reliability. We implemented a regression-based outlier detection method focusing on the fundamental relationship between EC and ion constituents, where points exceeding three standard deviations from the regression line were flagged as potential outliers. This analysis revealed minimal outliers across locations, as detailed in Table 1, with outlier percentages generally less than 3% across all constituents and locations. The strong coefficient of determination (R^2) values, particularly for chloride (0.928 at HRO, 0.688 at ONG, and 0.823 at TRP), indicated high data quality and validated our outlier detection approach. Figure 3 illustrates the relationship between EC and chloride concentrations at the three water intake locations, demonstrating both the effectiveness of our outlier detection method and the varying relationships at different locations.

Table 1. Summary of outlier analysis and regression relationships between EC and ion constituents at water intake locations.

Location	Constituent	Total Samples	Outliers (%)	R^2	Regression Equation
HRO	Chloride	1328	0.1	0.93	$Cl = 0.19 \times EC - 20.2$
HRO	Bromide	1214	1.1	0.78	$Br = 0.0007 \times EC - 0.1$
HRO	Sulfate	1328	0	0.53	$SO_4 = 0.06 \times EC + 2.2$
ONG	Chloride	2665	0.1	0.69	$Cl = 0.22 \times EC - 28.9$
ONG	Bromide	2620	0.4	0.57	$Br = 0.0007 \times EC - 0.1$
ONG	Sulfate	2783	1.3	0.33	$SO_4 = 0.08 \times EC - 3.4$
TRP	Chloride	1227	2.8	0.82	$Cl = 0.16 \times EC - 11.7$
TRP	Bromide	1181	1.4	0.67	$Br = 0.0006 \times EC - 0.05$
TRP	Sulfate	1233	0.6	0.58	$SO_4 = 0.06 \times EC + 2.5$

Following outlier detection, we implemented a comprehensive feature preprocessing pipeline using Scikit-learn's ColumnTransformer function to prepare the data for model training. Numerical predictors (EC and X2) were standardized using the StandardScaler function to transform them to zero mean and unit variance, ensuring consistent scale across features and preventing convergence issues during model training. Categorical variables (Location, WYT, Month) were transformed using one-hot encoding to convert them into binary features while avoiding multicollinearity. This preprocessing pipeline was saved using joblib to ensure consistent transformation of future data points. The standardization and encoding approach helps to prevent any single feature from dominating the model training process, ensures proper handling of categorical information, and facilitates stable model convergence during training.

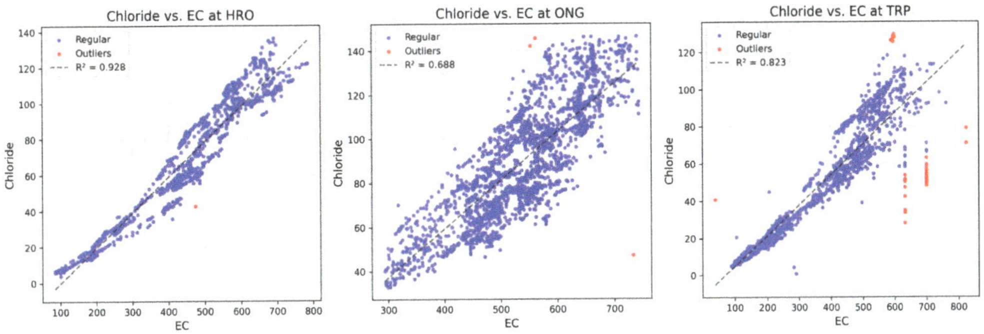

Figure 3. Relationship between EC and chloride concentrations at water intake locations (HRO, ONG, and TRP). Blue points represent regular observations, red points indicate identified outliers, and dashed lines show linear regression fit.

2.2.2. Machine Learning Approaches

Machine learning, a subset of artificial intelligence, enables computers to learn patterns from data without explicit programming [32]. In recent years, machine learning techniques have been increasingly applied to water quality modeling, due to their ability to capture complex, non-linear relationships in environmental systems. Numerous studies have demonstrated the effectiveness of machine learning in water quality applications, including the prediction of dissolved oxygen [33], suspended sediment [34], and salinity [35–38].

Artificial Neural Networks (ANNs) are particularly powerful machine learning models inspired by biological neural networks [39]. Among the various ANN architectures, we employed the Multi-Layer Perceptron (MLP), a feedforward neural network architecture well suited for regression problems like ion constituent prediction. The MLP architecture consists of multiple layers of interconnected nodes (neurons), with each connection having an associated weight that is adjusted during training. Each neuron applies a non-linear activation function to the weighted sum of its inputs, enabling the network to learn complex patterns in the data. This architecture has proven effective in various water quality applications, including prediction of chemical oxygen demand and analysis of heavy metal content.

Building upon our previous comparative analysis of machine learning algorithms [25], we employed ANNs for simulating ion constituent concentrations at water intake locations. The selection of ANNs was guided by their demonstrated superior performance in capturing non-linear relationships between environmental predictors and ion concentrations in the Delta system, particularly their ability to adapt to complex spatial and temporal patterns in water quality dynamics. We implemented an MLP architecture using TensorFlow [40]. The model training utilized the Adam optimizer with the mean squared error (MSE) loss function, chosen for its effectiveness in regression tasks and adaptability to varying gradients.

Hyperparameter optimization is a crucial step in developing neural networks, as these parameters significantly influence model performance, but cannot be learned directly from training data. Unlike model weights and biases that are updated during training, hyperparameters must be set before the training process begins. However, the complex interactions between different hyperparameters make it challenging to determine their optimal values theoretically, necessitating empirical optimization approaches.

MLPs have a number of hyperparameters, including network architecture (number of layers and neurons), activation functions, learning rate, batch size, optimizer settings, and regularization parameters. Optimizing all possible hyperparameters simultaneously would be computationally intensive and often impractical. For instance, considering just five options for ten hyperparameters would result in 5^{10} (approximately 9.7 million) possible combinations. Therefore, researchers typically focus on optimizing a subset of hyperparameters that are known to have the most significant impact on model performance.

In this study, we focused on three critical hyperparameters:

- Number of hidden layers (2 to 5 layers);
- Number of neurons per layer (10, 20, 30, or 40);
- Activation functions (ReLU, tanh, ELU, SELU, Leaky ReLU, and sigmoid) (Equations (1)–(6)).

Figure 4 illustrates these activation functions, grouped into two families. The ReLU family (Figure 4a) includes the following:

$$\text{ReLU: f(x)} = \max(0,\text{x}) \tag{1}$$

$$\text{ELU: f(x)} = \text{x if x} > 0 \text{ else } \alpha(\text{e}^{\text{x}} - 1), \tag{2}$$

where α is a positive constant

$$\text{Leaky ReLU: f(x)} = \text{x if x} > 0 \text{ else } \alpha\text{x}, \tag{3}$$

where α is a small positive constant

$$\text{SELU: f(x)} = \lambda\text{x if x} > 0 \text{ else } \lambda\alpha(\text{e}^{\text{x}} - 1), \tag{4}$$

where λ and α are learned parameters.

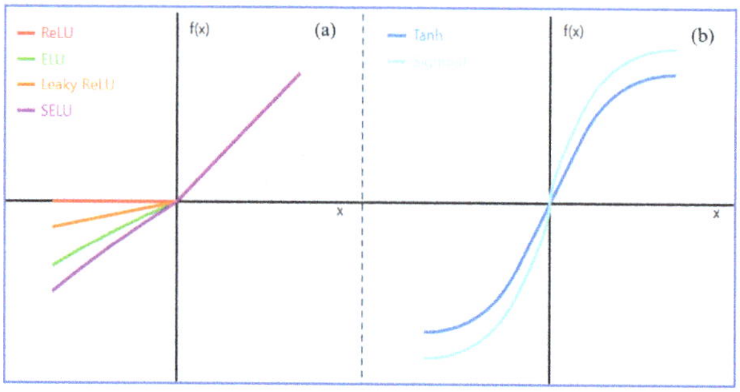

Figure 4. Common neural network activation functions: (**a**) ReLU family functions, showing different approaches to handling negative inputs; and (**b**) Sigmoid family functions with bounded outputs.

The sigmoid family (Figure 4b) includes the following:

$$\text{Tanh: f(x)} = \frac{e^x - e^{-x}}{e^x + e^{-x}} \tag{5}$$

$$\text{Sigmoid: f(x)} = \frac{1}{1 + e^{-x}} \tag{6}$$

This approach differs from our previous study [25] in several ways. In our earlier work, we fixed the number of hidden layers at 4, but allowed each layer to have different numbers of neurons (ranging from 20 to 44 in increments of 2) and different activation functions (choosing from ELU, ReLU, tanh, and sigmoid). This resulted in 13 possible neuron counts and 4 activation functions for each of the 4 layers, leading to $(13 \times 4)^4$ = 331,776 possible combinations. Given this large search space, we randomly sampled 100 combinations for evaluation.

In contrast, our current approach explores different network depths, while maintaining consistency across layers (same number of neurons and activation function for all hidden layers). We also expanded our activation function options by adding SELU and Leaky ReLU. This new approach results in 4 (layer options) × 4 (neuron options) × 6 (activation functions) = 96 possible combinations, allowing us to exhaustively evaluate all possibilities, rather than relying on random sampling.

We implemented a systematic grid search across these hyperparameters using TensorFlow. The model training utilized the Adam optimizer with the mean squared error (MSE) loss function, chosen for its effectiveness in regression tasks and adaptability to varying gradients. To prevent overfitting and ensure model generalization, we implemented early stopping with validation loss monitoring (patience = 50 epochs). The optimization process involved stratified data splitting (60% training, 20% validation, 20% testing), standardization of numerical features (EC, X2), and one-hot encoding of categorical variables (Location, WYT, Month). The preprocessing pipeline was preserved using joblib to ensure consistent transformation of future data points, which is crucial for model deployment and operational use.

2.2.3. Model Evaluation Metrics

Model evaluation in water quality prediction requires careful consideration of both prediction accuracy and generalization capability. We established an evaluation framework incorporating multiple complementary metrics and validation strategies to ensure robust assessment of model performance.

Our evaluation framework centered on three primary performance metrics, each capturing different aspects of model performance. The coefficient of determination (R^2), defined in Equation (7), quantifies the proportion of variance in the observed data explained by the model. While higher values indicate a better model fit, R^2 alone can be misleading, as it is sensitive to the range and distribution of the data. However, its widespread use in water quality modeling studies makes it valuable for comparing results across different research efforts and methodologies. The Mean Absolute Error (MAE, Equation (8)) provides a direct measure of prediction accuracy in the original units of measurement (mg/L), making it particularly valuable for operational applications. Unlike R^2, the MAE is less sensitive to outliers and offers more interpretable results. The magnitude of acceptable MAE varies by constituent, considering its typical concentration range and operational requirements. The third metric, Percentage Bias (PBIAS, Equation (9)), reveals systematic over- or under-prediction tendencies in the model. Positive values indicate overestimation, while negative values suggest underestimation. In water quality applications, consistent bias could significantly impact treatment processes and management decisions, making PBIAS an essential metric for operational reliability assessment.

We implemented a multi-layered validation approach to ensure robust evaluation of model performance. The dataset was divided into three subsets through random stratified sampling: 60% for training, 20% for validation during model development, and 20% reserved as an independent test set. This split ensured unbiased evaluation of final model performance on previously unseen data, while maintaining similar statistical distributions across all sets. The comparison of performance metrics between training and test sets served as a crucial check for potential overfitting, where small differences indicate good generalization, while large discrepancies would suggest model reliability issues requiring refinement.

The performance metrics are defined as follows:

$$R^2 = 1 - \frac{\sum\left(y_{true} - y_{pred}\right)^2}{\sum(y_{true} - \overline{y})^2} \quad (7)$$

$$MAE = \frac{\sum\left|y_{pred} - y_{true}\right|}{n} \quad (8)$$

$$\text{PBIAS} = 100 \times \frac{\sum\left(y_{pred} - y_{true}\right)}{\sum(y_{true})} \quad (9)$$

where y_{true} represents observed values, y_{pred} represents predicted values, $\bar{y}$ is the mean of observed values, and n is the number of observations.

To further assess model stability and generalization capabilities, we implemented a k-fold cross-validation analysis (k = 5). This approach systematically partitions the data into five equal folds, using four folds for training and one for testing, rotating through all possible combinations. We maintained consistent preprocessing across all folds using Scikit-learn's ColumnTransformer, which included standardization of numerical features (EC, X2) and one-hot encoding of categorical variables (Location, WYT, Month). For each fold, we loaded the previously trained models and preprocessors for each ion constituent, then evaluated their performance using the three metrics mentioned above (R^2, MAE, and PBIAS). This cross-validation implementation was performed independently for each ion constituent (chloride, bromide, and sulfate), allowing us to assess the stability of model performance across different subsets of data and different constituents. The analysis was automated using the TensorFlow and Scikit-learn libraries, ensuring consistent evaluation procedures across all folds and constituents. This systematic approach provides a more reliable estimate of model performance than single train-test splits, which was particularly important given the temporal and spatial variability in our water quality data.

2.3. Dashboard Development

Web-based dashboards offer significant advantages for water quality management tools, providing universal accessibility, real-time updates, and interactive visualizations without requiring local software installation. Our implementation leverages Microsoft Azure's Web Apps service, a cloud computing platform that provides scalable hosting infrastructure [41]. Azure Web Apps manage server maintenance, security updates, and load balancing, allowing the applications to efficiently handle multiple concurrent users. The system's computational resources are adjustable based on usage demands—if additional processing power is needed to handle more users or complex calculations, the cloud computing capacity can be increased by upgrading the subscription level. During periods of lower demand, capacity can be scaled down to optimize costs.

The development infrastructure uses GitHub (https://github.com/, accessed on 1 June 2022) as a version control and code hosting platform, integrated with Azure through continuous integration and deployment (CI/CD) pipelines. When code updates are pushed to GitHub, Azure automatically deploys these changes to the production environment after running automated tests, ensuring rapid updates while maintaining system stability. The cloud resources are configured based on anticipated computational needs and usage patterns, with subscription levels adjusted monthly to balance performance and cost.

All three dashboards share a common technical foundation, built using Python's scientific computing stack [42]. The visualization layer employs Panel for creating interactive web interfaces and HoloViews for generating dynamic plots [43]. These libraries were chosen for their ability to create responsive, interactive visualizations, while efficiently handling large datasets [30]. Bokeh serves as the underlying plotting library, enabling the creation of interactive plots that can be updated in real time as users adjust parameters [44]. The dashboards use a consistent set of UI components, such as sliders, dropdown menus, and buttons, to provide a familiar user experience across applications.

The Water Intake Locations Dashboard focuses on water intake locations, providing real-time predictions of ion concentrations based on user-specified parameters. Users can adjust parameters like EC, Sacramento X2, Water Year Type, and Month, with results displayed through interactive bar charts that compare different model predictions. By relying on pre-trained models, this dashboard delivers fast predictions without requiring the computational overhead of training models on-the-fly.

The Enhanced Interior Delta Dashboard extends our previous Interior Delta dashboard [25] by integrating Hutton et al.'s parametric regression equations alongside machine learning predictions [21]. This integration enables direct comparisons between four machine learning approaches—Regression Trees (RTs), Gradient Boosting (GB), Random Forest (RF), and Artificial Neural Networks (ANNs)—and the classical parametric method across three sub-regions. The dashboard also includes horizontal threshold lines for key ion constituents (chloride: 250 mg/L, sulfate: 250 mg/L, sodium: 60 mg/L, and total dissolved solids: 500 mg/L), allowing users to quickly identify when predicted concentrations exceed acceptable water quality standards [45].

The Model Interpretability Dashboard serves as an analytical tool to help users understand how different parameters influence ion predictions. Users can select any input parameter—such as EC, Sacramento X2, Water Year Type, or Month—as the x-axis to observe changes in ion concentrations across its full range, while keeping other variables constant. This functionality allows users to perform the following:

- Analyze the sensitivity of predictions to different input variables.
- Compare how different regions respond to parameter changes.
- Understand model behavior under various conditions.
- Assess model stability across different parameter ranges.
- Compare predictions from different modeling approaches (ML vs. classical methods).

The dashboard uses a cache-based system to optimize performance. For first-time parameter combinations, comprehensive calculations are performed, which may take a few seconds. However, these results are cached, enabling instant response times for subsequent queries with the same parameters. This caching mechanism is especially valuable for sensitivity analyses, where users often explore similar scenarios. The visualizations include interactive line plots for continuous variables and bar charts for categorical parameters, with options to display multiple regions and prediction methods simultaneously [43].

Each dashboard maintains its own logging system to track usage patterns and performance metrics, which informs future optimizations and resource allocation. Error handling and input validation are implemented to ensure robustness, and responsive design principles are applied to ensure usability across different devices and screen sizes.

2.4. *Addressing Model Limitations Through a Hybrid Approach*

While both machine learning and classical approaches have demonstrated strong predictive capabilities for ion constituents, they share two fundamental limitations. First, predictions are constrained to specific monitoring locations, lacking continuous spatial coverage across the Delta. Second, neither approach can independently evaluate potential future scenarios, limiting their utility for long-term planning and climate change impact assessment.

To overcome these limitations, we developed a hybrid water quality model that integrates the Delta Simulation Model 2 (DSM2) with our machine learning framework. DSM2 is a well-established hydrodynamic and water quality model used extensively for simulating water conditions in the Delta [45]. DSM2 can simulate EC across an extensive network of nodes throughout the Delta, providing spatial and temporal detail that can be leveraged to enhance prediction capabilities.

For this hybrid approach, we selected 162 points across three key sub-regions: 60 in the South Delta (green circles), 30 along the San Joaquin River corridor (red stars), and 72 in the Old–Middle River corridor (black squares), as illustrated in Figure 5. These points were carefully positioned to capture key water quality gradients and hydrodynamic features across the Delta system while maintaining computational efficiency. This strategic distribution ensures comprehensive coverage of critical areas and enables effective spatial interpolation between monitoring locations. The hybrid model operates in two stages:

- DSM2 simulation: DSM2 is used to generate EC values at all 162 points, ensuring extensive spatial and temporal coverage.
- ANN predictions: the simulated EC values are then used as inputs for our ANN model to predict ion concentrations at these points.

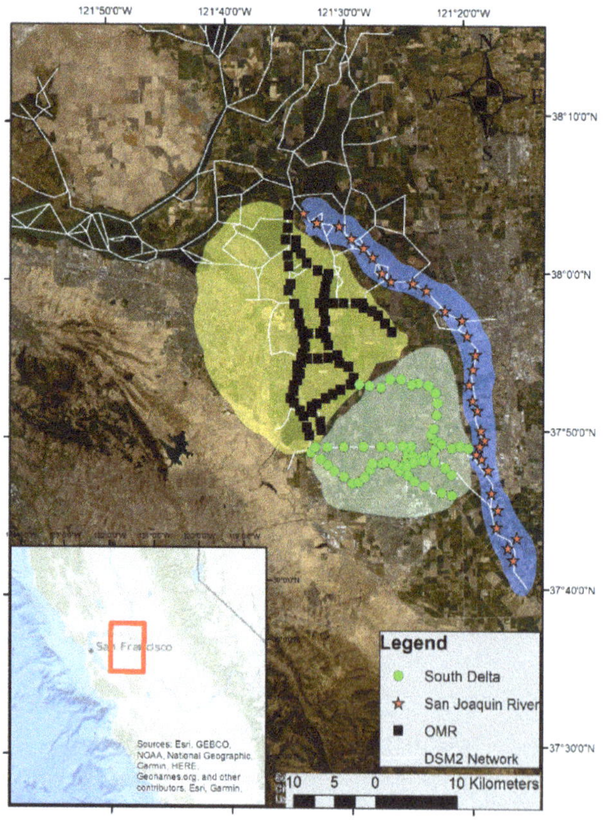

Figure 5. Distribution of 162 points across Sacramento–San Joaquin Delta for hybrid model implementation. Points are categorized by sub-region: South Delta (green circles), San Joaquin River corridor (red stars), and Old–Middle River corridor (black squares).

This hybrid model offers several advantages:

- Continuous spatial coverage: the use of DSM2 provides simulated EC values across much of the Delta's channel network, enabling spatial interpolation between the selected points to create a more continuous ion concentration map.
- Temporal flexibility: DSM2 can generate EC data for any desired time period, from historical conditions to future projections (for example in this project: 2000–2023), providing complete temporal coverage.
- Scenario analysis: the model's capability to modify DSM2 input parameters allows for evaluation of future scenarios, such as changes in climate, hydrology, or operational practices.
- Combining physics-based and data-driven insights: by integrating DSM2, a process-based model, with the ANN, a data-driven model, the hybrid system benefits from both detailed physical simulation and the flexibility of machine learning.

To make this hybrid model accessible, we developed a dedicated dashboard that visualizes spatial distributions of ion concentrations throughout the Delta. Users can select specific dates and ion constituents to view the spatial patterns on an interactive map. Due to the computational intensity of these calculations, particularly involving spatial interpolation, this dashboard is designed for local deployment, rather than cloud hosting. The full codebase is available through our GitHub repository, allowing users to conduct their analyses on local machines with their own computational resources.

This hybrid approach bridges the gap between localized predictions and broader spatial–temporal analysis, enabling scenario planning and improving water quality management throughout the Delta. The ability to produce continuous spatial distributions of ion concentrations represents a significant step forward in our capacity to understand and manage the complex dynamics of the Delta ecosystem.

3. Results

3.1. Model Performance at Water Intake Locations

The performance evaluation of our models began with identifying optimal hyperparameters for each constituent. Table 2 presents the optimal network architectures derived from our hyperparameter optimization process.

Table 2. Optimal neural network architectures for each constituent.

Constituent	Number of Layers	Neurons per Layer	Activation Function
Chloride	3	40	tanh
Bromide	4	40	relu
Sulfate	5	40	tanh

While all constituents achieved optimal performance with 40 neurons per layer, they required different network depths and activation functions. Chloride predictions were optimized with a relatively simple architecture (three layers), while sulfate required a deeper network (five layers), suggesting varying levels of complexity in the underlying relationships between input features and constituent concentrations.

Table 3 presents the detailed performance metrics for these optimized models on both the training and test datasets.

The models demonstrated strong performance across all constituents, with R^2 values above 0.96 for all test cases. The small differences between the training and test performance metrics ($\Delta R^2 \leq 0.007$) indicate good generalization capabilities. The minimal bias in predictions (all < 0.5%) further supports the reliability of these models for operational use. When evaluating the Mean Absolute Error (MAE), it is crucial to consider the distinct concentration ranges for each constituent at the water intake locations, as shown in Figure 2. Chloride concentrations typically range from 20 to 140 mg/L, while bromide varies between 0 and 0.5 mg/L, and sulfate ranges from 20 to 80 mg/L. Given these different scales, the MAE values (chloride: 2.672 mg/L, bromide: 0.011 mg/L, sulfate: 1.774 mg/L) should be evaluated relative to each constituent's typical concentration range, rather than compared directly across constituents. For example, the chloride model's MAE of 2.672 mg/L represents approximately 2.2% of its maximum observed concentration range, while the bromide model's MAE of 0.011 mg/L represents about 2.2% of its range, indicating comparable relative performance, despite very different absolute MAE values. To visualize the prediction accuracy across these different concentration ranges, Figure 6 presents scatter plots of the predicted versus observed values for all three constituents.

Table 3. Model performance metrics for water intake locations.

Constituent	Dataset	R^2	MAE (mg/L)	Percentage Bias (%)
Chloride	Training	0.988	2.017	−0.02%
	Test	0.981	2.672	−0.39%
Bromide	Training	0.975	0.01	−0.22%
	Test	0.971	0.011	−0.41%
Sulfate	Training	0.969	1.502	−0.24%
	Test	0.962	1.774	−0.01%

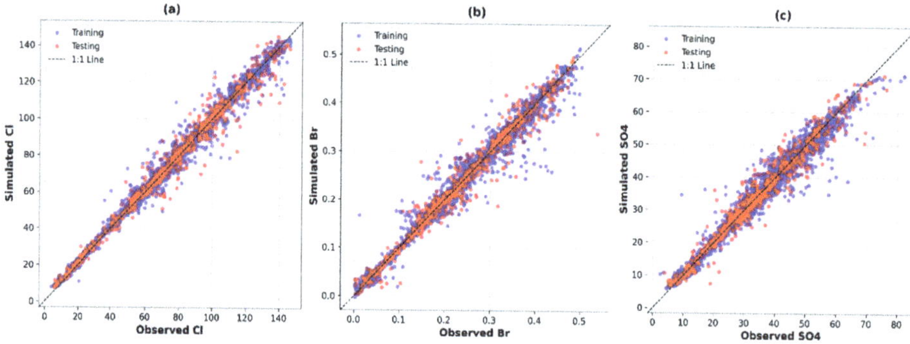

Figure 6. Scatter plots comparing predicted versus observed concentrations for (**a**) chloride [0–140 mg/L], (**b**) bromide [0–0.5 mg/L], and (**c**) sulfate [0–80 mg/L]. Blue points represent training data, red points represent test data, and dashed line represents perfect prediction (1:1 line).

The scatter plots reveal several important characteristics of the models' performance across the three constituents. For Chloride (Figure 6a), the model demonstrates excellent prediction accuracy across the full range of 0–146 mg/L, with particularly tight clustering around the 1:1 line at low-to-medium concentrations. While there is slightly increased scatter at higher concentrations (>100 mg/L), the model maintains overall accuracy. The nearly identical patterns between the training and test points indicate robust generalization capabilities.

Bromide predictions (Figure 6b) show notably high precision in the critical low-concentration range (0–0.2 mg/L) and maintain consistent performance across the entire range of 0–0.5 mg/L. Although some scattered points appear in the mid-range concentrations (0.2–0.3 mg/L), the model effectively maintains the overall trend and captures the relationship well, even in the upper concentration range.

The sulfate model (Figure 6c) exhibits strong performance across the observed range of 0–85 mg/L, with similar prediction patterns between the training and test datasets. Despite slightly increased scatter in the middle range (30–50 mg/L), the model maintains accuracy at both low and high concentration extremes.

Several common features are evident across all three constituents. The even distribution of points around the 1:1 line confirms the low bias shown in Table 3, while the consistent performance between the training and test datasets validates the models' generalization capabilities. Notably, there are no systematic deviations or patterns that would suggest model limitations, and all models demonstrate robust performance across the full range of environmentally relevant concentrations. These visualizations complement the numerical metrics in Table 3 by demonstrating that the models' high R^2 values and low MAEs are achieved consistently across all concentration ranges, not just in specific regions of the prediction space.

To evaluate the effectiveness of our location-specific approach, we conducted a fair comparison using test dataset performance against both the Interior Delta ML model and classical methods. While all three approaches achieved relatively high R^2 values (water intake location test: 0.962–0.981, Interior Delta test: 0.96–0.97, classical all data: 0.52–0.92), the MAE metric is a more discriminating metric for comparing model performance, particularly given its direct relevance to operational decision-making. Note that for classical methods, results are reported for the entire dataset, as this approach does not employ train–test splitting. Figure 7 presents a heatmap visualization of the percentage improvements in MAE achieved by our water intake locations model's test performance compared to the Interior Delta test performance and classical method results, ensuring a fair comparison basis across methods.

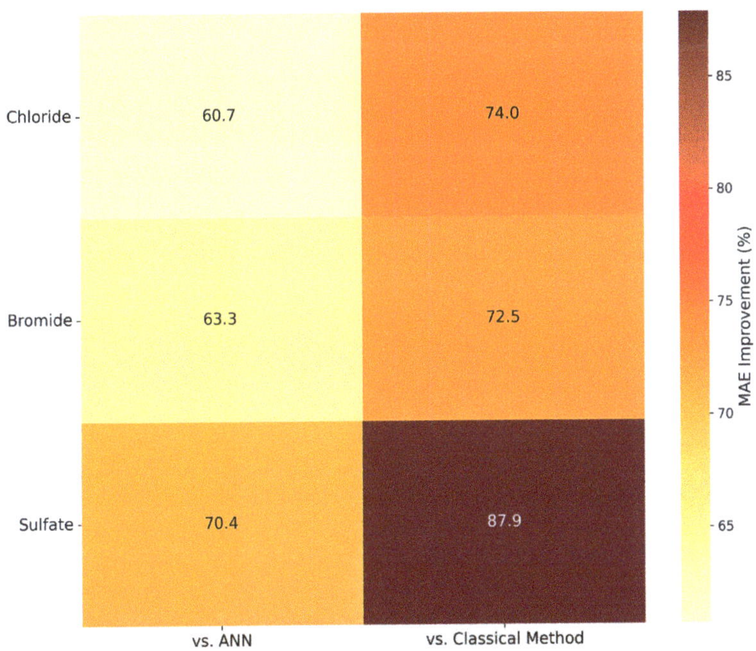

Figure 7. Heatmap showing percentage improvement in MAE achieved by water intake locations model compared to Interior Delta ML model and classical methods.

These results demonstrate that our location-specific optimization approach successfully captures the unique characteristics of water intake locations, leading to more precise predictions across all constituents. The magnitude of these improvements has significant implications for operational decision-making and water quality management at intake facilities.

3.2. Cross-Validation Results and Model Stability

The five-fold cross-validation analysis demonstrated strong and consistent performance across all folds, supporting the robustness of our models. Table 4 presents the comprehensive cross-validation results, showing the mean and standard deviation of performance metrics for each constituent.

Table 4. Cross-validation results showing mean ± standard deviation across 5 folds.

Constituent	R^2	MAE (mg/L)	PBIAS (%)
Chloride	0.983 ± 0.006	2.345 ± 0.301	−0.083 ± 0.316
Bromide	0.972 ± 0.004	0.011 ± 0.001	−0.324 ± 0.350
Sulfate	0.962 ± 0.001	1.671 ± 0.089	−0.585 ± 0.483

The cross-validation results reveal several important characteristics of model performance across the different constituents. For chloride predictions, the model achieved the highest R^2 values (0.983 ± 0.006) among all constituents, with the MAE ranging from 2.078 to 2.857 mg/L across folds. The model demonstrated consistent performance, with relatively small variations between folds, and PBIAS values close to zero (−0.083 ± 0.316%) indicate minimal systematic bias in predictions.

Bromide predictions maintained very strong performance, with an R^2 of 0.972 ± 0.004, and exhibited remarkably consistent MAE (0.011 ± 0.001 mg/L). While showing slightly higher relative bias (−0.324 ± 0.350%) compared to chloride, the bromide model maintained robust predictive capabilities across all folds. Sulfate predictions demonstrated strong and stable performance, with an R^2 of 0.962 ± 0.001; notably, they had the most consistent R^2 values, with the lowest standard deviation (±0.001). The MAE values showed good stability at 1.671 ± 0.089 mg/L, though the model exhibited slightly higher negative bias (−0.585 ± 0.483%) compared to other constituents.

The remarkably low standard deviations across all metrics, particularly for R^2 values ($\leq$0.006), demonstrate the models' robust generalization capabilities and stability across different data partitions. This consistent performance across folds suggests that the models successfully captured the underlying relationships between input features and ion concentrations, while avoiding overfitting to specific data subsets.

3.3. Dashboard Applications and Enhancements

3.3.1. Water Intake Locations Dashboard

The Water Intake Locations Dashboard employs ANN models to simulate ion constituents (chloride, bromide, sulfate) at three critical water intake locations. The interface features four predictors: EC (0–800 μS/cm), Sacramento X2 (40–90 km), Water Year Type, and Month. Due to the limited data availability period (2019–2023), the models for Harvey O. Banks (HRO) and the Tracy Pumping Plant (TRP) currently cover only three Water Year Types (D, W, and C), while the O'Neill Forebay (ONG), with its longer record (2012–2023), includes BN as well. The Above-Normal (AN) Water Year Type is excluded for all locations; users are advised to use Wet (W) as an approximation for AN conditions. The accuracy of predictions for HRO and TRP is expected to improve as future data collection encompasses more hydrological conditions.

Figure 8 demonstrates the dashboard's interface design and functionality using a specific example scenario where we simulate chloride concentrations under the following conditions: EC: 600 µS/cm, Sacramento X2: 80 km, Month: June, Water Year Type (WYT): Critical, ion constituent: chloride. In this particular example, when simulating chloride concentrations under the specified conditions, we observe distinct spatial variations: the O'Neill Forebay (ONG) and Harvey O. Banks (HRO) show similar predicted concentrations at around 90 mg/L, while the Tracy Pumping Plant (TRP) exhibits a lower predicted concentration of approximately 83 mg/L. These spatial differences in chloride concentrations under identical input conditions demonstrate how the complex hydrodynamics of the Delta system influence water quality patterns at different intake locations. Users can explore other scenarios by adjusting any of the input parameters, allowing them to investigate how ion concentrations vary under different operational and environmental conditions.

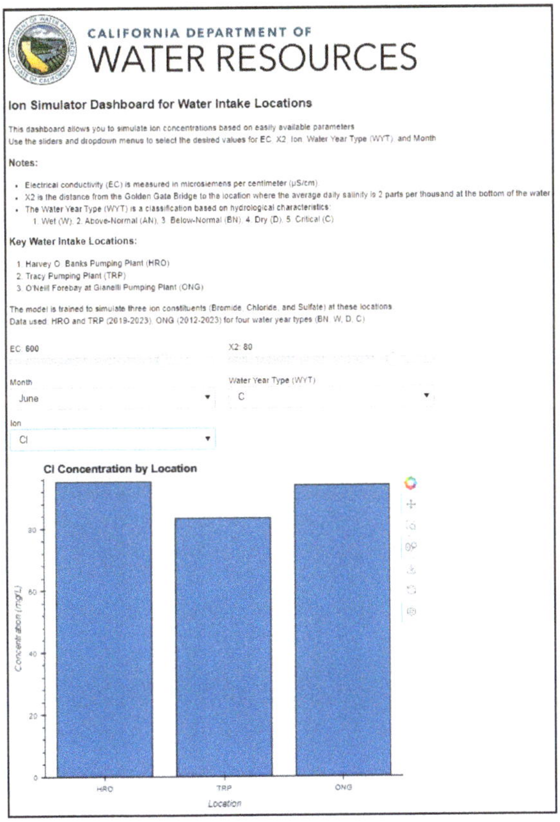

Figure 8. A screenshot of the Intake Locations Dashboard, showing the interface layout and chloride concentration predictions at water intake locations.

The dashboard's real-time visualization capabilities enable operators to compare constituent levels across locations simultaneously, evaluate the impacts of changing operational conditions, identify potential water quality concerns, and support rapid operational decision-making through scenario testing. The interface provides essential contextual information about measurement units, parameter definitions, and data coverage periods, ensuring proper interpretation of results. The visualization includes interactive tools for zooming, panning, and data export, represented by the icons on the right side of the plot.

3.3.2. Enhanced Interior Delta Dashboard

Building upon our previous work [25], the dashboard developed in that study has been significantly enhanced to provide comprehensive visualization and comparison capabilities for ion constituent predictions across the Delta. Key enhancements include the following:

Classical Method Integration: The most significant enhancement is the integration of the classical parametric regression method (PR), developed by [21], alongside machine learning predictions (RT, GB, RF, and ANN). This integration offers several benefits:

- Direct visual comparison between classical and ML approaches;
- Simplified access to ion constituent predictions without requiring manual coefficient calculations;
- Enhanced transparency in model comparison for stakeholders;
- Validation of ML predictions against established methodologies.

This unified platform allows water managers to make informed decisions by comparing multiple prediction methods simultaneously, while maintaining the user-friendly interface that characterized our original dashboard. The addition of the PR method particularly addresses stakeholder needs for comparative analysis between traditional and modern prediction approaches.

Spatial Comparison Capability: The dashboard now enables simultaneous visualization of predictions across all three sub-regions (OMR, SJR corridor, and South Delta), revealing important spatial patterns in ion constituent distributions. For instance, under identical predictor conditions (EC, X2, WYT, and Month), the ANN model demonstrates consistently higher chloride concentrations in the OMR region compared to other areas, reflecting its proximity to ocean influences—a primary source of chloride in the system.

Water Quality Standards Integration: A crucial addition is the incorporation of regulatory threshold indicators (shown as red dashed lines) for key constituents, based on guidelines from the United States Environmental Protection Agency (EPA) and the World Health Organization (WHO): chloride: 250 mg/L, sulfate: 250 mg/L, sodium: 60 mg/L, and total dissolved solids: 500 mg/L. These visual threshold indicators enable immediate assessment of water quality compliance across different regions and conditions.

Figure 9 demonstrates the dashboard's key capabilities through a typical usage case. In this example, chloride predictions are shown for specified conditions (EC: 1600 μS/cm, Sacramento X2: 75 km, Water Year Type: Above-Normal, Month: September) across all three sub-regions. The visualization clearly shows the spatial variation in chloride concentrations, with consistently higher levels in the OMR region across all models. This pattern aligns with the physical geography of the Delta, as the OMR region is situated closer to the San Francisco Bay and the Pacific Ocean, which serve as the primary sources of chloride through seawater intrusion. The other regions, being further inland, typically experience lower chloride concentrations, due to their greater distance from marine influences. The integrated EPA/WHO threshold line (250 mg/L) provides immediate context for water quality compliance assessment, while the side-by-side comparison of different models (RT, GB, RF, ANN, and PR) enables stakeholders to evaluate prediction consistency and model agreement.

3.3.3. Model Interpretability Dashboard

The Model Interpretability Dashboard was developed to provide deeper insights into model behavior and sensitivity for our Interior Delta predictions, focusing on comparing the machine learning models and the classical method presented in our previous study [21]. While our ion constituent simulator dashboard focuses on operational predictions across all locations, this interpretability dashboard specifically examines how different modeling approaches for the Interior Delta respond to variations in input parameters. The dashboard enables analysis of all nine ion constituents against any selected predictor variable, while holding other conditions constant, making it a powerful tool for understanding model dynamics and validating their physical consistency.

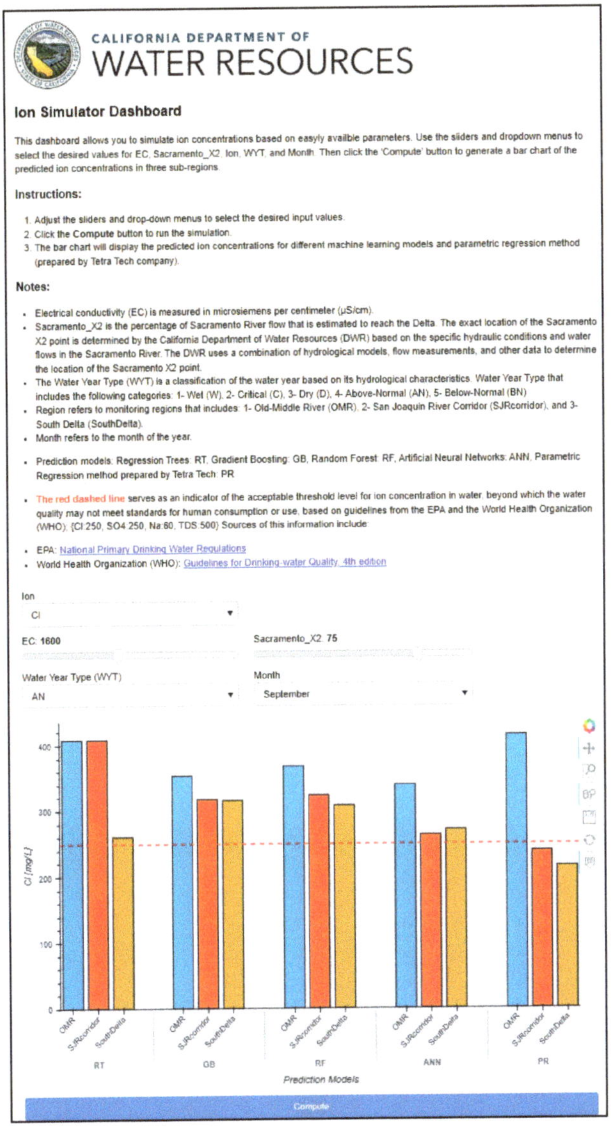

Figure 9. Screenshot of Enhanced Interior Delta Ion Constituent Prediction Dashboard.

Figure 10 demonstrates the dashboard's capability to analyze model behavior across the EC range for bromide predictions under specific conditions (Sacramento X2 = 80 km,

Water Year Type = Above-Normal, Month = September, South Delta region). The visualization reveals distinct behavioral patterns among different models, with the ANN model exhibiting notably smoother response curves compared to other ML approaches. A critical transition point occurs at EC ≈ 1200 μS/cm, where the relative predictions between ANN and the classical method (PR) reverse their relationship. Below this threshold, ANN predicts higher bromide concentrations, while above it, the classical method shows higher predictions. This smooth transition in the ANN model, compared to the more erratic behavior of other ML models, represents an additional advantage beyond its superior statistical performance metrics.

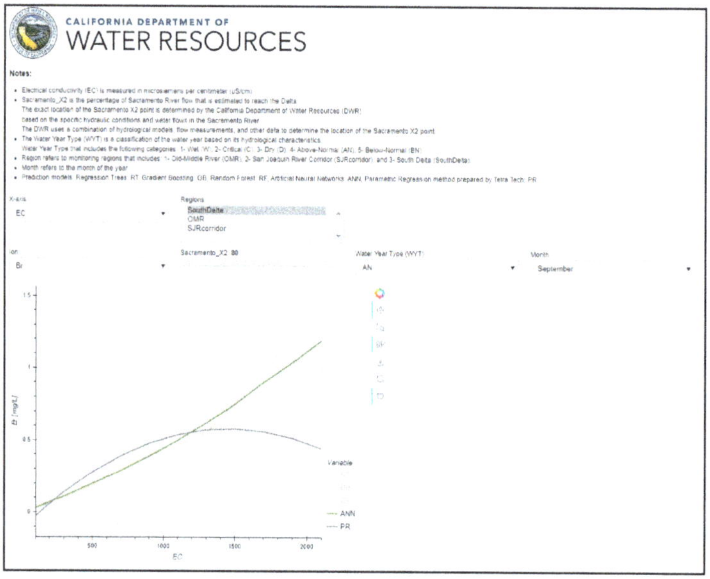

Figure 10. Model response comparison for bromide predictions across EC range in Model Interpretability Dashboard. Plot shows different model behaviors under specified conditions (Sacramento X2: 80 km, WYT: Above-Normal, Month: September) for South Delta region.

The dashboard's ability to analyze sensitivity to different predictors is further illustrated in Figure 11, which shows model responses to Sacramento X2 variations for chloride predictions (EC = 700 μS/cm, Month = September, Water Year Type = Dry, OMR region). This comparison reveals significant differences in how models incorporate the X2 influence. The classical method shows no sensitivity to X2 changes under these conditions, producing a flat response that suggests a potential limitation in capturing this important physical relationship. In contrast, the ANN model demonstrates smooth, physically plausible responses to X2 variations. Other ML models (RT, GB, and RF) exhibit high sensitivity to X2 changes with discontinuous jumps in predictions, suggesting potential overfitting to training data points.

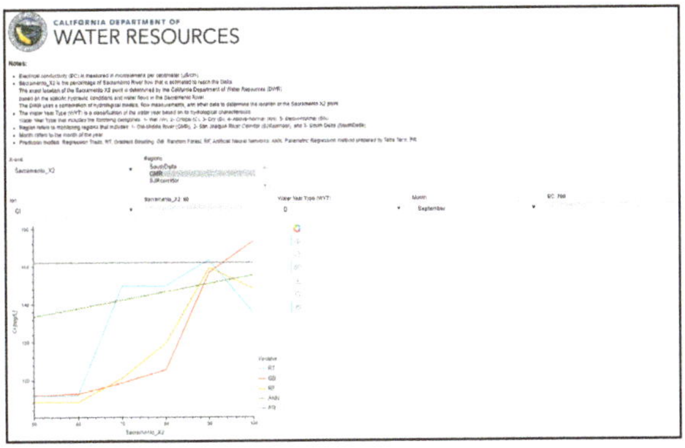

Figure 11. Analysis of model sensitivity to Sacramento X2 variations for chloride predictions using the Model Interpretability Dashboard. The comparison illustrates different model responses under fixed conditions (EC: 700 μS/cm, WYT: Dry, Month: September) for the OMR region.

The dashboard's capability to analyze categorical predictors is demonstrated in Figure 12, which examines model behavior across different Water Year Types for chloride predictions under consistent conditions (EC = 700 μS/cm, Sacramento X2 = 80 km, Month = September, OMR region). The classical method (PR) shows limited sensitivity to water year classification, consistently predicting the highest chloride concentrations (approximately 150 mg/L) across all WYTs. The ANN model demonstrates more nuanced and physically plausible behavior, showing a gradual decrease in chloride concentrations from Critical to Wet conditions, while maintaining predictions in the 120–140 mg/L range. Other ML models (RT, GB, and RF) show similar overall patterns, but with more variable responses and generally lower chloride concentration predictions compared to PR and ANN.

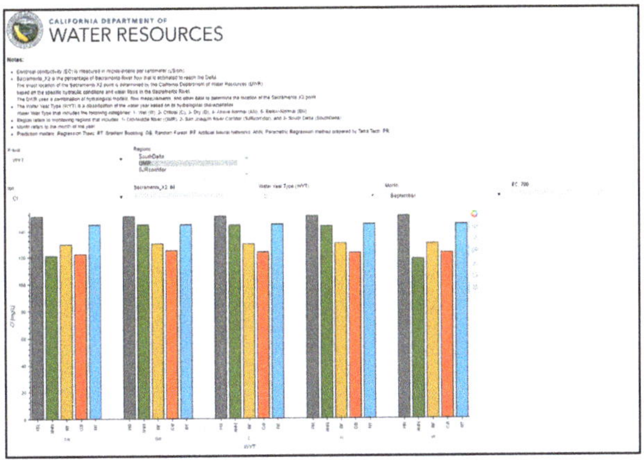

Figure 12. A model behavior comparison across Water Year Types (WYTs) for chloride predictions on the Model Interpretability Dashboard. The bar chart displays model predictions under consistent conditions (EC: 700 μS/cm, Sacramento X2: 80 km, Month: September) for the OMR region.

These visualizations collectively highlight the ANN model's superior characteristics beyond traditional performance metrics. Its smoother response surfaces across parameter ranges, more physically consistent behavior, balanced sensitivity to input parameters, and robust predictions without artificial discontinuities make it particularly reliable for operational applications. The Model Interpretability Dashboard proves especially valuable for stakeholders seeking to understand model reliability under different conditions, and for researchers validating model behavior against known physical relationships in the Delta system. This enhanced transparency in model behavior helps to build confidence in the predictions and provides crucial insights for water quality management decisions.

3.4. Hybrid Water Quality Dashboard

The Hybrid Water Quality Dashboard represents a novel integration of DSM2 hydrodynamic modeling with our ANN framework to provide comprehensive spatial coverage of ion constituent concentrations across the Interior Delta. This integration addresses a key limitation of standalone ANN models by enabling predictions at locations without monitoring stations. The dashboard's architecture combines DSM2-simulated EC values at 162 strategic points (60 in the South Delta, 30 along the San Joaquin River corridor, and 72 in the Old–Middle River corridor) with ANN predictions to generate spatially continuous constituent distributions.

Figure 13 demonstrates the dashboard's visualization capabilities through a bromide concentration map for 1 July 2015. The interface features a date selector and ion constituent dropdown menu, allowing users to explore temporal and chemical variations across the Delta. The example reveals distinct spatial patterns in bromide concentrations, with values ranging from 0.17 to 0.51 mg/L. A clear gradient is visible from the northern Delta (higher concentrations shown in purple/blue) to the southern regions (lower concentrations in green/yellow), reflecting the complex interplay of tidal influences and river inflows.

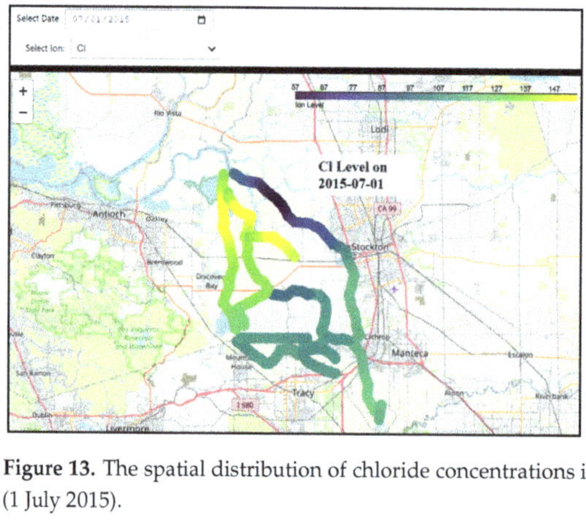

Figure 13. The spatial distribution of chloride concentrations in the Interior Delta during a dry year (1 July 2015).

Figures 13 and 14 demonstrate temporal variations in chloride concentrations by comparing Critical (2015) versus Wet (2017) water year conditions for July 1st. In the critical year (Figure 13), higher chloride concentrations are evident throughout the Delta, particularly in the OMR corridor, reflecting increased seawater intrusion due to reduced Delta outflow. In contrast, the Wet year (2017, Figure 14) shows notably lower chloride concentrations across all regions, as higher precipitation and river inflows create stronger seaward gradients that limit ocean-derived chloride intrusion.

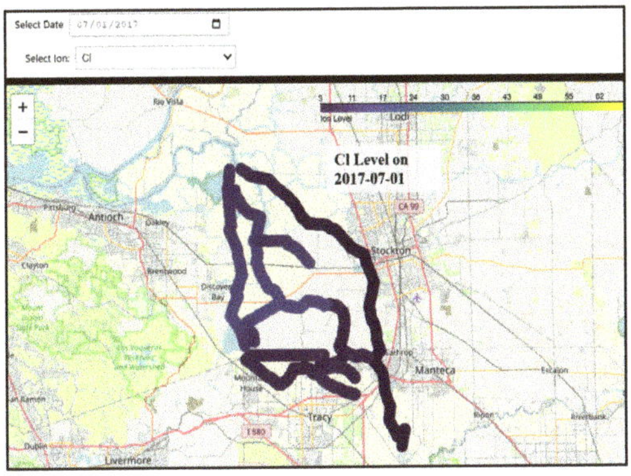

Figure 14. The spatial distribution of chloride concentrations during a wet year (1 July 2017).

Figure 15 illustrates spatial patterns of sulfate concentrations during the same period as Figure 13 (1 July 2015). The distribution reveals the highest sulfate levels along the San Joaquin River corridor and South Delta region, with concentrations decreasing notably in the OMR area. This pattern reflects the primary source of sulfate in the Delta system—agricultural drainage inputs predominant in the San Joaquin River and South Delta regions. The distinct spatial patterns between chloride (Figure 13) and sulfate (Figure 15) demonstrate how different sources—seawater intrusion versus agricultural drainage—create unique constituent distributions across the Delta.

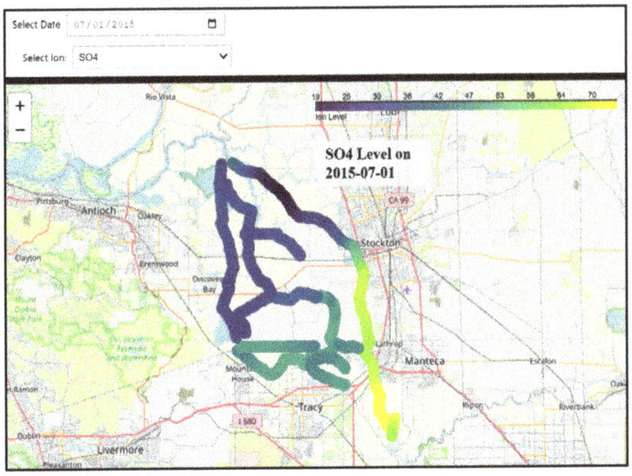

Figure 15. The spatial distribution of sulfate concentrations (1 July 2015).

The dashboard employs an intuitive color-coding scheme for concentration visualization, with a continuous scale that enables easy identification of spatial gradients and potential water quality concerns. Users can zoom and pan across the Delta region to examine specific areas of interest, while the base map provides geographical context through landmarks and waterway configurations. Unlike our Azure-hosted dashboards, this tool is available through GitHub for local deployment, allowing users to conduct intensive spatial analyses with their own computational resources.

This hybrid approach extends the temporal coverage from 2000 to 2023, providing historical context for water quality patterns. The combination of DSM2's hydrodynamic simulation capabilities with ANN's constituent prediction accuracy creates a powerful tool for understanding large-scale water quality dynamics in the Delta system. The spatial interpolation between the 162 monitoring points enables visualization of continuous concentration gradients, offering insights into water quality transitions that would be difficult to discern from point measurements alone.

For enhanced accessibility, we developed three Azure-hosted dashboards and one GitHub-hosted solution, as summarized in Table 5. The Azure-hosted dashboards (Interior Delta, Interpretability, and Water Intake Locations) are optimized for both desktop and mobile access, with QR codes provided for convenient mobile deployment. These QR codes enable stakeholders to quickly access the dashboards from their smartphones or tablets, facilitating real-time decision-making in the field. The mobile interface maintains all the functionality of the desktop version, while adapting to smaller screens.

Table 5. Summary of developed dashboards with web links.

Dashboards	Dashboard ID	Web Links
Interior Delta Dashboard	DD1	https://dwrdashion.azurewebsites.net/Dashboard (accessed on 1 March 2025)
Interpretability Dashboard	DD2	https://dwrdashionsensitivity.azurewebsites.net/Sensitivity_IonStudy (accessed on 1 March 2025)
Water Intake Locations Dashboard	DD3	https://dwrdashionintake.azurewebsites.net/Intake_dashboard (accessed on 1 March 2025)
Hybrid Water Quality Dashboard	DD4	https://github.com/PeymanHNamadi/Hybrid_WaterQualityDashboard (accessed on 1 March 2025)

4. Discussion

The development and implementation of machine learning approaches for ion constituent prediction in the Sacramento–San Joaquin Delta represents a significant advancement in water quality monitoring and management. Our results demonstrate both the technical advantages of these new methods and their practical utility for operational decision-making. This discussion examines the implications of our findings across several key dimensions: model performance and reliability, operational utility, and broader implications for water quality management.

4.1. Model Performance and Reliability

The superior performance of our location-specific ANN models compared to both the Interior Delta ML model and classical methods highlights several important considerations in water quality modeling. The substantial improvements in Mean Absolute Error (MAE)—ranging from 24% for TDS to 59% for sulfate—demonstrate that tailoring models to specific monitoring locations can significantly enhance prediction accuracy. This improvement likely stems from the models' ability to capture location-specific relationships between Electrical Conductivity and ion constituents, which vary considerably across the Delta system.

The cross-validation results, showing remarkably low standard deviations across all metrics (e.g., R^2 variations ≤ 0.006), confirm the robust generalization capabilities of our models. This stability is particularly noteworthy given the complex and dynamic nature of the Delta system, where multiple factors influence water quality simultaneously. The consistent performance across different data partitions suggests that the models have successfully captured underlying physical relationships, rather than simply fitting to training data patterns.

However, it is important to acknowledge certain limitations in our approach. The relatively short data record for some locations (2019–2023 for HRO and TRP) means that the models have not been exposed to the full range of possible hydrological conditions. While performance metrics remain strong, continued validation against new data will be crucial for ensuring long-term reliability. Additionally, the current exclusion of Above-Normal Water Year Types from some location-specific models represents a gap that should be addressed as more data become available.

4.2. Operational Utility and Dashboard Implementation

The Water Intake Locations Dashboard's (DD3) focus on critical infrastructure points demonstrates how location-specific optimization can enhance operational relevance. The real-time visualization capabilities and mobile accessibility through QR codes reflect a practical understanding of how these tools will be used in the field. However, the dashboard's current limitations regarding Water Year Types highlight the ongoing need for data collection and model refinement.

The development of three complementary dashboards, each serving different user needs, represents a significant step forward in making complex modeling tools accessible to water managers. The Enhanced Interior Delta Dashboard's (DD1) integration of classical methods alongside machine learning predictions provides a valuable bridge between traditional and modern approaches, helping to build user confidence through direct comparison. The addition of regulatory threshold indicators directly addresses operational needs by providing immediate visual feedback on water quality compliance.

The Model Interpretability Dashboard addresses (DD2) a crucial challenge in the adoption of machine learning methods: the perceived "black box" nature of these models. By enabling users to visualize how models respond to different predictors and operating conditions, this tool helps to build trust in ML predictions. The smooth response curves demonstrated by the ANN model, particularly in comparison to other ML approaches, suggest that it has captured physically meaningful relationships, rather than arbitrary patterns in the training data.

Our hybrid approach, combining DSM2 hydrodynamic modeling with ANN predictions, represents a significant advancement in addressing spatial coverage limitations inherent in point-based monitoring systems. The ability to generate continuous spatial distributions of ion concentrations provides water managers with unprecedented insight into water quality dynamics across the Delta. This capability is particularly valuable for understanding how different operational decisions might affect water quality throughout the system, not just at monitoring locations.

The dashboard implementation strategy, using both cloud-hosted and locally deployed solutions, balances accessibility with computational requirements. The Azure-hosted dashboards provide widespread access to essential prediction tools, while the GitHub-hosted hybrid model enables more intensive analyses when needed. This dual approach ensures that different stakeholder needs can be met effectively, while managing computational resources efficiently.

5. Conclusions

This research introduces new machine learning tools for simulating ion constituents in California's Sacramento–San Joaquin Delta. Our approach advances water quality monitoring through improved prediction accuracy and accessibility. Our work makes three significant contributions:

First, we demonstrated that location-specific ANN models substantially improve prediction accuracy for ion constituents at critical water intake locations (cross-validated $R^2 > 0.96$, $\sigma \leq 0.006$).

Second, we developed four complementary web-based tools that make these modeling capabilities accessible to water managers: (1) an Enhanced Interior Delta Dashboard integrating classical and machine learning approaches; (2) a Model Interpretability Dashboard visualizing model behavior; (3) a Water Intake Locations Dashboard providing targeted predictions; and (4) a Hybrid Water Quality Dashboard enabling spatial coverage.

Third, our hybrid modeling approach combines DSM2 hydrodynamic simulations with ANN predictions, enabling continuous spatial prediction of ion constituents across the Delta and addressing the fundamental limitations of point-based monitoring.

Future Research

Our current models effectively simulate present conditions (nowcasting), but future work will focus on forecasting capabilities. We plan to extend these models to predict ion constituent levels 1–7 days in advance, developing an early warning system for water managers. Additionally, we aim to incorporate climate change scenarios into our hybrid model to assess long-term impacts on Delta water quality. As computational capabilities and data collection improve, we envision expanding this approach to model additional water quality parameters and implementing real-time monitoring integration for more responsive management systems.

Author Contributions: Conceptualization, P.N., M.H. and P.S.; methodology, P.N., M.H. and P.S.; software, P.N.; validation, P.N., M.H. and P.S.; formal analysis, P.N.; investigation, P.N.; writing—original draft preparation, P.N.; writing—review and editing, P.N., M.H. and P.S.; visualization, P.N. All authors have read and agreed to the published version of the manuscript.

Funding: This work was supported by the Municipal Water Quality Investigations Program.

Data Availability Statement: Data used in this study are available at https://github.com/CADWRDeltaModeling/Ion_Study_Dashboard/tree/main (accessed on 1 March 2025) and https://github.com/PeymanHNamadi/Hybrid_WaterQualityDashboard (accessed on 1 March 2025).

Acknowledgments: The authors thank their colleague Brad Tom for generating the model simulations used in the hybrid modeling portion of the study. The views expressed in this study are those of the authors, and do not necessarily reflect those of their employer.

Conflicts of Interest: The authors declare no conflicts of interest.

References

1. California Legislative Analyst's Office. Achieving State Goals for the Sacramento–San Joaquin Delta. 2015. Available online: https://lao.ca.gov/reports/2015/res/Delta/sac-sj-delta-011515.aspx (accessed on 11 March 2025).
2. Hartman, R.; Knowles, N.; Fencl, A.; Ekstrom, J. Drought in the Delta: Socio Ecological Impacts, Responses, and Tools. *San Fr. Estuary Watershed Sci.* **2025**, *23*, Art. 3. [CrossRef]
3. Grossman, G. Predation on fishes in the Sacramento–San Joaquin Delta: Current knowledge and future directions. *San. Fr. Estuary Watershed Sci.* **2016**, *14*, 1–29. [CrossRef]
4. Madani, K.; Lund, J.R. California's Sacramento–San Joaquin Delta Conflict: From Cooperation to Chicken. *J. Water Resour. Plan. Manag.* **2012**, *138*, 90–99. [CrossRef]
5. Denton, R. *Delta Salinity Constituent Analysis*; Richard Denton & Associates: Concord, CA, USA, 2015.
6. Chow, A.T.; Dahlgren, R.A.; Harrison, J.A. Watershed sources of disinfection by-product precursors in the Sacramento and San Joaquin Rivers, California. *Environ. Sci. Technol.* **2007**, *41*, 7645–7652. [CrossRef]
7. Hutton, P.H.; Roy, S.B.; Krasner, S.W.; Palencia, L. The municipal water quality investigations program: A retrospective overview of the program's first three decades. *Water* **2022**, *14*, 3426. [CrossRef]
8. Richardson, S.D.; Thruston, A.D.; Rav-Acha, C.; Groisman, L.; Popilevsky, I.; Juraev, O.; Glezer, V.; McKague, A.B.; Plewa, M.J.; Wagner, E.D. Tribromopyrrole, brominated acids, and other disinfection by-products produced by disinfection of drinking water rich in bromide. *Environ. Sci. Technol.* **2003**, *37*, 3782–3793. [CrossRef]

9. Weston, D.P.; Lydy, M.J. Urban and agricultural sources of pyrethroid insecticides to the Sacramento–San Joaquin Delta of California. *Environ. Sci. Technol.* **2010**, *44*, 1833–1840. [CrossRef]
10. Lund, J.R. California's Agricultural and Urban Water Supply Reliability and the Sacramento–San Joaquin Delta. *San Fr. Estuary Watershed Sci.* **2016**, *14*, 6. [CrossRef]
11. Cloern, J.E.; Jassby, A.D. Drivers of Change in Estuarine–Coastal Ecosystems: Discoveries from Four Decades of Study in San Francisco Bay. *Rev. Geophys.* **2012**, *50*, RG4001. [CrossRef]
12. Fulton, S.G.; Stegen, J.C.; Kaufman, M.H.; Dowd, J.; Thompson, A. Laboratory Evaluation of Open-Source and Commercial Electrical Conductivity Sensor Precision and Accuracy: How Do They Compare? *PLoS ONE* **2023**, *18*, e0285092. [CrossRef]
13. Forhad, H.M.; Uddin, M.R.; Chakrovorty, R.S.; Ruhul, A.M.; Faruk, H.M.; Kamruzzaman, S.; Sharmin, N.; Jamal, A.S.I.M.; Haque, M.M.U.; Morshed, A.M. IoT-Based Real-Time Water Quality Monitoring System in Water Treatment Plants (WTPs). *Heliyon* **2024**, *10*, e40746. [CrossRef]
14. Taylor, M.; Elliott, H.A.; Navitsky, L.O. Relationship between Total Dissolved Solids and Electrical Conductivity in Marcellus Hydraulic Fracturing Fluids. *Water Sci. Technol.* **2018**, *77*, 1998–2004. [CrossRef] [PubMed]
15. Cox, R.A.; Culkin, F.; Riley, J.P. The electrical conductivity/chlorinity relationship in natural sea water. *Deep Sea Res. Oceanogr. Abstr.* **1967**, *14*, 203–220. [CrossRef]
16. Marion, G.M.; Babcock, K.L. Predicting specific conductance and salt concentration in dilute aqueous solutions. *Soil Sci.* **1976**, *122*, 181–187. [CrossRef]
17. Thirumalini, S.; Joseph, K. Correlation between electrical conductivity and total dissolved solids in natural waters. *Malays. J. Sci.* **2009**, *28*, 55–61. [CrossRef]
18. Guivetchi, K. Salinity Unit Conversion Equations. California Department of Water Resources Interoffice Memorandum. 1986. Available online: https://www.waterboards.ca.gov/waterrights/water_issues/programs/bay_delta/california_waterfix/exhibits/docs/petitioners_exhibit/dwr/dwr_316.pdf (accessed on 11 March 2025).
19. Jung, M. Revision of Representative Delta Island Return Flow Quality for DSM2 and DICU Model Runs, prepared for the CALFED Ad-Hoc Workgroup to Simulate Historical Water Quality Conditions in the Delta, Consultant's Report to the CDWR MWQI Program. December 2000. Available online: www.rtdf.info (accessed on 11 March 2025).
20. Suits, B. Relationships between Delta water quality constituents as derived from grab samples. In *Methodology for Flow and Salinity Estimates in the Sacramento–San Joaquin Delta and Suisun Marsh: 23rd Annual Progress Report*; California Department of Water Resources: Sacramento, CA, USA, 2002.
21. Hutton, P.H.; Sinha, A.; Roy, S.B.; Denton, R.A. A simplified approach for estimating ionic concentrations from specific conductance data in the San Francisco Estuary. *San Fr. Estuary Watershed Sci.* **2023**, *21*, art6. [CrossRef]
22. Monismith, S.G.; Kimmerer, W.; Burau, J.R.; Stacey, M.T. Structure and flow-induced variability of the subtidal salinity field in northern San Francisco Bay. *J. Phys. Oceanogr.* **2002**, *32*, 3003–3019. [CrossRef]
23. He, M.; Sandhu, P.; Namadi, P.; Reyes, E.; Guivetchi, K.; Chung, F. Protocols for water and environmental modeling using machine learning in California. *Hydrology* **2025**, *12*, 59. [CrossRef]
24. Namadi, P.; He, M.; Sandhu, P. Salinity-constituent conversion in South Sacramento–San Joaquin Delta of California via machine learning. *Earth Sci. Inform.* **2022**, *15*, 1749–1764. [CrossRef]
25. Namadi, P.; He, M.; Sandhu, P. Modeling ion constituents in the Sacramento–San Joaquin Delta using multiple machine learning approaches. *J. Hydroinform.* **2023**, *25*, 2541–2560. [CrossRef]
26. California Department of Water Resources. *Emulation of DWRDSM Using Artificial Neural Networks and Estimation of Sacramento River Flow: 12th Annual Progress Report*; California Department of Water Resources: Sacramento, CA, USA, 1996.
27. Rusydi, A.F. Correlation between conductivity and total dissolved solid in various types of water: A review. *IOP Conf. Ser. Earth Environ. Sci.* **2018**, *118*, 012019. [CrossRef]
28. Jassby, A.D.; Kimmerer, W.J.; Monismith, S.G.; Armor, C.; Cloern, J.E.; Powell, T.M.; Schubel, J.R.; Vendlinski, T.J. Isohaline position as a habitat indicator for estuarine populations. *Ecol. Appl.* **1995**, *5*, 272–289. [CrossRef]
29. Kimmerer, W.J. Effects of freshwater flow on abundance of estuarine organisms: Physical effects or trophic linkages? *Mar. Ecol. Prog. Ser.* **2002**, *243*, 39–55. [CrossRef]
30. Dettinger, M.D.; Cayan, D.R. Large-scale atmospheric forcing of recent trends toward early snowmelt runoff in California. *J. Clim.* **1995**, *8*, 606–623. [CrossRef]
31. California Department of Water Resources. Water Year Classification Indices. 2020. Available online: https://cdec.water.ca.gov/reportapp/javareports?name=WSIHIST (accessed on 11 March 2025).
32. Goodfellow, I.; Bengio, Y.; Courville, A. *Deep Learning*; MIT Press: Cambridge, MA, USA, 2016.
33. Mosavi, A.; Ozturk, P.; Chau, K.W. Flood prediction using machine learning models: Literature review. *Water* **2018**, *10*, 1536. [CrossRef]

34. Kim, H.S.; He, M.; Sandhu, P. Suspended sediment concentration estimation in the Sacramento–San Joaquin Delta of California using long short-term memory networks. *Hydrol. Process.* **2022**, *36*, e14694. [CrossRef]
35. He, M.; Zhong, L.; Sandhu, P.; Zhou, Y. Emulation of a process-based salinity generator for the Sacramento–San Joaquin Delta of California via deep learning. *Water* **2020**, *12*, 2088. [CrossRef]
36. Roh, D.M.; He, M.; Bai, Z.; Sandhu, P.; Chung, F.; Ding, Z.; Qi, S.; Zhou, Y.; Hoang, R.; Namadi, P.; et al. Physics-informed neural networks-based salinity modeling in the Sacramento–San Joaquin Delta of California. *Water* **2023**, *15*, 2320. [CrossRef]
37. Rath, J.S.; Hutton, P.H.; Chen, L.; Roy, S.B. A hybrid empirical–Bayesian artificial neural network model of salinity in the San Francisco Bay–Delta estuary. *Environ. Model. Softw.* **2017**, *93*, 193–208. [CrossRef]
38. Qi, S.; He, M.; Bai, Z.; Ding, Z.; Sandhu, P.; Chung, F.; Namadi, P.; Zhou, Y. Novel salinity modeling using deep learning for the Sacramento–San Joaquin Delta of California. *Water* **2022**, *14*, 3628. [CrossRef]
39. Rumelhart, D.E.; Hinton, G.E.; Williams, R.J. Learning representations by back-propagating errors. *Nature* **1986**, *323*, 533–536. [CrossRef]
40. Abadi, M.; Barham, P.; Chen, J.; Chen, Z.; Davis, A.; Dean, J.; Devin, M.; Ghemawat, S.; Irving, G.; Isard, M.; et al. TensorFlow: A System for Large-Scale Machine Learning. In Proceedings of the 12th USENIX Symposium on Operating Systems Design and Implementation (OSDI 16), Savannah, GA, USA, 2–4 November 2016; pp. 265–283.
41. Microsoft. Azure Web Apps Overview. 2023. Available online: https://azure.microsoft.com (accessed on 11 March 2025).
42. Oliphant, T.E. Python for Scientific Computing. *Comput. Sci. Eng.* **2007**, *9*, 10–20. [CrossRef]
43. Yang, S.; Madsen, M.S.; Bednar, J.A. HoloViz: Visualization and interactive dashboards in Python. In Proceedings of the 28th ACM SIGKDD Conference on Knowledge Discovery and Data Mining, Washington, DC, USA, 14–18 August 2022; pp. 4846–4847. [CrossRef]
44. Bokeh Development Team. Bokeh: Python Library for Interactive Visualization. 2018. Available online: https://bokeh.org (accessed on 11 March 2025).
45. California Department of Water Resources. *Water Quality Standards for the Delta*; California State Publishing: Sacramento, CA, USA, 2022.

water

MDPI

Article

Analysis of Prediction Confidence in Water Quality Forecasting Employing LSTM

Pan Fang [1,2,†], Yonggui Wang [3], Yanxin Zhao [4] and Jin Kang [5,6,*,†]

1 Institute for Advanced Study, China University of Geosciences, Wuhan 430078, China; fangpan@cestc.cn
2 CECloud Computing Technology Co., Ltd., Wuhan 430056, China
3 Hubei Key Laboratory of Regional Ecology and Environmental Change, School of Geography and Information Engineering, China University of Geosciences, Wuhan 430074, China; wangyg@cug.edu.cn
4 Center of Eco-Environment of the Yangtze River Economic Belt, Chinese Academy of Environmental Planning, Beijing 100014, China; zhaoyx@caep.org.cn
5 Hubei Provincial Academy of Eco-Environmental Sciences (Provincial Ecological Environment Engineering Assessment Center), Wuhan 430072, China
6 Hubei Key Laboratory of Pollution Damage Assessment and Environmental Health Risk Prevention and Control, Wuhan 430072, China
* Correspondence: kangjin314@163.com; Tel.: +86-15102789737
† These authors contributed equally to this work as co-first authors.

Abstract: Water quality prediction serves as an important foundation for risk control and the proactive management of the aquatic environment, and the Long Short-Term Memory (LSTM) network has gained recognition as an effective approach for achieving high-precision water quality predictions. However, despite its potential, there is a significant gap in the literature regarding the confidence analysis of its prediction accuracy and the underlying causes of variability across different water quality indicators and basins. To address this gap, the present study introduces a novel confidence evaluation method to systematically assess the performance of LSTM in predicting key water quality parameters, including ammonia nitrogen (AN), biochemical oxygen demand (BOD), chemical oxygen demand (COD), dissolved oxygen (DO), hydrogen ion concentration (pH), and total phosphorus (TP). This evaluation was conducted across three basins with distinct geographical, climatic, and water quality conditions: the Huangshui River Basin (HSB), the Haihe River Basin (HRB), and the Yangtze River Basin (YRB). The results of the confidence evaluation revealed that LSTM exhibited higher credibility in the Haihe River Basin compared to the Yangtze River Basin. Additionally, LSTM demonstrated greater accuracy and stability in predicting total phosphorus (TP) compared to other water quality indicators in both basins, with median *NSE* values of 0.71 in the HRB and 0.73 in the YRB. Additionally, the research demonstrated a linear relationship between the ability of LSTM models to predict the water quality and temporal autocorrelation as well as the cross-correlation coefficients of the water quality parameters. The coefficients of determination (R^2) ranged from 0.59 to 0.85, with values of 0.59 and 0.79 for the YRB and 0.85 and 0.80 for the HRB, respectively. This finding underscores the importance of considering these correlation metrics when evaluating the reliability of LSTM-based predictions.

Keywords: confidence analysis; forecasting water quality; long short-term memory (LSTM) models; Yangtze River basin; Haihe River basin; Huangshui River basin

1. Introduction

Over the past few decades, artificial intelligence (AI) has emerged as an effective alternative approach for modeling complex nonlinear systems [1]. Extensive research has been conducted on river water quality prediction, monitoring, and management using various AI models, including artificial neural networks (ANNs) [1–4], adaptive neuro-fuzzy inference systems (ANFISs) [5–7], and support vector machines (SVMs) [8,9], among others. However, recurrent neural networks (RNNs) are particularly well suited for sequential data due to their inherent structure [10], and they have been utilized for water quality prediction [11,12]. The Long Short-Term Memory (LSTM) network addresses the vanishing or exploding gradient issues associated with RNNs [13], making it even more effective for handling sequential data, such as water quality information [10,14]. Recently, LSTM-based models have demonstrated promising performance in water quality prediction [14–20].

Recent studies have demonstrated that Long Short-Term Memory (LSTM) networks hold significant value in many time-series predictions and exhibit excellent simulation capabilities. For instance, in the field of runoff prediction, LSTM models have been shown to provide accurate and reliable results, especially when combined with time-frequency analysis methods. However, it is crucial to recognize that LSTM models have their own limitations and are not universally applicable. They are designed to handle specific types of sequential data and may not perform well in all scenarios. For example, the Root Mean Square Errors (RMSEs) of LSTM models for dissolved oxygen (DO) differ across various basins: 0.0396 in Jiangsu Wuxi of the Yangtze River, China; 0.55 in the Prespa Basin in southeastern Europe; 0.07 in Tai Lake, China; and 0.067 in Victoria Bay, China [14,15,21]. Similarly, within the same basin, LSTM models exhibit different performances for various indicators. In the Beilun Estuary of the Guangxi Autonomous Region, China, the Mean Absolute Errors (MAEs) for predicting the weekly pH, DO, chemical oxygen demand (COD), and ammonia nitrogen (AN) were 0.58, 0.51, 0.71, and 0.68, respectively [20]. Moreover, the performance of LSTM is influenced by various factors, including data characteristics, model parameters, and training methods. Despite these considerations, there is a notable gap in the literature regarding the credibility analysis of LSTM predictions. While numerous studies have focused on comparing different models or incrementally improving existing ones, there is a lack of in-depth research on evaluating the confidence level of LSTM predictions in diverse applications [1,18]. For example, in water quality prediction, LSTM models show varying performances across different basins and indicators. The Root Mean Square Errors (RMSEs) of LSTM models for dissolved oxygen (DO) differ significantly across various basins: 0.04 in Jiangsu Wuxi of the Yangtze River, China; 0.545 in the Prespa Basin in southeastern Europe; 0.07 in Tai Lake, China; and 0.067 in Victoria Bay, China [22,23]. Similarly, within the same basin, LSTM models exhibit different performances for various indicators. In the Beilun Estuary of the Guangxi Autonomous Region, China, the Mean Absolute Errors (MAEs) for predicting the weekly pH, DO, chemical oxygen demand (COD), and ammonia nitrogen (AN) were 0.58, 0.51, 0.71, and 0.68, respectively [23–25]. However, the underlying causes of these variations remain largely unexplored. There is a striking deficiency in the literature regarding the feasibility and applicability of data-driven models for various water quality indicators and across diverse basins. Although it is evident that LSTM performance fluctuates with different indicators and basins, the root causes of these discrepancies remain largely unexplored. The absence of comprehensive credibility analysis significantly hampers the reliable application and broader adoption of LSTM models in practical scenarios.

The Yangtze River Basin (YRB), the Haihe River Basin (HRB), and the Huangshui River Basin (HSB) are three different basins with distinctive features in China, located in the southern, northern, and northwestern regions, respectively. The YRB is characterized by its humid climate, abundant rainfall, and rich water resources, whereas the HRB is relatively dry, with scarce water resources and frequent water shortages [18,26–29]. It is worth highlighting that the Haihe River Basin (HRB) experiences the most critical levels of water scarcity and pollution when compared to the other six major river basins in China [30]. Accurate water quality forecasts are essential for early warning and timely intervention to mitigate pollution in these basins [31]. The water shortage problem is more serious in the HSB; the water quantity is scarce, and the water pollution level is serious [32]. Various models, including process-based and machine learning models, have been applied to predict water quality in the HRB [18,33–37] and YRB [21,38–41]. However, LSTM-based models have been underutilized in these regions [18,21]. Moreover, LSTM applications have been limited to predicting the dissolved oxygen (DO) in both basins, while other critical pollution indicators, such as the biochemical oxygen demand (BOD) and total phosphorus (TP), urgently need to be assessed. Further research is needed to analyze the prediction accuracy of LSTM for these additional water quality indicators. Additionally, comparative analysis of LSTM performance in areas with different water resource availability and pollution conditions is essential to evaluate its applicability.

Comparing LSTM performance between the HRB and YRB can not only reveal the characteristics and influencing factors of LSTM in different regions but also provide valuable insights for similar studies worldwide. As a result, an analytical comparison of the effectiveness of water quality forecasting using LSTM models was conducted across the Yangtze and Haihe River Basins. Our aims are twofold: (1) to demonstrate the diverse performance of LSTM in predicting water quality across different basins and elucidate the underlying causes and (2) to establish the connection between LSTM performance and the data characteristics of the modeled region.

2. Methodology

2.1. Study Area

The Yangtze River Basin (YRB), Haihe River Basin (HRB), and Huangshui River Basin (HSB) were selected as the study area (shown in Figure 1).

The YRB is the largest river in China and the third longest river in the world [42–44]. Most of the YRB is in the subtropical monsoon climate zone [29] and is characterized by a warm/humid climate but with an irregular distribution of annual average temperature and precipitation in both the spatial and temporal dimensions. The Haihe River is the largest river catchment in Northern China [45].

The Haihe River Basin (HRB) has about 1.05×10^5 m^3/km^2 (less than one-fifth of that in the YRB) of water resources, covering an area of about 2.6×10^5 km^2. Located in a continental monsoon climate zone, the HRB belongs to semi-humid and semi-arid region [46]. With the rapid urbanization and economic development, the HRB receives large amount of sewage and waste discharged from varied origins [45,47]. Accordingly, the HRB suffers the most water shortages and water pollution among all seven river basins in China.

The Huangshui River Basin (HSB) plays a crucial role in the socio-economic advancement of Qinghai Province, China. However, due to its typical alpine and arid climate, the basin faces significant challenges, including drought, water shortages, and soil erosion. The average annual precipitation in the HSB ranges from 300 to 500 mm, while the average annual evaporation rate is as high as 800 to 1500 mm, indicating a severe water scarcity situation [30].

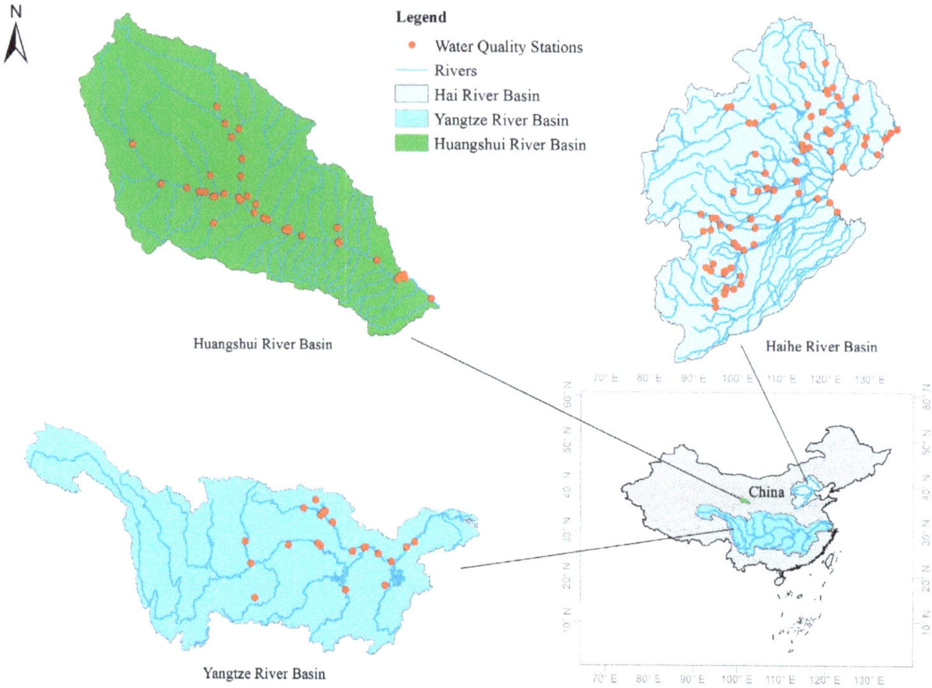

Figure 1. Geographical regions and placement of water quality stations.

2.2. Data Sources

The water quality dataset encompasses monthly records for 23 monitoring sites within the Yangtze River Basin (YRB) spanning from 2003 to 2018, 76 monitoring sites in the Huai River Basin (HRB) from 2010 to 2020, and 68 monitoring sites in the Huangshui River Basin (HSB) between 2011 and 2021 (Figure 1). The monthly concentrations for biochemical oxygen demand (BOD), chemical oxygen demand (COD), dissolved oxygen (DO), ammonia nitrogen (AN), total phosphorus (TP), and hydrogen ion concentration (pH) were collected.

The mean, minimum, maximum, standard deviation (SD), and coefficient of variation (CV) were used for a comparative evaluation of model performance in the three basins, as shown in Table 1.

Table 1. Basic statistical analysis for water quality indicators of three basins.

Basins	Indicators	Unit	Mean	Minimum	Maximum	SD	CV
YRB	AN	mg/L	0.178	0.025	1.340	0.184	1.035
	BOD	mg/L	1.100	0.500	2.500	0.400	0.300
	COD	mg/L	2.200	0.500	4.100	0.500	0.200
	DO	mg/L	8.530	4.400	13.10	1.510	0.180
	pH	-	7.970	6.930	8.920	0.330	0.040
	TP	mg/L	0.072	0.005	0.250	0.051	0.706
HRB	AN	mg/L	8.104	0.012	122.00	14.554	1.796
	BOD	mg/L	12.300	0.200	220.00	24.700	2.000
	COD	mg/L	11.000	0.600	127.00	16.00	1.400
	DO	mg/L	6.750	0.020	18.80	3.500	0.520
	pH	-	7.890	6.420	8.990	0.380	0.050
	TP	mg/L	0.730	0.005	8.880	1.243	1.703

Table 1. *Cont.*

Basins	Indicators	Unit	Mean	Minimum	Maximum	SD	CV
HSB	AN	mg/L	0.572	0.011	10.80	0.929	1.622
	BOD	mg/L	2.100	0.200	24.00	1.700	0.800
	COD	mg/L	2.100	0.200	13.00	1.000	0.500
	DO	mg/L	8.190	3.220	12.700	1.200	0.150
	pH	-	8.220	6.490	9.290	0.310	0.040
	TP	mg/L	0.081	0.005	1.190	0.103	1.270

The Maximal Information Coefficient (MIC) can be utilized as a versatile tool for detecting and quantifying complex dependencies between variables, going beyond traditional correlation measures, which often fail to capture nonlinear relationships [48]. The MIC was computed and employed to determine both temporal autocorrelation and cross-correlation coefficients for water quality indicators across various lag periods, ranging from 1 to 12 months, as illustrated in Figure 2. The measurement MIC is symmetric and normalized into a range $[0, 1]$. A high MIC value indicates a strong dependency between the investigated variables, whereas MIC = 0 describes the relationship between two independent variables.

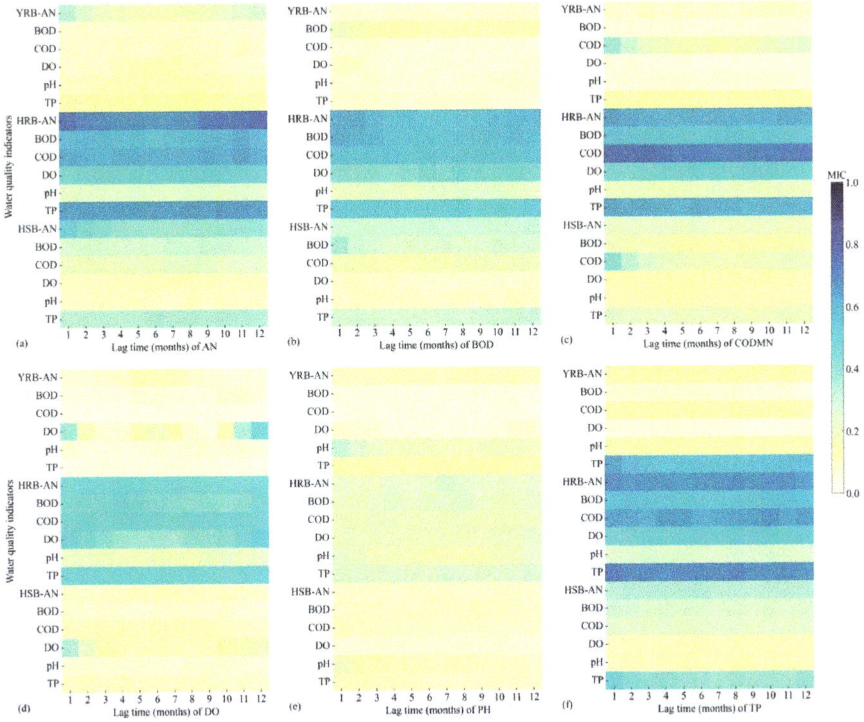

Figure 2. Temporal autocorrelation and cross-correlation coefficients based on Maximal Information Coefficient (MIC) of each water quality indicator in the same basin with a lag time from 1 to 12 months. The abbreviations mean: biochemical oxygen demand (BOD), chemical oxygen demand (COD), dissolved oxygen (DO), ammonia nitrogen (AN), total phosphorus (TP), and hydrogen ion concentration (pH), the Huangshui River Basin (HSB), the Haihe River Basin (HRB), and the Yangtze River Basin (YRB). Figure (**a–f**) mean temporal autocorrelation and cross-correlation coefficients of AN, BOD, CODMN, DO, PH and TP, respectively.

2.3. Model Development Based on LSTM Models

2.3.1. Principle of the Model

As an advanced recurrent neural network (RNN), Long Short-Term Memory (LSTM) has a specialized neuron structure. This neuron structure has a selective memory function [14,49] by means of a special model structure design, a memory block, for information filtering and conversion [16,50]. A memory block is composed of a forget gate, an input gate, a memory cell, and an output gate [13,51], with its state at a given time depicted in Figure 3.

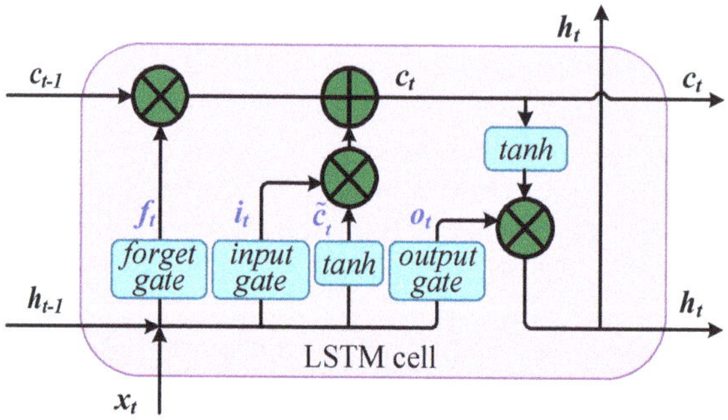

Figure 3. Conceptual illustration of memory block in LSTM.

In the Long Short-Term Memory (LSTM) model, at the preceding time step $(t-1)$, the memory block retains crucial information in the form of both the cell state (C_{t-1}) and the output (h_{t-1}). The initial values for the cell state (C_0) and the hidden state (h_0) are typically set to zero, serving as the starting point for the LSTM cell's memory. As it progresses to time step t, the current inputs (X_t) become accessible. The process begins with deriving the hidden state (h_t) through a nonlinear transformation. This transformation utilizes the output gate and the newly updated cell state, as described in Equation (1). This hidden state (h_t) encapsulates the information learned from the previous time steps and the current input, forming the basis for subsequent calculations. Next, four multi-layer perceptrons (MLPs) are employed to compute key gates and states within the LSTM cell. Specifically, these MLPs are formulated as Equations (2)–(4) and (6), and they calculate the forget gate (f_t), the candidate cell state $\left(\widetilde{c_t}\right)$, the input gate (i_t), and the output gate (o_t). These calculations are based on the previous hidden state (h_{t-1}) and the current input (X_t). The forget gate (f_t) decides what information to discard from the previous cell state, the input gate (i_t) determines what new information to add, and the candidate cell state $\left(\widetilde{c_t}\right)$ represents the potential new values for the cell state. The output gate (o_t) controls the output based on the updated cell state. Finally, the updated cell state (C_t) is calculated using the relationship outlined in Equation (5). This new cell state incorporates the forgotten information, the new input, and any other modifications decided by the gates, effectively updating the LSTM cell's memory for the current time step.

$$h_t = o_t * tanh(c_t) \tag{1}$$

$$f_t = \sigma\left(W_{xf}X_t + W_{hf}h_{t-1} + b_f\right) \tag{2}$$

$$\widetilde{c_t} = tanh(W_{xc}X_t + W_{hc}h_{t-1} + b_c) \tag{3}$$

$$i_t = \sigma(W_{xi}X_t + W_{hi}h_{t-1} + b_t) \tag{4}$$

$$c_t = f_t * c_{t-1} + i_t * \widetilde{c_t} \tag{5}$$

$$o_t = \sigma(W_{xo}X_t + W_{ho}h_{t-1} + b_O) \tag{6}$$

In Equations (2)–(4) and (6), W denotes the matrices of weights for the gates or cells with the corresponding subscripts, and b represents learnable biases. Additionally, σ and $tanh$ denote the sigmoid function and the tanh function, respectively.

Long Short-Term Memory (LSTM) networks are powerful tools in the realm of deep learning, distinguished by their ability to retain and manipulate information over extended temporal periods. By incorporating specialized memory cells, LSTM networks can "remember" the most pertinent input values, making them exceptionally well suited for sequential learning tasks [49]. This capability enables LSTM models to capture long-term dependencies and patterns within data sequences, which is crucial for accurate predictions and understanding of temporal dynamics.

In the context of the present study, LSTM models were meticulously developed to forecast future one-month values of a specific water quality indicator. To achieve this, two distinct configurations of input variables were employed:

(1) Univariate inputs: In this configuration, the LSTM models solely rely on historical data of the targeted water quality indicator. This approach is straightforward and focuses on understanding the temporal evolution of the single variable of interest.

(2) Multivariate inputs: Alternatively, the LSTM models were also trained using historical data from all available water quality indicators. This multivariate approach provides a more comprehensive view of the water quality system, potentially capturing interdependencies and correlations between different indicators.

For both input configurations, the time steps (k) for the model inputs were varied from 1 to 12 months. This means that the LSTM models were fed with time-series data spanning from the previous 1 to k months.

2.3.2. Model Training and Testing

The data from each station were randomly divided into three subsets—80% for training, 10% for validation, and 10% for testing—to facilitate model training. And same kind of subsets from the same basin were combined to form the training, validation, and test dataset. Then, all water quality indicators were standardized according to Equation (1) to ensure that the model's input variables remained on the same scale and to guarantee a stable convergence of parameters in LSTM [52].

$$\tilde{x}_i = \frac{x_i - \overline{x}}{std} \tag{7}$$

Here, $\overline{x}$ and std signify the average and standard deviation of the training dataset, respectively; x_i defines every single value of the raw data, while $\tilde{x}_i$ represents the standardized value.

2.3.3. Model Optimization

For each LSTM model, Keras Tuner (https://keras-team.github.io/keras-tuner/, accessed on 31 March 2025) was employed to optimize the model hyperparameters via a 150-time random search (10 repetitive times training for each random search) from the searching space of hidden layers and neurons [53]. Other hyperparameters of LSTM were determined using trial and error. The batch size and the dropout rates were 256 and 0.5, respectively. The maximum number of epochs for model training was established at 200, signifying that the training phase would terminate when convergence was reached or when the 200 predefined epochs had elapsed. Furthermore, an initial learning rate of 0.1 was assigned to facilitate the learning process. This rate undergoes automatic adjustment downwards if there is no notable improvement in the loss function on the validation dataset for five straight epochs.

During the training phase, models are developed using the training dataset, with Keras Tuner being instrumental in pinpointing the best model architectures through thorough performance assessments on the validation dataset. After identifying these optimal architectures, the LSTM models are fine-tuned through a retraining process and subsequently assessed on both training and testing datasets. To mitigate the variability caused by the random initialization of model parameters, the entire cycle of training, optimization, retraining, and evaluation is carefully repeated 100 times, ensuring the creation of robust, reliable, and generalizable LSTM models.

2.4. *Confidence Analysis of LSTM Models*

In the field of deep learning and machine learning, confidence is an important measure of the predictive reliability of a model and a key factor in determining whether a model is usable. Confidence usually refers to the degree to which a model is certain of its predictions, and this degree of confidence can be quantified. The applicability of the LSTM model is a key issue that needs to be evaluated for practical application scenarios, such as watershed water quality prediction. Although LSTM performs well in many time-series prediction tasks, its scope of application is limited and cannot be arbitrarily applied to all scenarios. Therefore, this paper constructs a set of model confidence evaluation methods.

2.4.1. Model Accuracy Calculation

By comparing the predicted value and the real value, the accuracy of the model base on the Nash–Sutcliffe Efficiency (*NSE*; Equation (8)) [50] will be judged. The Nash–Sutcliffe Efficiency (*NSE*; Equation (8)) [50], a widely recognized metric for hydrological model evaluation [2,51], is employed as the model assessment criterion in the current study:

$$NSE = 1 - \frac{\sum_{i=1}^{n}\left(y_{m,i} - y_{p,i}\right)^2}{\sum_{i=1}^{n}\left(y_{m,i} - \overline{y_m}\right)^2} \quad (8)$$

where n is the number of observations; $y_{m,i}$ and $y_{p,i}$ define the ith observations and the corresponding values predicted using LSTM, respectively; and $\overline{y_m}$ represents the average of observations. In a lot of studies, the *NSE* ranges from $-\infty$ to 1. The model is acceptable when the $NSE \geq 0.60$ [54] and good when the $NSE \geq 0.75$ [2,55].

2.4.2. Confidence Analysis

The confidence level and confidence interval were used for confidence analysis based on the *NSE*. The confidence value under confidence level Z is:

$$c_Z = \overline{X} \pm \frac{Z \times \sigma}{\sqrt{n}} \quad (9)$$

The confidence coefficient, defined as δ, can be simulated as:

$$\delta = max\left(0, \overline{X} - d_Z\right) \quad (10)$$

$$d_Z = \frac{max(c_Z) - min(c_Z)}{max(c_Z)} \quad (11)$$

where $\overline{X}$ is the mean of the *NSE*, σ is the standard deviation, n is the sample size, Z is the coefficient under the confidence level (usually 95%), and d_Z is the confidence width of the confidence interval under Z. According to the accuracy evaluation basis of the NSE, the classification criteria are divided into four levels and determined as follows:

$$L\delta = \begin{cases} L1 & bad,\ \delta < 0.4 \\ L2 & acceptable,\ 0.4 \le \delta < 0.6 \\ L3 & good, 0.6 \le \delta < 0.7 \\ L4 & excellent, \delta \ge 0.7 \end{cases} \tag{12}$$

3. Results and Discussion

3.1. Model Accuracy Evaluation of Water Quality Prediction with Different Indexes

The accuracy of the model for various water quality indicators across different basins was examined and compared, considering the input variables and their respective lag times, as illustrated in Figures 4 and 5.

For models with different input conditions (shown in Figure 4), when inputting more variables, most of the median NSEs of the YRB decreased by nearly 0.25 (Figure 4b), while, except for the pH, there was little deterioration in the performances of the models of the HRB (Figure 4a). However, LSTM models to predict the DO in the HRB performed better after inputting data for more water quality indicators (Figure 4a). Moreover, with more time-step data inputs, most LSTM models improved to some extent. Models for predicting the AN and TP of the YRB with multivariate inputs improved significantly when inputting more time-step data.

Figure 5 shows the differences in the performance of LSTM in the HRB and in the YRB under different inputs conditions. In the univariate input condition, LSTM models for the pH in the YRB are better than those for the pH in the HRB, while little difference between the YRB and HRB for predicting the DO or TP was found; the performances for predicting the AN, BOD, and COD in the HRB are much better than those for the YRB. As for the multivariate input condition, LSTM models for all water quality indicators of the HRB perform significantly better than those for the YRB. Nevertheless, the prediction accuracy gap between the HRB and YRB for the AN or TP was narrowed with increasing input time steps.

Table 2 presents the average Nash–Sutcliffe Efficiency (NSE) values of the LSTM model for five key water quality indicators across three different basins: the Yangtze River Basin (YRB), the Haihe River Basin (HRB), and the Huangshui River Basin (HSB).

This table provides a comprehensive comparison of the model's performance in predicting these indicators within each basin, highlighting the variability in prediction accuracy across different regions and parameters. The NSE values in the Haihe River Basin (HRB) are generally higher than those in the Yangtze River Basin (YRB) and the Huangshui River Basin (HSB). For instance, the HRB achieves NSE values of 0.773 for BOD, 0.785 for COD, 0.623 for DO, 0.631 for NH_3-N, and 0.644 for TP, indicating relatively high prediction accuracy. In contrast, the YRB shows much lower NSE values, such as 0.060 for BOD, 0.041 for COD, and 0.375 for DO, suggesting that the LSTM model performs less effectively in this basin. The HSB has intermediate NSE values, with 0.523 for BOD, 0.420 for COD, and 0.540 for DO. Notably, the HRB has the highest NSE for NH_3-N (0.631) and TP (0.644), while the YRB has the lowest NSE for most indicators, except for TP, where the HSB has the lowest value (0.478). The significant differences in NSE values across the three basins highlight the importance of considering regional characteristics when applying the LSTM model for water quality prediction. The higher NSE values in the HRB suggest that the model may be more suitable for this region due to specific environmental conditions or data quality. Conversely, the lower NSE values in the YRB indicate potential limitations of the model in this basin, possibly due to more complex hydrological processes or higher data variability. These findings emphasize the need for further investigation into the factors influencing model performance and the importance of tailoring model applications to specific basin conditions. Overall, while the LSTM model shows promise in some regions, its applicability and reliability vary significantly, necessitating careful evaluation and adaptation for different environments.

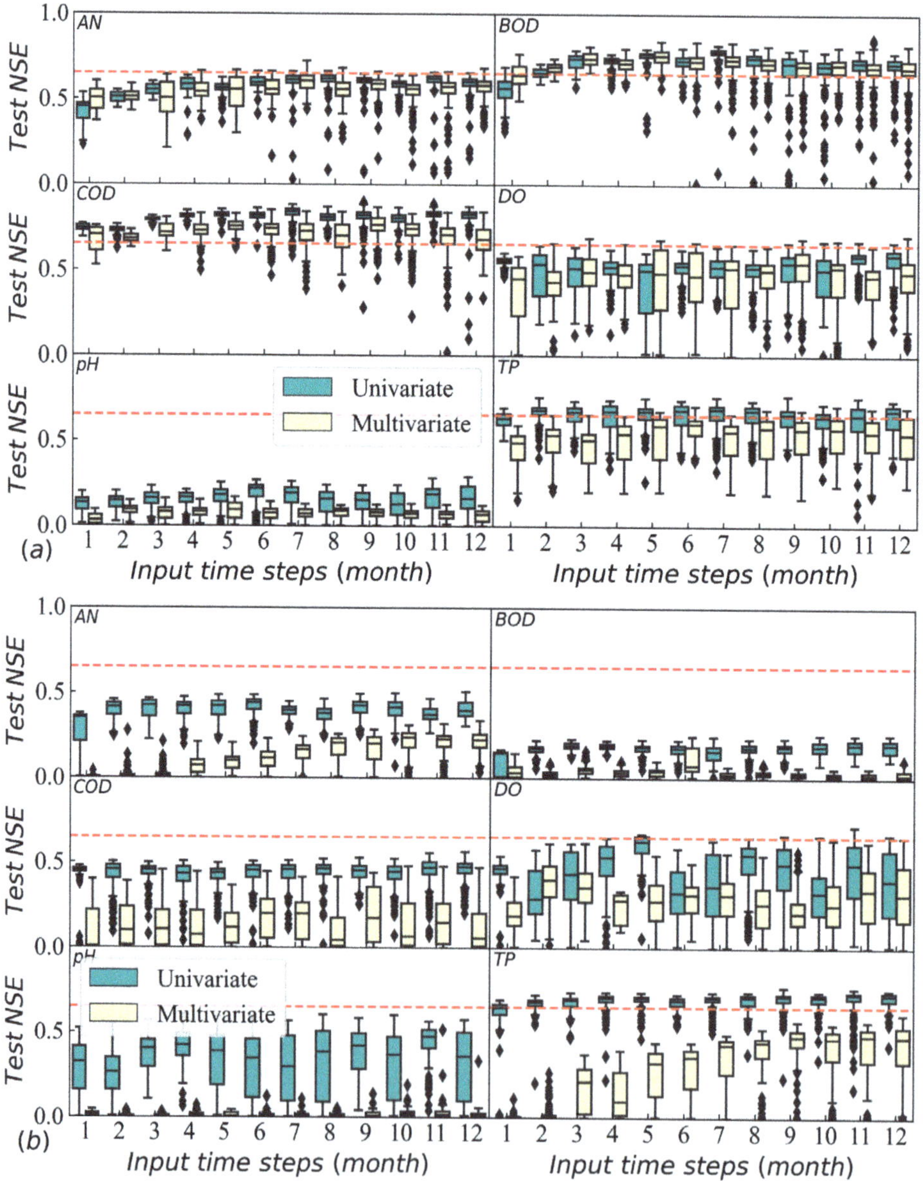

Figure 4. Comparison of distribution of *NSE* values for water quality indicators of (**a**) the HRB and (**b**) YRB predicted using different input conditions. The red dotted lines represent the threshold of acceptable model performance (*NSE* = 0.65).

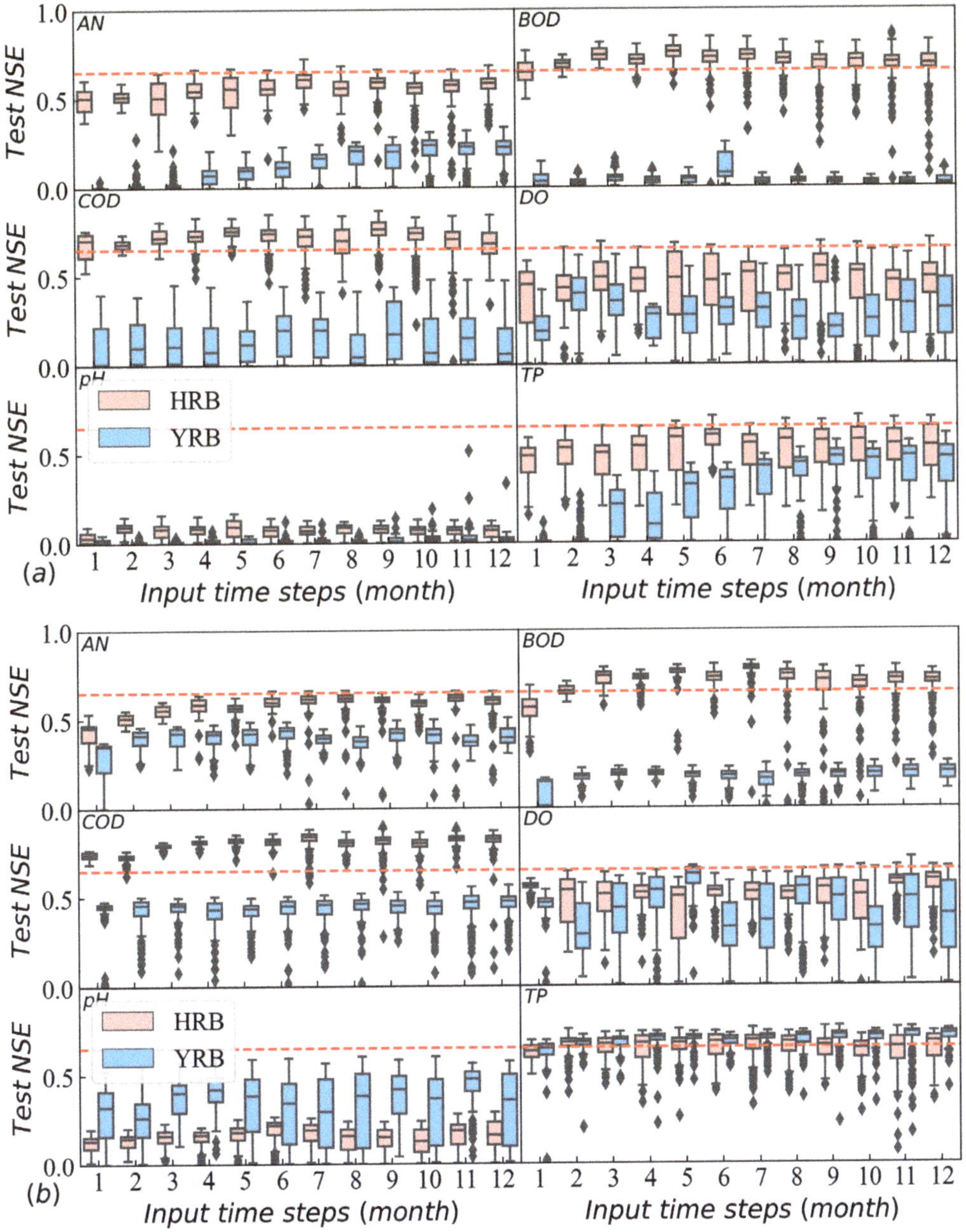

Figure 5. A comparison of the distribution of *NSE* values for each water quality indicator across different basins, predicted using (**a**) univariate inputs and (**b**) multivariate inputs. The red dashed lines indicate the threshold for acceptable model performance (*NSE* = 0.65).

Table 2. Average ($\overline{X}$) and standard deviation (σ) of Nash–Sutcliffe Efficiency (*NSE*) values with the LSTM model.

Index	YRB		HRB		HSB	
	$\overline{X}$	σ	$\overline{X}$	σ	$\overline{X}$	σ
BOD	0.060	0.038	0.773	0.070	0.523	0.167
COD	0.041	0.047	0.785	0.047	0.420	0.166
DO	0.375	0.174	0.623	0.055	0.074	0.110
NH_3-N	0.233	0.038	0.631	0.082	0.706	0.043
TP	0.463	0.033	0.644	0.056	0.478	0.048
Ph	0.382	0.108	0.471	0.092	0.421	0.103

3.2. Confidence Analysis of LSTM for Water Quality Prediction in the Three Basins

We calculated the mean and standard deviation of the Nash–Sutcliffe Efficiency (*NSE*) for the predictions of BOD, COD, DO, NH_3-N, TP, and pH across all models in three watersheds. Based on these values, we determined the 95% confidence intervals. Subsequently, we obtained the widths of these confidence intervals under different confidence levels. The results are presented in Table 3.

Table 3. Confidence interval (C_i) and confidence width d_Z of three basins under confidence level of 95%.

Index	YRB		HSB		HRB	
	C_i	d_Z	Ci	d_Z	C_i	d_Z
BOD	[0.058, 0.062]	0.071	[0.769, 0.777]	0.011	[0.514, 0.532]	0.035
COD	[0.038, 0.044]	0.123	[0.782, 0.788]	0.007	[0.411, 0.429]	0.043
DO	[0.365, 0.385]	0.053	[0.620, 0.626]	0.010	[0.068, 0.080]	0.156
NH_3-N	[0.231, 0.235]	0.019	[0.626, 0.636]	0.015	[0.704, 0.708]	0.007
TP	[0.461, 0.465]	0.008	[0.641, 0.647]	0.010	[0.475, 0.481]	0.011
Ph	[0.376, 0.388]	0.033	[0.465, 0.477]	0.023	[0.415, 0.427]	0.028

For all water quality indicators, as the confidence level increases, the width of the confidence intervals gradually expands, indicating increased prediction uncertainty at higher confidence levels. As shown in the table, at a 95% confidence level, the HRB demonstrates high prediction certainty for multiple key indicators. Particularly for biochemical oxygen demand (BOD), chemical oxygen demand (COD), dissolved oxygen (DO), and pH, the HRB exhibits the narrowest confidence intervals, implying relatively more stable prediction results. In contrast, the YRB shows lower prediction certainty for BOD, COD, and DO, with wider confidence intervals, while the HSB has relatively lower certainty in predicting DO and ammonia nitrogen (NH_3-N). Analyzing the specific performance for different water quality indicators, the HRB performs best in terms of the BOD, COD, and DO, with high prediction certainty and narrow confidence intervals. For the prediction of NH_3-N, the HSB performs optimally, with the smallest confidence interval width. Regarding the prediction of total phosphorus (TP), the YRB also demonstrates relatively high certainty. Overall, the HRB exhibits the best prediction stability, surpassing that of the HSB and YRB. Further analysis of the mean *NSE* values and prediction reliability levels for the three watersheds is presented in the Figure 6.

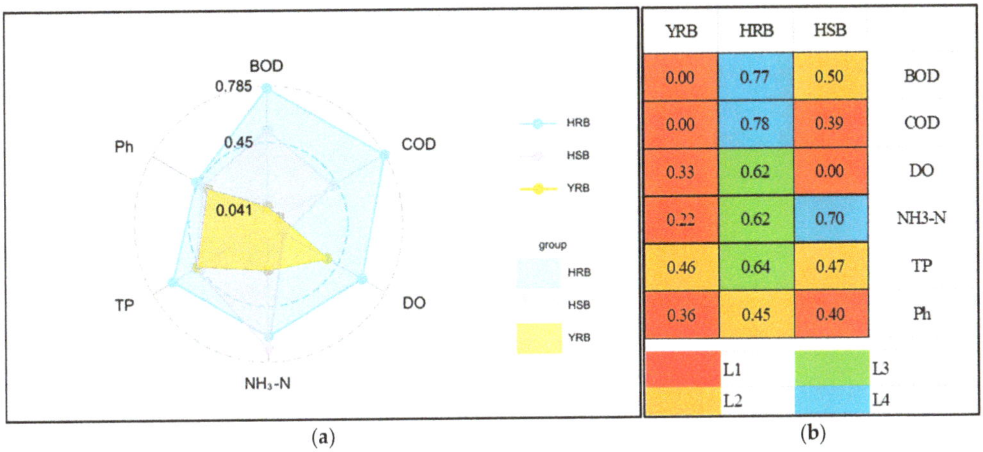

Figure 6. Simulation reliability grades of LSTM model for different indicators in three watersheds. (**a**) Mean *NSE*; (**b**) confidence levels.

From Figure 6a, it can be seen that in terms of model simulation accuracy, the HRB performs the best, followed by the HSB, with the YRB performing the worst. Further analysis of the simulation reliability grades of the LSTM model for different indicators in the three watersheds in Figure 6b reveals that in the YRB, except for TP, the simulation reliability of the other indicators is at grade L1, indicating poor simulation performance and suggesting that the LSTM model may not be suitable for water quality prediction in the YRB; whereas in the HRB, the simulation performance of BOD and COD reaches grade L4 (excellent), and the simulation reliability of indicators such as DO and NH_3-N is also at grade L3, indicating good simulation performance and suggesting that the LSTM model is highly suitable for water quality prediction in the HRB; as for the HSB, the simulation performance of NH_3-N is good, and the simulation performance of other indicators is acceptable, indicating that with certain optimizations, the LSTM model can also be used for water quality prediction in the HSB watershed, and the prediction results are basically reliable.

3.3. *Influencing Factors of Model Performance in Different Basins*

Most of these phenomena in Sections 3.1 and 3.2 can be associated with temporal autocorrelation and cross-correlation coefficients, as mentioned in Section 2.2. For water quality indicators of the three basins, their autocorrelation coefficients are generally bigger than their cross-correlation coefficients with other indicators. In the YRB, the cross-correlation coefficients are much smaller (most MICs smaller than 0.2). As the volume of input data expanded in the LSTM model, water quality variables that exhibited low correlation with the output variables introduced significant amounts of redundant information. This influx of redundant data acted as a barrier, impeding the model's capacity to discern meaningful and coherent patterns among the variables. Consequently, the prediction performance of the model suffered, as it struggled to differentiate between relevant and irrelevant information, leading to a decline in its overall accuracy and reliability [56–58]. Therefore, models with multivariate inputs of the YRB will be worse than those with univariate inputs of the YRB, which is consistent with the results. Similarly, TP has the biggest autocorrelation coefficients (MICs more than 0.6, 0.75, and 0.7 in the YRB, HRB, and HSB, respectively) among water quality indicators. However, except for the pH, water quality indicators in the HRB have bigger cross-correlation coefficients (MICs generally bigger than 0.5). In the YRB, LSTM performs the best in predicting the TP. And in the HRB, the prediction performance for the pH is the worst.

Therefore, the correlations between LSTM performances of water quality prediction and temporal autocorrelation/cross-correlation coefficients of water quality indicators warrant further exploration (as shown in Figure 7). In all scenarios, model performances are linearly dependent on the maximum values of temporal autocorrelation/cross-correlation coefficients. In addition, the linear relation is more stable and significant with the coefficients of determination for MICs ranging from 0.59 to 0.85. Different scenarios have almost the same slope in Figure 7. Significant variations are observed in the intercepts of the models, which can be attributed to various factors. One possible reason is the inherent differences across different basins, such as geographical characteristics, climate conditions, and water flow patterns. Another contributing factor could be the distinct input patterns used in the models, which may reflect variations in data collection methods, preprocessing steps, or the specific variables included in the analysis. These differences in intercepts highlight the complexity and variability inherent in modeling water quality across diverse basins and underscore the importance of considering basin-specific characteristics and input patterns when developing and applying such models.

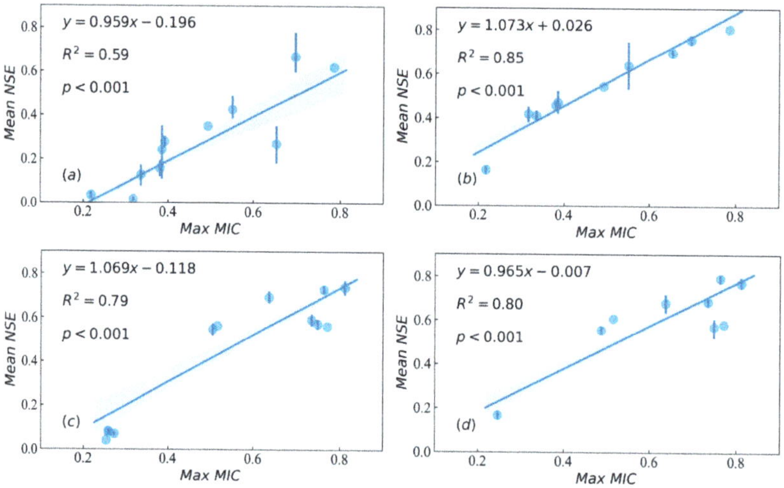

Figure 7. Scatter plots illustrating the relationship between model performance and input–output correlations, as calculated using the MIC, are presented for the following cases: (**a**) the YRB with multivariate inputs, (**b**) the YRB with univariate inputs, (**c**) the HRB with multivariate inputs, and (**d**) the HRB with univariate inputs. In these plots, R^2 represents the coefficient of determination, while p denotes the p-value from the statistical significance test of linear regression.

Furthermore, pollutant concentrations of the HRB and HSB are much higher than those of the YRB (Table 1), which could be attributed to large geographical variations [1]. Additionally, the Haihe River Basin is a typical sluice-controlled river. The construction and operation of sluice dams change the flow and other hydrological factors of rivers, which have adverse effects on water [30]. The high pollutant concentration and poor fluidity of water in the HRB and HSB may be one of the reasons why water quality indicators in these two basins are more related than those in the YRB.

Thus, the relationship between concentration and variability of water quality indicators (described by the mean value and coefficient of variation (CV), respectively) and prediction performance (exhibited by the median *NSE*) has been investigated, as shown in Figure 8. The results revealed that little linear correlation exists between the mean concentration and prediction performance of water quality indicators with coefficients of determination (R^2) as low as 0.10. Additionally, there is a certain correlation between model performances and CVs of water quality indicators with an R^2 of 0.44 and p-value < 0.05. Thus, the performance of the model demonstrates a pronounced and robust linear dependence on the maximum values of the temporal autocorrelation and cross-correlation coefficients of the water quality indicators. This suggests that the model's predictive accuracy and reliability are significantly influenced by the strength and patterns of temporal relationships among the water quality indicators. Specifically, as the maximum values of these correlation coefficients increase, the model's performance tends to improve, indicating that it is better able to capture and utilize the underlying temporal dynamics of the water quality data. This finding underscores the importance of considering temporal autocorrelation and cross-correlation in the development and optimization of water quality models.

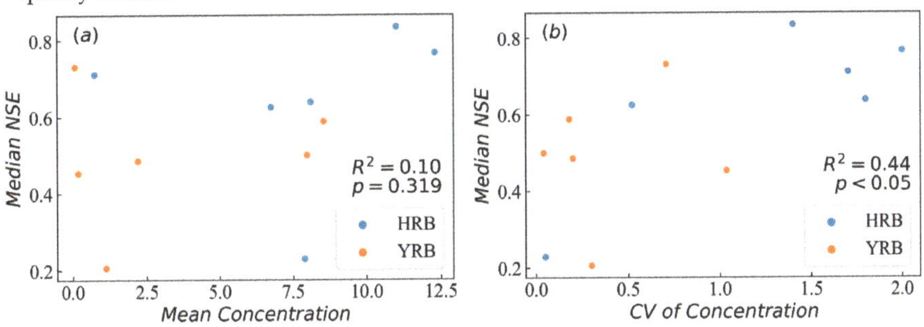

Figure 8. Scatter plot demonstration of the relationship between (**a**) the mean concentration and (**b**) coefficient of variation of water quality indicators and prediction performance.

In this study, forecasts for water quality in the YRB, a large, water-rich basin, were less accurate than those for the HRB and HSB, which are smaller, water-scarce basins. Considering that the HRB and HSB have smaller watershed areas and more intensive monitoring networks than the YRB, more data are available. The variation in water quality is small, which makes the autocorrelation and cross-correlation of water quality variables bigger. This makes the LSTM model more applicable in both basins. It also shows, from the other hand, that increasing the density of the water quality monitoring network and the amount of available water quality data for training can significantly improve the surface water quality prediction performances of the learning models [23].

4. Conclusions

In this study, we developed a comprehensive confidence evaluation method for assessing the performances of LSTM models in water quality prediction. This method was applied to three major river basins in China. By comparing the prediction performance of the LSTM model across these basins for various water quality indicators, including the AN, BOD, COD, DO, pH, and TP, we found that the LSTM model exhibited superior performance in the Haihe River Basin and the Huangshui River Basin. Specifically, the prediction accuracy was higher in these two basins, and the confidence intervals for water quality prediction were also wider, indicating greater reliability. Our analysis revealed that for indicators with higher temporal autocorrelation, such as TP and COD, the prediction accuracy was significantly higher. Further analysis revealed a linear association between the predictive accuracy of the LSTM model and the temporal autocorrelation as well as the inter-variable correlation coefficients of the water quality indicators. This finding underscores the importance of considering the specific data characteristics of the target indicators when applying LSTM models for water quality prediction. While the LSTM model is a powerful tool for time-series prediction, its application must be tailored to the specific characteristics of the basin and the data features of the target indicators. A thorough analysis of the data features and the underlying temporal relationships of the water quality indicators is essential to maximize the effectiveness of the LSTM model in different river basins.

Author Contributions: Resources, Y.Z. and J.K.; Writing—original draft, P.F.; Writing—review & editing, Y.W.; Funding acquisition, J.K. All authors have read and agreed to the published version of the manuscript.

Funding: This work was supported by the Hubei Provincial Natural Science Foundation and Three Gorges Innovative Development Foundation of China (2024AFD371), National Key R&D Program of China (No. 2022YFC3203502), Open Fund State Key Laboratory of Hydraulic Engineering Simulation and Safety Tianjin University (HSEE-2311), and Science and Technology Major Project of Hubei Province, China (2023BCA003).

Data Availability Statement: Restrictions apply to the availability of these data. Data were obtained from Chinese Academy of Environmental Planning and are available from the authors with the permission of Chinese Academy of Environmental Planning (http://www.caep.org.cn/).

Acknowledgments: The authors wish to express their gratitude to the Hebei Provincial Academy of Ecological and Environmental Science, China (http://www.hebhky.cn/), for providing the data used in this study. Additionally, thanks are extended to the Changjiang Water Resources Commission of the Ministry of Water Resources, China (http://www.cjw.gov.cn/).

Conflicts of Interest: Author Pan Fang was employed by the CECloud Computing Technology Co., Ltd. The remaining authors declare that the research was conducted in the absence of any commercial or financial relationships that could be construed as a potential conflict of interest.

References

1. Ahmed, A.N.; Othman, F.B.; Afan, H.A.; Ibrahim, R.K.; Fai, C.M.; Hossain, M.S.; Ehteram, M.; Elshafie, A. Machine learning methods for better water quality prediction. *J. Hydrol.* **2019**, *578*, 124084. [CrossRef]
2. Tung, T.M.; Yaseen, Z.M. A survey on river water quality modelling using artificial intelligence models: 2000–2020. *J. Hydrol.* **2020**, *585*, 124670. [CrossRef]
3. Kim, H.G.; Hong, S.; Jeong, K.-S.; Kim, D.-K.; Joo, G.-J. Determination of sensitive variables regardless of hydrological alteration in artificial neural network model of chlorophyll a: Case study of Nakdong River. *Ecol. Model.* **2019**, *398*, 67–76. [CrossRef]
4. Kim, S.E.; Seo, I.W. Artificial Neural Network ensemble modeling with conjunctive data clustering for water quality prediction in rivers. *J. Hydro-Environ. Res.* **2015**, *9*, 325–339. [CrossRef]
5. Ahmed, A.A.M.; Shah, S.M.A. Application of adaptive neuro-fuzzy inference system (ANFIS) to estimate the biochemical oxygen demand (BOD) of Surma River. *J. King Saud Univ.-Eng. Sci.* **2017**, *29*, 237–243. [CrossRef]
6. Mahmoodabadi, M.; Arshad, R.R. Long-term evaluation of water quality parameters of the Karoun River using a regression approach and the adaptive neuro-fuzzy inference system. *Mar. Pollut. Bull.* **2018**, *126*, 372–380. [CrossRef]
7. Yi, H.-S.; Park, S.; An, K.-G.; Kwak, K.-C. Algal Bloom Prediction Using Extreme Learning Machine Models at Artificial Weirs in the Nakdong River, Korea. *Int. J. Environ. Res. Public Health* **2018**, *15*, 2078. [CrossRef] [PubMed]
8. Fan, J.; Li, M.; Guo, F.; Yan, Z.; Zheng, X.; Zhang, Y.; Xu, Z.; Wu, F. Priorization of River Restoration by Coupling Soil and Water Assessment Tool (SWAT) and Support Vector Machine (SVM) Models in the Taizi River Basin, Northern China. *Int. J. Environ. Res. Public Health* **2018**, *15*, 2090. [CrossRef] [PubMed]
9. Ji, X.; Shang, X.; Dahlgren, R.A.; Zhang, M. Prediction of dissolved oxygen concentration in hypoxic river systems using support vector machine: A case study of Wen-Rui Tang River, China. *Environ. Sci. Pollut. Res.* **2017**, *24*, 16062–16076. [CrossRef]

10. Kratzert, F.; Klotz, D.; Brenner, C.; Schulz, K.; Herrnegger, M. Rainfall-runoff modelling using Long Short-Term Memory (LSTM) networks. *Hydrol. Earth Syst. Sci.* **2018**, *22*, 6005–6022. [CrossRef]
11. Antanasijević, D.; Pocajt, V.; Povrenović, D.; Perić-Grujić, A.; Ristić, M. Modelling of dissolved oxygen content using artificial neural networks: Danube River, North Serbia, case study. *Environ. Sci. Pollut. Res.* **2013**, *20*, 9006–9013. [CrossRef]
12. Li, L.; Jiang, P.; Xu, H.; Lin, G.; Guo, D.; Wu, H. Water quality prediction based on recurrent neural network and improved evidence theory: A case study of Qiantang River, China. *Environ. Sci. Pollut. Res.* **2019**, *26*, 19879–19896. [CrossRef]
13. Hochreiter, S.; Schmidhuber, J. Long Short-Term Memory. *Neural Comput.* **1997**, *9*, 1735–1780. [CrossRef]
14. Zhou, J.; Wang, Y.; Xiao, F.; Wang, Y.; Sun, L. Water Quality Prediction Method Based on IGRA and LSTM. *Water* **2018**, *10*, 1148. [CrossRef]
15. Barzegar, R.; Aalami, M.T.; Adamowski, J. Short-term water quality variable prediction using a hybrid CNN-LSTM deep learning model. *Stoch. Environ. Res. Risk Assess.* **2020**, *34*, 415–433. [CrossRef]
16. Liang, Z.; Zou, R.; Chen, X.; Ren, T.; Su, H.; Liu, Y. Simulate the forecast capacity of a complicated water quality model using the long short-term memory approach. *J. Hydrol.* **2020**, *581*, 124432. [CrossRef]
17. Liu, P.; Wang, J.; Sangaiah, A.; Xie, Y.; Yin, X. Analysis and Prediction of Water Quality Using LSTM Deep Neural Networks in IoT Environment. *Sustainability* **2019**, *11*, 2058. [CrossRef]
18. Song, C.; Yao, L.; Hua, C.; Ni, Q. A novel hybrid model for water quality prediction based on synchro squeezed wavelet transform technique and improved long short-term memory. *J. Hydrol.* **2021**, *603*, 126879. [CrossRef]
19. Luo, Q.; Peng, D.; Shang, W. Water quality analysis based on LSTM and BP optimization with a transfer learning model. *Environ. Sci. Pollut. Res.* **2023**, *30*, 124341–124352. [CrossRef]
20. Zou, Q.; Xiong, Q.; Li, Q.; Yi, H.; Yu, Y.; Wu, C. A water quality prediction method based on the multi-time scale bidirectional long short-term memory network. *Environ. Sci. Pollut. Res.* **2020**, *27*, 16853–16864. [CrossRef]
21. Ueda, F.; Tanouchi, H.; Egusa, N.; Yoshihiro, T. A Transfer Learning Approach Based on Radar Rainfall for River Water-Level Prediction. *Water* **2024**, *16*, 607. [CrossRef]
22. Chang, F.-J.; Tsai, Y.-H.; Chen, P.-A.; Coynel, A.; Vachaud, G. Modeling water quality in an urban river using hydrological factors—Data driven approaches. *J. Environ. Manag.* **2015**, *151*, 87–96. [CrossRef]
23. Chen, K.; Chen, H.; Zhou, C.; Huang, Y.; Qi, X.; Shen, R.; Liu, F.; Zuo, M.; Zou, X.; Wang, J.; et al. Comparative analysis of surface water quality prediction performance and identification of key water parameters using different machine learning models based on big data. *Water Res.* **2020**, *171*, 115454. [CrossRef]
24. Nemati, S.; Fazelifard, M.H.; Terzi, Ö.; Ghorbani, M.A. Estimation of dissolved oxygen using data-driven techniques in the Tai Po River, Hong Kong. *Environ. Earth Sci.* **2015**, *74*, 4065–4073. [CrossRef]
25. Parmar, K.S.; Makkhan, S.J.S.; Kaushal, S. Neuro-fuzzy-wavelet hybrid approach to estimate the future trends of river water quality. *Neural Comput. Appl.* **2019**, *31*, 8463–8473. [CrossRef]
26. Wang, C.; Shan, B.; Zhang, H.; Zhao, Y. Limitation of spatial distribution of ammonia-oxidizing microorganisms in the Haihe River, China, by heavy metals. *J. Environ. Sci.* **2014**, *26*, 502–511. [CrossRef]
27. Zhu, S.; Luo, X.; Yuan, X.; Xu, Z. An improved long short-term memory network for streamflow forecasting in the upper Yangtze River. *Stoch. Environ. Res. Risk Assess.* **2020**, *34*, 1313–1329. [CrossRef]
28. Zhu, Y.; Drake, S.; Lü, H.; Xia, J. Analysis of temporal and spatial differences in eco-environmental carrying capacity related to water in the Haihe river basins, China. *Water Resour. Manag.* **2010**, *24*, 1089–1105. [CrossRef]
29. Xu, J.; Liu, R.; Ni, M.; Zhang, J.; Ji, Q.; Xiao, Z. Seasonal variations of water quality response to land use metrics at multi-spatial scales in the Yangtze River basin. *Environ. Sci. Pollut. Res.* **2021**, *28*, 37172–37181. [CrossRef]
30. Wang, Y.; Ding, X.; Chen, Y.; Zeng, W.; Zhao, Y. Pollution source identification and abatement for water quality sections in Huangshui River basin, China. *J. Environ. Manag.* **2023**, *344*, 118326. [CrossRef]
31. Maguire, J.; Cusack, C.; Ruiz-Villarreal, M.; Silke, J.; McElligott, D.; Davidson, K. Applied simulations and integrated modelling for the understanding of toxic and harmful algal blooms (ASIMUTH): Integrated HAB forecast systems for Europe's Atlantic Arc. *Harmful Algae* **2016**, *53*, 160–166. [CrossRef] [PubMed]
32. Salacinska, K.; El Serafy, G.Y.; Los, F.J.; Blauw, A. Sensitivity analysis of the two dimensional application of the Generic Ecological Model (GEM) to algal bloom prediction in the North Sea. *Ecol. Model.* **2010**, *221*, 178–190. [CrossRef]
33. Li, R. Water quality forecasting of Haihe River based on improved fuzzy time series model. *Desalination Water Treat.* **2018**, *106*, 285–291. [CrossRef]
34. Liang, N.; Zou, Z.; Wei, Y. Regression models (SVR, EMD and FastICA) in forecasting water quality of the Haihe River of China. *Desalination Water Treat.* **2019**, *154*, 147–159. [CrossRef]
35. Liu, X.B.; Peng, W.Q.; He, G.J.; Liu, J.L.; Wang, Y.C. A Coupled Model of Hydrodynamics and Water Quality for Yuqiao Reservoir in Haihe River Basin. *J. Hydrodyn.* **2008**, *20*, 574–582. [CrossRef]

36. Zhang, L.; Zou, Z.H.; Zhao, Y.F. Application of chaotic prediction model based on wavelet transform on water quality prediction. *IOP Conf. Ser. Earth Environ. Sci.* **2016**, *39*, 012001. [CrossRef]
37. Zhang, X.; Jiang, H.L.; Zhang, Y.Z. The Hybrid Method to Predict Biochemical Oxygen Demand of Haihe River in China. *Adv. Mater. Res.* **2012**, *610–613*, 1066–1069. [CrossRef]
38. Chen, S.; Fang, G.; Huang, X.; Zhang, Y. Water Quality Prediction Model of a Water Diversion Project Based on the Improved Artificial Bee Colony–Backpropagation Neural Network. *Water* **2018**, *10*, 806. [CrossRef]
39. Deng, W.; Wang, G.; Zhang, X. A novel hybrid water quality time series prediction method based on cloud model and fuzzy forecasting. *Chemom. Intell. Lab. Syst.* **2015**, *149*, 39–49. [CrossRef]
40. Di, Z.; Chang, M.; Guo, P. Water Quality Evaluation of the Yangtze River in China Using Machine Learning Techniques and Data Monitoring on Different Time Scales. *Water* **2019**, *11*, 339. [CrossRef]
41. Zhou, C.; Gao, L.; Gao, H.; Peng, C. Pattern Classification and Prediction of Water Quality by Neural Network with Particle Swarm Optimization. In Proceedings of the 2006 6th World Congress on Intelligent Control and Automation, Dalian, China, 21–23 June 2006; pp. 2864–2868. [CrossRef]
42. Gao, Y.; Zhang, W.; Li, Y.; Wu, H.; Yang, N.; Hui, C. Dams shift microbial community assembly and imprint nitrogen transformation along the Yangtze River. *Water Res.* **2021**, *189*, 116579. [CrossRef] [PubMed]
43. Hu, M.; Liu, Y.; Zhang, Y.; Shen, H.; Yao, M.; Dahlgren, R.A.; Chen, D. Long-term (1980–2015) changes in net anthropogenic phosphorus inputs and riverine phosphorus export in the Yangtze River basin. *Water Res.* **2020**, *177*, 115779. [CrossRef]
44. Liu, X.; Beusen, A.H.W.; Van Beek, L.P.H.; Mogollón, J.M.; Ran, X.; Bouwman, A.F. Exploring spatiotemporal changes of the Yangtze River (Changjiang) nitrogen and phosphorus sources, retention and export to the East China Sea and Yellow Sea. *Water Res.* **2018**, *142*, 246–255. [CrossRef] [PubMed]
45. Dang, B.; Mao, D.; Xu, Y.; Luo, Y. Conjugative multi-resistant plasmids in Haihe River and their impacts on the abundance and spatial distribution of antibiotic resistance genes. *Water Res.* **2017**, *111*, 81–91. [CrossRef]
46. Bao, Z.; Zhang, J.; Wang, G.; Fu, G.; He, R.; Yan, X.; Jin, J.; Liu, Y.; Zhang, A. Attribution for decreasing streamflow of the Haihe River basin, northern China: Climate variability or human activities? *J. Hydrol.* **2012**, *460–461*, 117–129. [CrossRef]
47. Zheng, M.; Zheng, H.; Wu, Y.; Xiao, Y.; Du, Y.; Xu, W.; Lu, F.; Wang, X.; Ouyang, Z. Changes in nitrogen budget and potential risk to the environment over 20 years (1990–2010) in the agroecosystems of the Haihe Basin, China. *J. Environ. Sci.* **2015**, *28*, 195–202. [CrossRef] [PubMed]
48. Reshef, D.N.; Reshef, Y.A.; Finucane, H.K.; Grossman, S.R.; McVean, G.; Turnbaugh, P.J.; Lander, E.S.; Mitzenmacher, M.; Sabeti, P.C. Detecting novel associations in large data sets. *Science* **2011**, *334*, 1518–1524. [CrossRef] [PubMed]
49. Zhang, J.; Zhu, Y.; Zhang, X.; Ye, M.; Yang, J. Developing a Long Short-Term Memory (LSTM) based model for predicting water table depth in agricultural areas. *J. Hydrol.* **2018**, *561*, 918–929. [CrossRef]
50. Panfeng, B.; Songlin, Z.; Hongyu, C.; Caiwei, L.; Pengtao, W.; Lichang, Q. Structural monitoring data repair based on a long short-term memory neural network. *Sci. Rep.* **2024**, *14*, 9974. [CrossRef]
51. Bennett, N.D.; Croke, B.F.W.; Guariso, G.; Guillaume, J.H.A.; Hamilton, S.H.; Jakeman, A.J.; Marsili-Libelli, S.; Newham, L.T.H.; Norton, J.P.; Perrin, C.; et al. Characterising performance of environmental models. *Environ. Model. Softw.* **2013**, *40*, 1–20. [CrossRef]
52. Sadiki, N.; Jang, D.W. Estimation of Hydraulic and Water Quality Parameters Using Long Short-Term Memory in Water Distribution Systems. *Water* **2024**, *16*, 3028. [CrossRef]
53. Gachloo, M.; Liu, Q.; Song, Y.; Wang, G.; Zhang, S.; Hall, N. Using Machine Learning Models for Short-Term Prediction of Dissolved Oxygen in a Microtidal Estuary. *Water* **2024**, *16*, 1998. [CrossRef]
54. Ritter, A.; Muñoz-Carpena, R. Performance evaluation of hydrological models: Statistical significance for reducing subjectivity in goodness-of-fit assessments. *J. Hydrol.* **2013**, *480*, 33–45. [CrossRef]
55. Moriasi, D.N.; Arnold, J.G.; Van Liew, M.W.; Bingner, R.L.; Harmel, R.D.; Veith, T.L. Model evaluation guidelines for systematic quantification of accuracy in watershed simulations. *Trans. ASABE* **2007**, *50*, 885–900. [CrossRef]
56. Galelli, S.; Humphrey, G.B.; Maier, H.R.; Castelletti, A.; Dandy, G.C.; Gibbs, M.S. An evaluation framework for input variable selection algorithms for environmental data-driven models. *Environ. Model. Softw.* **2014**, *62*, 33–51. [CrossRef]
57. Maier, H.R.; Jain, A.; Dandy, G.C.; Sudheer, K.P. Methods used for the development of neural networks for the prediction of water resource variables in river systems: Current status and future directions. *Environ. Model. Softw.* **2010**, *25*, 891–909. [CrossRef]
58. Maier, H.R.; Dandy, G.C. Neural networks for the prediction and forecasting of water resources variables: A review of modelling issues and applications. *Environ. Model. Softw.* **2000**, *15*, 101–124. [CrossRef]

MDPI AG
Grosspeteranlage 5
4052 Basel
Switzerland
Tel.: +41 61 683 77 34

Water Editorial Office
E-mail: water@mdpi.com
www.mdpi.com/journal/water